MOVING IMAGE TECHNOLOGY

MOVING IMAGE TECHNOLOGY
from zoetrope to digital

Leo Enticknap

WALLFLOWER PRESS
LONDON & NEW YORK

First published in Great Britain in 2005 by
Wallflower Press
4th Floor, 26 Shacklewell Lane, London E8 2EZ
www.wallflowerpress.co.uk

A catalogue for this book is available from the British Library

ISBN 1-904764-06-1 (pbk)
ISBN 1-904764-07-X (hbk)

Book design by Elsa Mathern

Printed in Turin, Italy by Grafiche Dessi s.r.l.

contents

acknowledgements vii

introduction — affecting a knowledge of antiques 1

chapter one — film 4

chapter two — cinematography and film formats 29

chapter three — colour 74

chapter four — sound 98

chapter five — cinema exhibition 132

chapter six — television and video 159

chapter seven — archival preservation and restoration 187

chapter eight — new moving image technologies 202

notes 232
chronology 247
glossary 252
bibliography 264
index 273

acknowledgements

Throughout the two and a half years this book has been in preparation I have incurred debts to a great many individuals and organisations. The research has involved using a number of specialist research libraries, two of which deserve a special mention. The staff at the British Film Institute's National Library have, as always, been helpful, knowledgeable and prepared to go an extra mile to find what was in some cases quite obscure material. Dr Luke McKernan at the British Universities Film and Video Council (BUFVC) very kindly arranged extensive access to the unrivalled collection of technical books and periodicals formerly held by the British Kinematograph, Sound and Television Society, and also offered extensive and extremely useful feedback on the initial draft of the manuscript.

At my workplace, the University of Teesside/Northern Film and Television Archive, a number of colleagues have offered logistical support and comments on earlier drafts of the text. To this end I am especially grateful to Dr Joe Kember, Simon Popple and Professor Paul Wells. The History Research Group at the University of Teesside provided financial assistance for research visits and the production and copyright clearance of some of the illustrations. Other colleagues and contacts in the film archive world have been helpful in answering queries and pointing me in the direction of relevant material. I would also like to record my thanks to those members of the Association of Moving Image Archivists (AMIA), too numerous to mention individually, who have answered queries and brought relevant information and issues to my attention (often unintentionally) through this organisation's annual conferences and e-mail discussion list, AMIA-L. Darren Briggs of City Screen Ltd gave me very useful feedback on an early draft of chapter five, and I am also grateful to him for permission to publish some of the photographs showing cinema projection equipment. Special thanks must go to a number of people who have offered extensive and valuable feedback on the initial draft without whose help and support this book would have gone to press with a number of significant errors, omissions and misconceptions: in particular Charles Hopkins, Stephen Herbert, David Pierce, Brian Pritchard and Barry Salt. Those errors and omissions which remain are, of course, entirely my own.

This book was completed entirely in my own time and without any research leave. To this end many friends and relatives had to put up with me seeming to disappear off the face of the planet, often for several weekends in succession. The greatest acknowledgement of all, however, must be reserved for the publisher, Yoram Allon of Wallflower Press. He stoically put up with a year and a half of delays and setbacks caused by my trying to juggle writing this book with my 'day job', to the extent that I am sure many others would have pulled the plug. I hope that the result justifies the wait.

introduction | **affecting a knowledge of antiques**

Images don't move on their own. They have to be made to move – or *appear* to move – by technology, invented and developed by human beings with the purpose of tricking the human brain into thinking that it sees a continuously moving two-dimensional (or in some cases, three-dimensional) picture. No technology = no movement. The technology in question has two components: the images themselves and the processes by which they are made to appear to move. With one or two small-scale exceptions, such as stop-motion animation, the overwhelming majority of human activity in this area has used photographic images as the raw material for movies; probably because for most people photos are the most realistic-looking form of graphical illustration. They are understood to represent 'real life' in a way which no other medium can, hence the phrase 'the camera cannot lie'. Actually it can lie. It has done so throughout its history and with a vengeance, from the days when Stalin's spin doctors airbrushed out the latest functionary to be declared *persona non grata* to the digital era of the 1990s, when the verb 'Photoshop' evolved into an euphemism meaning to digitally manipulate a photographically-originated image to make it represent something it was never intended to.

But that's beside the point. The popular association of photography with realism – it is usually understood to be a medium rather than an art form – made photographs the raw material of choice for the scientists and engineers who wanted images to move, and on an industrial scale. And so we arrive at the close of the nineteenth century, at which point George Eastman, W. K. L. Dickson, the Lumières and many others besides arrive on the scene, bringing with them film, printers, projectors, large-scale financial investment, technical standards and all the other factors that would result in moving images becoming a multi-million dollar industry in today's money, within the space of around 15 years. The story of those 15 years and of the 95 or so which followed is the purpose of this book, but before embarking on it one key issue and its widespread implications need to be acknowledged.

Because moving images are primarily a medium rather than an art form in themselves, the overwhelming bulk of critical attention, by which I mean everything from a group of friends casually discussing a film they have just seen in the pub to academic monographs which apply obscure sociological theory to analyse the representation of characters in TV soap operas, tends to focus on the 'artistic' use of that medium as distinct from the aesthetic characteristics of the medium itself. Here is an example. When I was a film archiving Master's student in the mid-1990s, we shared one film history seminar with a group of film *studies* (i.e. whose programme was purely academic and contained no significant technical or vocational element) students. In one of these sessions the lecturer showed us an excerpt from a low-budget British crime melodrama from the 1930s. In it a male detective interrogates his female suspect while leaning over the chair in which the latter is sitting, appearing to shout at her. The sequence is lit in such a way that the protagonist appears to cast a long, menacing shadow over the subject of the interrogation. After stopping the video the lecturer invited comments from his class as to why the scene had been shot and staged as it had. His request immediately elicited an impassioned analysis from one of my film studies colleagues. She was (and as far as I know, still is) an enthusiastic advocate of 'feminist film theory', a body of research which holds that economic control of the Hollywood film industry can be interpreted through the psychoanalysis as proposed by Freud to encourage representations of women which are dominated by nasty male desires, and therefore undesirable.[1] Needless to say she had a field day describing the domination metaphor as applied to the images she had just seen.

It eventually emerged that the reason for our hero towering over his petite murder suspect was somewhat more pragmatic. The microphone recording the dialogue, we were told, had been concealed in a vase of flowers behind the chair in which she was sitting. Furthermore, being a primitive 1930s microphone, its sensitivity was limited, thereby requiring the detective to position his mouth as close as he could to it (hence leaning over the occupied chair) and speaking as loudly as possible (hence the menacing voice). In this instance an understanding of the electromagnetic properties of microphone technology would have been rather more useful than the Freudian claptrap which was inflicted on us that afternoon. Sometimes a candlestick really is just a candlestick. In fact, this is quite a mild example of the ways in which humanities and social sciences academics habitually misrepresent the role of engineering and technology in our everyday lives and culture. An extreme one would be the experiment carried out by the American physicist Alan Sokal. In 1996 he submitted an essay to the editorial board of *Social Text*, a 'cultural studies' journal which enjoyed an eminent reputation among academics in that field, which consisted of a satirical parody.[2] He was particularly concerned that humanities academics seemed to be trying to find ways of disputing scientific phenomena which had been proven and were demonstrable through empirical research and experiments, mainly for political and ideological reasons. As Sokal wrote, 'fair enough – anyone who believes that the laws of physics are mere social conventions is invited to try transgressing those conventions from the window of my twenty-first-floor flat'. Astonishingly *Social Text* published the essay, and in doing so revealed the 'self-perpetuating academic subculture that ignores (or disdains) reasoned criticism from outside'.[3]

If evidence that such a subculture exists within the academic fields of research which involve attempting to understand the role of moving images in society were needed, the mere fact that those who argue for an informed understanding of the role of technology (rather than one which is skewed to fit a dubious ideological agenda) should be central to this activity are habitually considered 'outsiders' provides it. In the introduction to what is beyond any reasonable doubt the standard history of film technology, Barry Salt refers to the continuing resistance of the academic humanities community to objective knowledge and empirical historiography in understanding film technology and other things.[4] To borrow a line from *The Big Sleep*, his allegation is that they 'affect a knowledge of antiques, but haven't any'.

Just how crucial that knowledge is to avoiding fundamental misunderstandings can be summed up in the unique nature of moving images relative to virtually any other cultural artefact one cares to mention. A spoken narrative can be replayed and duplicated using human memory and word of mouth, music can be sung and stage performances can take place in nothing more than an open space. In order to view (analogue) still photographic images only the physical medium on which they are recorded is necessary. Recorded moving images and sound, however, require technology in order to be perceived as such to the viewer or listener.

This book is not intended and certainly will not succeed in fundamentally changing the ways in which students and academics working in the humanities think about and understand moving images. Rather, I am trying to offer an accessible and coherent way in to what the state of the various arts involved was at any given time, their opportunities, characteristics and limitations. To this end the book is divided into eight chapters, which cover what I would argue are the key forms of technological research and development which comprise the origination, manipulation, distribution, reproduction and preservation of moving images. Each chapter consists essentially of a linear narrative explaining what happened, when and why, and the wider impact of each development on other related technologies and their usage.

The use of technologically specific terminology, or 'jargon' to put it brutally, is a bullet which cannot avoid being bitten. In an attempt to minimise the extent to which it could potentially obstruct the reader, two principles have been followed throughout the book: the use of such terms sequentially, and the provision of a glossary for reference purposes. Whenever such a term is used for the first time in the main text, the process or phenomenon it describes is explained in as close as possible to plain English as I can make it. As an *aide memoir* an etymological explanation is also given in the case of terms (a surprising number, given that moving image technology is essentially a twentieth-century phenomenon) with Greek or Latin origins. Thereafter the glossary should serve as a point of reference if needed, especially for readers who are going directly to individual sections of the book rather than reading it as a whole (by chapter eight the acronyms are flowing thick and fast!).

I hope, therefore, that this book will be able to function both as a quick reference point for readers seeking answers or explanations relating to specific technical issues, and also as a broader narrative for readers looking for a historical overview of the role of technology within the economic, industrial, political and cultural roles of moving images.

chapter one | film

> It was meant to be a showcase for Britain's electronic prowess – a computer-based, multimedia version of the Domesday Book. But 16 years after it was created, the £2.5 million BBC Domesday Project has achieved an unexpected and unwelcome status: it is now unreadable.[1]

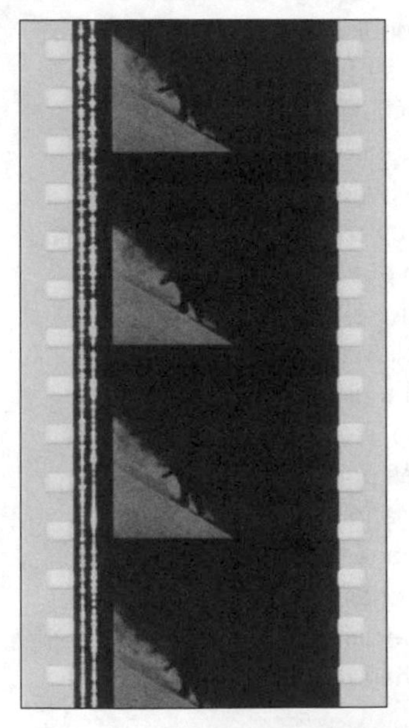

Fig. 1.1 A 35mm cinema release print. With the exception of the optical soundtrack (left), this photographic film format has been in constant use for recording and reproducing moving images since the early 1890s, making it the longest lasting form of motion picture media by a margin of several decades.

In 1987 the British Broadcasting Corporation (BBC) created a computer-based multimedia resource which was intended to serve as a snapshot of life and work in the UK, consisting of text, still photographs, sound and full-motion video. As *The Observer*'s reporter notes, it was supposed to be a reworking of the Domesday Book of 1086; a reworking both terms of its content (the original was basically a glorified asset register and as such was never intended to record any details of England's social, cultural or political life) and more importantly to this discussion, its form. The BBC's 'Domesday Disc' was recorded on 12" laserdiscs, a consumer and semi-professional video format used between the late 1970s and early 1990s. The BBC had adapted laserdiscs to hold digital content by means of a dedicated hardware interface and software designed to work on a non-standard type of computer, one which was only used on any significant scale in British secondary schools. Both the hardware and software are now obsolete, and unlike its Norman predecessor, the 'Domesday Disc' of 1986 has had to undergo 'format migration' (i.e. be copied onto a medium which is supported by current technologies) in order to remain accessible in 2002.

This is an example of a computer-based multimedia format that went from being in common use to unreadable within less than two decades. By contrast, the use of photographic film – that is, a flexible, transparent solid which supports a light-sensitive chemical layer – to record moving images and sound dates from the earliest successful commercial exploitation of this technology in 1889. Furthermore, it is still the most widely used format for feature film production, for television production in cases when image quality is considered important and budgets allow, for the application of special effects, for the mass-duplication of copies and for cinema exhibition. It is therefore the one and only form of moving image media to have remained in continuous, mainstream use for over a century. That fact in itself will not guarantee its long-term future. Many in the industry believe that the technical and economic climates are changing, and that digital technologies will surely supersede film for all of these uses just as soon as the processor speeds and storage capacity in the computers needed to drive them become cheaper than exposing and processing photographic film stock (these issues will be discussed at greater length in chapter eight). There is after all a precedent for this: digital technology replaced the gramophone record for recording sound and distributing it in the commercial music industry with the introduction of compact discs in 1982, just under a century after Emile Berliner produced the first shellac records in 1887. But whether or not this book will turn out to have been written in the closing years of film, the evolution of this technology is of key importance to the development of the moving image history and culture during the period covered here; so much so that it takes the year in which film was first used commercially, 1889, as its starting point.

This longevity is a key reason why film was the most widely used technology for recording moving images during the twentieth century. Assuming that chemical decomposition or physical damage is not an issue (this is covered in chapter seven), a moving image film manufactured in the 1890s could be viewed and copied using equipment in widespread use in 2005, with only minor modifications. Promoters of digital alternatives (notably the producer and director George Lucas) would therefore argue that the industry remains saddled with all the inherent faults of a Victorian technology. Defenders of film believe that it offers the best of both worlds: a tried and tested technology which needs relatively little cascaded expenditure on research and dev-elopment by its users, but one which is also capable of developing and improving the image quality it can deliver:

> Some existing technologies wither and disappear quickly in the face of new technologies. Some make subtle changes and adapt, becoming better. Some reach a stage of development that achieves a kind of stasis. The piano has not altered much since the days of Chopin, 150 years ago.[2]

The history of motion picture film, therefore, is the history of a very subtle industrial and economic combination of continuity, evolution and outright change. The remainder of this chapter will attempt to identify the technological specificities of the earliest and most widely used of all moving image media, and to show how these specificities shaped the industries, institutions and cultures which used it.

Moving images before 1889: establishing the physical properties and performance requirements of motion picture film

There is no such thing as a moving image. No means has ever been devised of continually recording the sequence of changing light over an extended period of time as it is perceived by the human optical and nervous system, and then reproducing it in a way that is perceived identically to the original source. The 'moving image' technologies we have today are all, without exception, based on the discovery during the mid-nineteenth century that a sequence of still images, photographed or created in rapid succession, will, when projected or otherwise mechanically displayed in equally rapid succession, be perceived by the human brain as a continuously moving image. Two processes need to take place in order for this to be achieved. The first is that the camera has to record the sequence of still images, or 'frames', at a fast enough rate that a distinct difference between two individual frames cannot be perceived by the viewer when the sequence is reproduced at a similar speed. The second is that the transition between each photograph in the moving image sequence has to be achieved in a way which is invisible to the viewer, i.e. without a perceptible fluctuation in light level, or 'flicker', when it is reproduced by a displaying device (e.g. a projector or TV monitor). For well over a century both of these processes were thought to have been achieved by a single phenomenon, known as the 'persistence of vision' effect. This held that the human brain continues to 'see' a projected image momentarily even after it has ceased to be projected, thereby enabling a seamless transition from one image to the next to be perceived. But in the second half of the twentieth century researchers identified some serious flaws in this theory. For example why do we perceive motion reproduced by methods which *do not* have a 'black' period between each frame, e.g. an interlaced scanning TV monitor (see chapter 6) or DLP projection (see chapter 8), in the same way as we perceive motion as reproduced by a film projector? One explanation is that of 'visual maksing', a human thought process which 'is said to occur when two visual presentations are made sequentially and one renders the other invisible.'[3] But the bottom line is that we don't know the whole truth of why the illusion of movement is achieved once these two conditions have been fulfilled. For example Stephen Herbert suggests that the periods of darkness in between each frame being displayed by a film projector 'are sufficiently short that they do not register in our brain':[4] yet if that is the case why did early TV engineers have to interlace the scanning sequence in a cathode ray tube display to get rid of the perceived intermittent 'darkness' which resulted from progressive scanning? All we know is that the impression of movement can be created in this way – and has been, on an industrial scale, ever since the discovery was made.

This knowledge was initially exploited several decades before the images could be created photographically in a series of toys which appeared from the mid-1830s onwards. The Phenakistiscope, demonstrated by Joseph Plateau in 1833, was a cardboard disc with sixteen evenly-spaced notches cut out of the edge and sixteen sequential drawings or paintings placed between each notch. The viewer held the painted side of the disc facing a mirror and rotated it while looking through the notches at the reflection. This gave the effect of a moving image. The Zoetrope, described by the Englishman William Horner in 1834 but not built or mass-marketed until the 1860s, consisted of a perforated drum mounted horizontally on a spindle. A paper strip containing a sequence of drawings or paintings was placed around the inner edge of the drum, and by looking through the perforations as the drum was rotated, the appearance of movement could be perceived. It could be argued that the Zoetrope was one of the earliest forms of moving image mass-media: the 'software', in the form of printed paper strips, sold for less than the price of a newspaper and remained popular throughout the late nineteenth century. A whole genre of optical toys based on the idea proliferated during this period, a process which culminated in Émile Reynaud's *Théâtre O ptique* of 1892, which projected a linear sequence of hand-painted lantern slides onto a large screen in front of a theatrical audience. Embodying principles both of the Victorian optical toys and the modern cinema projector, the *Théâtre Optique* consisted of slides mounted horizontally on a linear band, with each joint perforated. The complete sequence of between 500 and 700 images was fed between spools and passed in front of a light source. However, the system of rotating mirrors which enabled the transition from each slide to appear seamless and the mechanical tolerances of the mechanism could not even approach the speed needed to project a perceived moving image to the quality achieved when the first public demonstration of moving image film projection took place a little over three years later.[5]

In the half-century before 1889, therefore, the optical toy industry established that the perception of continuous movement could be induced mechanically. Two other technologies were also developed during this period which, when combined with the illusion of continuous movement, resulted in the film stock first sold by George Eastman in 1889. These were photography and the film base itself (i.e. the physical support on which a photographic image is held).

Photography – the creation of a permanent record of the existence of light in a given place and at a given moment in time – also had its origins in the early nineteenth century. A photographic image is created through the use of substances in which chemical change is induced by exposure to light, and in which that change can be permanently recorded and perceived by the human eye. In order to create a photograph two technologies are needed: the photosensitive medium, and the device used to achieve the exposure, i.e. a camera. The mechanical and optical properties of cameras are discussed at length in chapter two. As far as photographic technology is concerned, this chapter will concentrate on the chemical processes needed to create a photographic image.

Like the mechanical processes which exploited the persistence of vision effect, photochemical technology also has its origins in the nineteenth century. In its crudest form a photographic image is created by exposing a silver-based halide (a compound of halogen with a metal or radical, such as bromide) to light. This causes a chemical reaction to take place which is initially invisible (known as the *latent image*), but which can be made visible and permanent by a two-stage chemical process. This procedure, known as *processing*, consists firstly of *development* – immersing the photochemical substance in a chemical which converts the exposed halide into a pure metallic silver, which is visible. The developed image is then *fixed* by making the undeveloped (i.e. still photosensitive) silver halide soluble in water by sodium thiosulphate and then washing it away. This leaves only the metallic silver dye in place, and this is visible to the naked eye as a black-and-white photograph.

As well as the photochemical layer itself, a surface is needed to support it during exposure, processing and subsequent viewing of the processed image. When referring to motion picture film the flexible, transparent support is known as the *base* whilst the photochemical layer is termed the *emulsion*. The earliest bases, however, were not flexible or transparent, and were quite unsuitable for use in any motion picture process. Experiments took place involving paper, leather, canvas, glass and copper since Thomas Wedgwood saturated a leather canvas with a solution of silver salts in the early 1800s.

The very early photographic emulsions were only used successfully for producing images of silhouettes (for example leaves, which were used in Wedgwood's canvas images) or of existing images on a transparent base such as stained-glass windows. They were unusable for creating photographs of real-life subjects due to their very low *speed*. The speed of a photographic emulsion measures the intensity of light and length of exposure needed to produce the latent image. Nowadays the speed of film is generally specified using the *exposure index* (EI) scale. The film used in 35mm still cameras today usually has an EI rating between 100 and 400. Film rated at EI100, for example, would require an exposure of 1/125 second to photograph a typical street scene in bright sunlight.[6] By comparison, Wedgwood's sensitised canvas would have needed several tens of hours. Given that the perception of continuous movement requires an effective minimum of 16 images per second, it is clear that the creation of moving images photographically was at that stage a long way off. Emulsion speeds gradually improved during the Victorian period as techniques were increased to improve the concentration and grain structure of the silver halide. Two important breakthroughs happened in the 1830s: the discovery by Louis Daguerre that the sensitivity of silver halides could be increased by exposure to iodine, and the announcement in 1839 by Henry Fox-Talbot that he had successfully invented a means of copying photographic images. This would, of course, be of crucial importance to the application of photography for moving images, and for two reasons.

Fig. 1.2 Negetive (above) and positive photographic images

When initially exposed and developed, the silver-based photographic emulsion yields an image which is known as a *negative*. That is to say, the greater the intensity of light in the original subject, the more opaque the developed area of the photographic image is, and vice-versa. The negative photograph thus looks like an inverse of the actual scene in real life. In order to produce a *positive* image – a photograph which actually resembles the original subject – it is necessary to create a *print*. This is done by passing light through the negative image onto a second layer of photosensitive emulsion. When developed, this second photograph yields a negative image of the original negative, thus making a positive print. It is possible to make multiple prints from the same negative, which, when photography began to be used for moving images, was the means by which many hundreds of prints could be made of a single film for distribution to cinemas. The *reversal* process, originally described in 1899 by the Italian chemist Rodolfo Namias,[7] was a process for developing a photographic image directly to a positive which consisted, in effect, of processing the image twice in one procedure. The image is developed in the same way as a negative, after which the converted halides are bleached away leaving the unexposed silver halides to form a positive image of the original negative latent image. This residue is then exposed to light (or treated with a chemical which achieves the same effect) and the second latent image is redeveloped and fixed, thus creating a negative of the negative – in other words, a positive image of the original subject. The reversal process was eventually used in two significant areas of moving image technology: amateur filmmaking, in which the original camera negative could be used for projection, thus avoiding the expense of making additional copies, and films intended for television use, where speed of processing is crucial but the need to make multiple film copies is not.

By the late nineteenth century the sensitivity of photographic emulsions had improved to reduce exposure times to significantly under a second and the techniques for coating them onto various different bases had evolved into an efficient mass-production process. Crucial to the latter was the invention of the gelatine bromide process in 1871.[8] Gelatine is a transparent semi-liquid adhesive derived from albumin. By suspending particles of cadmium bromide and silver nitrate in gelatine, it was found possible to produce an emulsion which was many times faster than any of its predecessors (exposures of $^{1}/1000$ second were now possible) and which could easily be coated onto any surface – initially paper and glass plates, but eventually film. Though the gelatine bromide chemistry was originally demonstrated by the British chemist Richard Leach Maddox, a patent for the automated process of coating it onto paper was applied for by George Eastman (the founder of Eastman Kodak) in 1884, and eventually granted in 1890.[9] Eastman photographic paper rolls went on sale to the public in 1885.

By the mid-1880s it was possible to change the perception of continuous movement mechanically and to create photographic images at the quality and frequency needed for use in a moving image device. The only remaining issue was the production of a physical support – the base – that would allow photographs to be exposed and projected as moving images. It was established relatively early on that the most suitable method for passing large numbers of still photographs through a mechanism at high speed was to place them consecutively on a continuous strip of material.

Working from a rented workshop in a suburb of Leeds, the French inventor Louis Augustin Le Prince built a working camera in which paper roll film was exposed and advanced intermittently in 1888.[10] But because paper was opaque it could not be used for projection. Projected still photographs on a glass base were by then an accepted and growing part of magic lantern performances, and the *Théâtre Optique* would later establish the technique of feeding multiple frames through a projection mechanism in roll form. For the new photography/moving image hybrid, projection on a screen before a theatrical audience was the goal, but for that a transparent base was essential. In the (as yet) absence of film, Le Prince attempted to build a projector which advanced individually-mounted glass slides using a mechanism that worked on similar principles to that of a modern Carousel slide projector. It achieved a speed of approximately seven frames per second – less than half the rate that would be needed to display a flicker-free image with fluent motion.[11] Meanwhile the use of paper-base photographic material in optical toys began to be established. In 1876 the Englishman Wordsworth Donisthorpe patented the Kinesigraph,[12] a single-lens camera similar to Le Prince's projector, which exposed glass-plate negatives at eight frames per second. It was used to produce photographic images which were then printed on paper and viewed in a Phenakistiscope.[13] But projected moving images remained stubbornly out of reach.

The research which would eventually lead to the production of a flexible, transparent film base began slightly later than the development of persistence of vision devices and photography. Cellulose nitrate – a liquid formed by dissolving cellulose (a wood derivative) in nitric acid, is believed to have been discovered in Germany around 1845–46.[14] During the following three decades processes were invented for refining this substance into a flexible, transparent solid.[15] The end result was a patent granted to the Eastman Kodak company on 10 December 1889, for transparent sheets of celluloid: 'a mixture of methyl alcohol, camphor, nitrocellulose, amyl acetate and fusel oil, dried on a polished support, then taken off and coated with the photographic emulsion'.[16] It was widely reported that the first cellulose nitrate film supplied for motion picture research was received by William Kennedy Laurie Dickson on 2 September 1889. It would seem that the Edison company (Dickson's employers) was the first to successfully use the new material as a moving image film base, though many of the inventors who had previously worked in the field, including Le Prince and William Friese-Greene, experimented with celluloid but failed, initially, to make it work in a camera or projector.[17] But by 1889 the three essential ingredients of the first mass-produced form of moving image technology were in place: the ability to induce the perception of continuous movement effect mechanically, photographic emulsions which were fast enough to produce the images needed for these devices and a strong, flexible and transparent film base to support them on.

The remainder of this chapter divides the history of photographic film in moving image technology into two sections: the period when cellulose nitrate was used as the principal film base within the moving image industry, which lasted almost exactly the first half of the twentieth century, and the period following its replacement in 1948– 50 by acetate and polyester bases. In doing this I will argue that the use of nitrate and its eventual obsolescence was a fundamental influence in the industrial evolution of moving image technology in general and of film in particular. The simplicity, reliability and low cost of manufacturing nitrate film coupled with its high tensile strength made it an almost ideal medium for originating, distributing and projecting moving images. Almost, because another characteristic of the medium – its high volatility and inflammability – necessitated extensive health and safety precautions wherever it was used or transported. A further attribute of nitrate is that it is prone to long-term chemical decomposition which, while not an issue when the base was in everyday industrial use, most certainly is an issue for archivists attempting to preserve it decades later, one which is covered in chapter seven. The development and evolution of monochrome (black-and-white), silver-based emulsions will also be considered here alongside film bases, though colour film technologies are covered separately in chapter three.

Film technology during the nitrate period: 1889–1950

In the decade following 1889 film-based moving image technology developed quickly into a mass-medium. Historians have identified a number of economic, cultural and political reasons for the unusually rapid growth in the film industry, from the experiments of a small number of engineers and scientists to the provider of a popular and expanding leisure activity. These were helped and influenced by the evolution of the technology underpinning it – the film itself. The production and exhibition of films was at first very closely linked to that of magic lantern slides. During the mid-Victorian period these had started to be produced photographically as distinct from images drawn or painted onto slides. The glass slides used for the still images in magic lanterns were an ideal base for photographic emulsions for almost thirty years before a flexible equivalent was available for moving image projection, and so the companies and individuals which produced lantern slides were already used to working with one of the key technologies used in moving image film. In fact, many early motion picture projectors were supplied in the form of mechanisms which used the same light source as a magic lantern, thus enabling exhibitors to use both still and moving images interchangeably. The relatively short length of most early films ensured that they could easily be assimilated into the existing leisure industries of the period, notably Music Hall performances and fairground attractions. And the economic climate in which the motion picture pioneers operated, one in which the service sector economy underwent rapid expansion on both sides of the Atlantic, also worked in the new technology's favour.

The characteristics of film itself and the ability of that technology to be adapted for compatibility with existing cultural practices were also important factors in the rapid growth in its use. But initial progress was slow. In the period between 1889 and 1895 the equipment and processes began to be developed which enabled and then extended the activities of moving image photography and film duplication. Among the improvements made during these years were variations to the nitrate base formulation which made it easier to perforate and more resistant to tearing when stressed by sprocket teeth engaging the perforations, a slower emulsion designed specifically for printing and the emergence of methods for cutting and joining together separate pieces of film.[18] But projection remained a problem, and without the means of showing a moving image on a large screen in front of a paying audience, the economic potential of film remained severely limited, not least because it was incompatible with most pre-existing forms of leisure activity. The only commercially marketed means of viewing moving image film between 1889 and 1895 was the Edison Kinetoscope, a device hastily designed by Dickson in 1891 to exploit the Edison company's film, which at that stage represented a major research and development expenditure without any means of exploiting it. The Kinetoscope was a wooden cabinet housing an endless loop of 35mm film transported continuously through a series of rollers under a rotating shutter illuminated by an incandescent filament bulb. A magnifying glass built into the top of the cabinet enabled the viewer to observe, through the shutter, a moving image sequence lasting about twenty seconds. 'Kinetoscope parlours' featuring a number of the coin-operated machines sprung up in New York and America's major cities in the early 1890s, but the Kinetoscope turned out to be a purely transitional form of film exhibition. When the mass-audience problem was finally overcome, which is generally accepted to have happened when Louis and Auguste Lumière demonstrated a working projector in Paris on 28 December 1895, the stage was set for moving image film to become second only to paper as the software medium on which the mass-communication industry was based.

Crucial to the growth of the medium was the means of editing and duplication. Editing, or the technique of cutting lengths of film supplied by manufacturers and joining them together in different configurations after exposure and processing, was possible almost as soon as moving image film started to be used. The method which Dickson recalled using during his early film experiments with Edison involved 'a clamp with steady pins to fit the punch holes, to use in joining the films with a thin paste of the base dissolved in amyl acetate which, I suppose, is still [in 1933] commonly used'.[19] This is now known as cement splicing, and involves the use of a chemical compound which dissolves a thin layer of film base on two facing surfaces, which are then pressed together under considerable pressure to form an adhesive seal. It remained the sole method of joining film until the late 1960s. In fact Dickson's clamp was relatively sophisticated even compared to common practice two decades later: as late as the 1920s, the routine method of producing splices in studios, laboratories and projection rooms was still by hand using a razor blade, without any form of mechanisation.[20] It was not until 1918 that the Bell and Howell company marketed an automatic splicer in which the film perforations were held by registration pins as the emulsion was scraped off, in order to produce a splice of a pre-set width and which was guaranteed not to be visible in projection. The 1918 version was only suitable for use with original camera negatives, as the splices it made were strong enough for printing but not for projection. In 1922 Bell and Howell introduced a modified version suitable for use on release prints.[21] Before the introduction of mechanised splicing, however, film editing was a slow and laborious process. It is interesting to note that genres and individual filmmakers whose work relies on the use of complex editing techniques, for example the Hollywood continuity system or Soviet montage, did not become firmly established until the mid-1920s, after the widespread introduction of automated splicing. Another labour-saving innovation from Bell and Howell was the introduction of automatic film perforators in 1908; before then stock was generally supplied by the manufacturer unperforated, leaving the customer to perforate the stock for use in the sprockets of a camera, printer and/or projector.[22] Bell and Howell's perforators were subsequently used by Eastman Kodak to perforate their raw stock before coating, thus automating another aspect of the manufacture and use of moving image film. Laboratory technology – the ability to duplicate film, manipulate the visual qualities of the photographic image in the course of so doing and produce large numbers of release prints for showing in cinemas – improved and expanded rapidly in the years following the discovery and early use of nitrate film as a photographic base in 1889. Methods of film printing fall into two categories. In *continuous* printing, the processed original containing the image to be copied and the unexposed film stock which is receiving the copy are transported over a light source at a constant speed. This technique can be further subdivided into continuous contact printing, in which the emulsion surfaces of the source and destination film elements are placed in physical contact with each other as they pass the light source, and continuous optical printing, in which the light source is used to 'project' an image of the moving source film through a lens onto the surface of the receiving film. In *step* printing, each individual frame is held stationary for the duration of an exposure and the film is then advanced intermittently as in a camera or projector. Nowadays both step contact and step optical printing are used, though continuous contact printing was used exclusively in the period between 1889 (the invention of film) and 1895 (the first successful demonstration of an intermittent mechanism). Optical printing was not used on any significant scale until the 1910s, though the early British film pioneer Cecil Hepworth used a modified projector and camera to produce trick special effects around the turn of the century. As optical printing produces a higher contrast, less sharp duplicate than contact printing (because the lens introduces imperfections in the way it refracts the flow of light), it is only used for film duplication which cannot be achieved by contact printing. The three main uses of contact printing are enlargement and reduction between gauges (e.g. making a 35mm print from a 16mm negative), copying damaged originals, usually for archival preservation, and introducing special visual effects. During the first two decades in which moving image film was used as a mass-medium there was only one format in widespread use (35mm), and all the film in existence was relatively new and undamaged. Special effects were generally produced in front of the camera rather than by manipulating the image during the process of duplication.

The basic technology needed to carry out continuous and step contact printing was developed in the early 1890s. W. K. L. Dickson designed a continuous printer which he used for striking release prints for the Kinetoscope parlours.[23] Step printing began with the Cinématogràphe used by the Lumière brothers for their public film show in December 1895. It was in fact an integrated device which performed all three of the mechanical and optical functions needed to use photographic film as a mass-medium: it was a combined camera, printer and projector. Its printing functions were carried out by the same film transport mechanism used for projection, which advanced the film intermittently. Purpose-built step printers started appearing in 1896, and were thereafter the most widely used method of film duplication for the next two decades. The fact that both the source and destination film stocks were held stationary during exposure ensured more accurate registration, thus producing a sharper copy than any of the continuous printers available at the time could achieve: as Barry Salt notes, 'it was realised almost immediately [in 1895] that the only type of mechanism that gave good registration between the positive and negative was that with an intermittent claw pull-down'.[24]

Together with the basic chemistry of black-and-white film processing, the technique of step contact printing represented the effective limit of what could be achieved in a moving image film laboratory until World War One. Film stock and laboratory technology were refined slightly during the 1900s. With the introduction by the French Gaumont company of continuous processing machines in 1907, the first viable alternative was available to the method devised by Dickson in 1899, in which lengths of exposed film were wound on a cylindrical wooden frame which was transferred manually between chemical baths. The Gaumont system, the basic principle of which is used in processing machines today, transported the exposed film stock through a sequence of steel tubes which directed it through immersion in developer and fixer sequentially. In modern processing machines non-abrasive recessed rollers have replaced the tubes in order to minimise the risk of introducing dirt or scratching while the film is being processed, of which a typical unit contains several hundred. In 1911 Bell and Howell introduced a continuous contact printer that offered vastly superior film registration and consistency of exposure than could be obtained from any previous design. Continuous contact printing would eventually become the mainstay of release print production, because it offers the image quality advantage of step contact printing but can run at much higher speeds than a step printer (modern continuous printers are capable of running at speeds up to 1,000 feet per minute). The need to produce many hundreds of release prints of a given title did not become systematic until the vertical integration of the global film industry, which began to take shape in the aftermath of World War One. This economic model, in which the same company owned studios (the means of production), distribution infrastructure and cinemas, required individual films to be shown in vast numbers of cinemas. This in turn necessitated the technology to produce hundreds, and sometimes thousands, of release prints from a single original camera negative. Continuous printing and developing was one of the ways in which this was accomplished, but it was soon found to be impossible to strike more than a few tens of prints by contact printing from a camera negative. An early demonstration of this came with Cecil Hepworth's 1905 film *Rescued by Rover*. This was a five-minute chase narrative in which a baby is kidnapped and then rescued by her family's dog. It proved so popular that the cut and spliced camera negative was, in Hepworth's words, 'worn out' through repeated continuous step printing to the point at which it was no longer possible to strike further prints. Hepworth's solution was simple: he

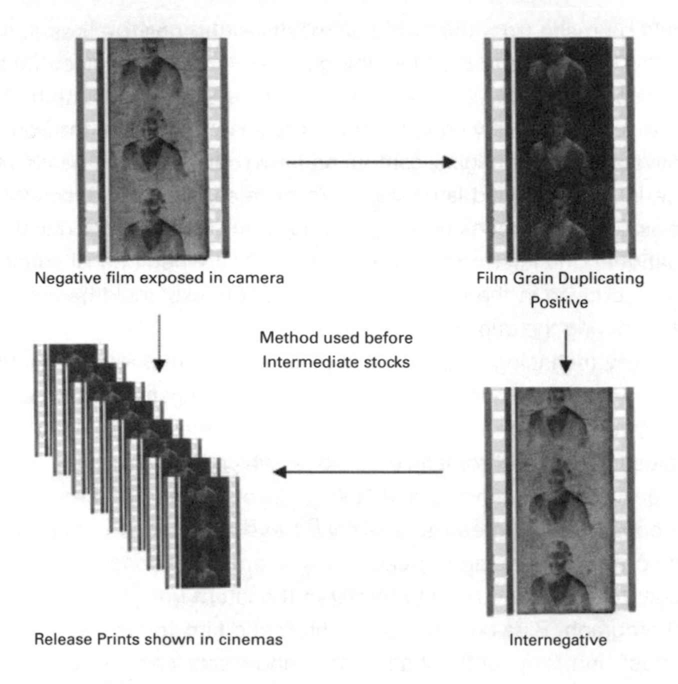

Fig. 1.3 The basic stages in film duplications

reassembled the actors and reshot the film – twice. In the final version, the 'baby' has quite clearly outgrown the pram from which she is kidnapped![25] For the mass copying of films, therefore, Hepworth had discovered that cellulose nitrate would only withstand repeated mechanical pressure up to a given point before perforation damage and contamination of the film surface made it unprintable. The solution was the introduction of intermediate elements.

Like optical printing, widespread use of intermediate elements, or as it was called at the time, 'duplicating stock', did not take place until the 1920s, although it is likely that intermediate printing using standard negative and print stocks took place much earlier for films where large numbers of release prints were required. Separate emulsions intended for negative and positive printing were marketed by Eastman Kodak and the Blair company in the UK from 1895; the positive stocks were extremely fine-grain, low-speed (unlike with camera stocks, speed did not matter in film intended for step printing, which could simply be exposed for as long as was needed) and with a slightly higher contrast and density under uniform exposure than most negative stocks available. This was in order to minimise the duplication of grain from the original. The principle was further improved with the introduction of Eastman Cine-Positive stock in 1916, which with minor modification remained the standard release print stock used by Western film industries for over two decades. Eastman Kodak introduced the first purpose-designed duplicating film in 1926. This was a very slow, fine-grain emulsion which was used to produce a contact duplicate from an original camera negative with very little loss of quality. A further duplicate negative would be made from the fine grain positive (the positive was sometimes called a 'lavender print', due to the blue-tinted base used to reduce contrast), and it was from this second negative that release prints were struck. This method offered several advantages over simply contact-printing release prints straight from the cut camera negative. Firstly, the original camera negative only had to be passed through a printer once in order to yield large numbers of release prints. Secondly, optical effects such as dissolves, fades and superimposition could be introduced by optical printing between the intermediate generations. And when optical sound-on-film started to be introduced in the late 1920s, the soundtracks could be synchronised with the picture during the duplication process.

The other key technological advance in black-and-white photochemistry which took place during the nitrate period was the introduction of panchromatic emulsions. The earliest film emulsions to be used for moving image photography were only sensitive to blue light (orthochromatic emulsions, which were sensitive to both blue and green, were introduced in the mid-1910s). In other words, areas of a photograph which appear red or green to the naked eye (or shades which have a red and/or green component) would not cause any chemical changes to the emulsion during exposure, and thus would not register in the latent image or be visible in the developed photograph. Blue-only and orthochromatic film tends therefore to have a higher contrast and less subtle shades of shadow and contrast than the black-and-white photographic images we are used to nowadays. Two particular problems which cinematographers in the 1900s and 1910s frequently raised were that clouds were usually invisible – skies tended to reproduce as uniform shades of grey – and the hair of blonde actresses appeared a lot darker than in real life. The latter became a particular problem during the early 1920s due to the fashion of using hydrogen peroxide (bleach) to lighten the appearance of women's hair, as the hair of a 'peroxide blonde' film star would frequently turn out to be less than flattering in the final print! The reproduction of light by a blue-only emulsion can be seen to the naked eye by looking through a blue-tinted piece of transparent material, a practice that was frequently used by cinematographers on location.

As is becoming increasingly apparent, a pattern emerges whereby a siginificant technological advance, be it optical printing, continuous developing, or as we shall see in subsequent chapters sound, colour and widescreen, tends to happen in two stages: the research and development which makes the process technically viable, and the changes to economic and industrial practice which enables its widespread commercial use. The introduction of panchromatic emulsions – ones which are uniformly sensitive to the entire red, blue and green areas of the colour spectrum – is another example of this, one which happened gradually over a 15-year time scale. C. A. Kenneth Mees notes that there was 'no great difficulty' in sensitising emulsion panchromatically right from the outset, but that difficulties in its manufacture and use initially inhibited the technology.[26] Unprocessed panchromatic stock has to be handled in total darkness, whereas blue-only and orthochromatic can be handled by cinematographers and laboratory technicians under a red 'safelight' to which the emulsion is insensitive. The earliest known use of panchromatic emulsions for moving image use was by Léon Gaumont in France, in the short-lived additive three-strip 'Chronochrome' colour process (see chapter 3) first marketed in 1908. Eastman Kodak supplied panchromatic film on an experimental basis from 1913, and following the successful production of what is believed to be the first panchromatic feature, the Will Rogers horror spoof *The Headless Horseman* (1922, dir. Edward D. Venturini), introduced it as a regular product the following year.[27]

The widespread introduction of panchromatic film into industry use did not happen for a further five years. The stock itself was much more expensive than orthochromatic and laboratories, which had to process it in total darkness, also charged more. Intermediate and release prints could continue to be made (and black-and-white prints still are made) on blue-only stock, as the image on a panchromatic negative is of course black-and-white (i.e. it does not contain any red or green which a blue-only monochromatic stock could not reproduce) and the light source used in the printer is of a uniform colour temperature, with the only adjustment being to its intensity. The eventual catalyst was the launch of a much cheaper panchromatic stock by Eastman Kodak in 1926, another panchromatic stock by Kodak's main US rival, Du Pont, in 1928, and a systematic, planned change in the lighting technology and practices used by the major Hollywood studios. The method of lighting used in film studios at the time was either carbon-arc or mercury vapour discharge lamps. In carbon-arc lighting, illumination is produced by burning carbon by passing a low-volt-age, high-current electrical signal across a gap between two carbon rods. Carbon-arc illumination was also used extensively in cinema projectors, and as such is discussed further in chapter 5. Mercury vapour discharge lighting ignited a gas to similar effect, and required the use of two highly inflammable substances. Both were especially suited to orthochromatic film, as the colour temperature of the light they produced was concentrated in those areas of the visible spectrum to which orthochromatic emulsions were most sensitive. Exposed under artificial light, however, panchromatic film required a form of illumination that gave a more even output across the entire range of colour temperatures in the visible spectrum to produce an even exposure. The solution came in the form of tungsten incandescent lighting. This consists of a conductive filament encased in an airtight glass bulb filled with a gas (tungsten at first, later halogen) which emits light at a high temperature. A high-voltage, low-current charge is passed through the filament in order to illuminate the gas. The resulting light output was far more even across the visible spectrum than either arc or mercury discharge lighting, and also offered several further advantages. It was powered by a conventional mains supply, whereas arc and discharge lighting needed rectifiers to produce the type of current required. Running costs were cheaper and studios lit by incandescent bulbs needed far fewer technicians to operate them. A series of organised tests took place at the Warner studios early in 1928 in order to determine the incandescent lighting requirements of panchromatic stock, its sensitivity and exposure characteristics and the likely cost implications of a wholesale conversion. It was found that the savings achieved with the new lighting technology more than offset the increased cost of panchromatic film stock, so much so that nine major studios expressed the belief that incandescent lighting had reduced their maintenance and electricity costs by half.[28]

Thus panchromatic film, tungsten lighting and synchronised sound all converged in 1928–32 to become key technologies in studio production practice. By the end of the decade Kodak, Du Pont and the German Aktien-Gelleschaft für Analin Fabrikation (Agfa) company were all marketing panchromatic film, which became the standard for studio and location use. Thereafter orthochromatic emulsions gradually dropped out of use for motion picture imaging. There were few significant developments in nitrate film base and black-and-white emulsions apart from the introduction of stocks intended specifically for optical sound recording (see chapter four) and for Technicolor separations (chapter three), though 1938 saw the introduction by Kodak of a faster camera negative stock. The next major technology which would bring about significant industrial change was the introduction of cellulose triacetate 'safety' film base in 1948.

As has been noted above, one major problem with cellulose nitrate was its inflammability. Once ignited nitrate film burns fiercely, generates highly toxic fumes and cannot be extinguished. This is because the combustion process itself generates oxygen, which makes the process autocatalytic. Even a roll of nitrate film immersed in water will continue to burn, and the only effective safety procedure with a nitrate fire is to contain it and evacuate everyone from the immediate vicinity. This unfortunate property of the film base necessitated elaborate health and safety precautions wherever it was used, ones which had a wide-ranging impact on virtually every aspect of film industry activity during the 1889–1948 period. In the US, a series of highly publicised nitrate fires during the 1900s and 1910s led to the gradual introduction of state legislation which determined the necessary fire precautions in studios, cinemas, labs and distributors' depots (or 'exchanges'). In Britain a similar process took place about five years earlier, although the dangers of nitrate had been well known ever since a fire at a charity bazaar in a temporarily constructed venue in Paris had killed 121 theatre-goers on 4 May 1897.[29] As most early film exhibition took place in music halls, fairgrounds and later in halls hastily converted from other premises such as shops which were known as 'penny gaffs', the projector was usually positioned in the middle of the auditorium rather than in a projection box separated by a partition wall, as it is today. This meant, of course, that if the film did ignite the consequences were likely to be serious. After a number of fires which resulted in deaths and serious injuries, mainly from smoke inhalation, the 1909 Cinematograph Act was passed. It stated that nitrate films could only be shown to the paying public on premises which had been licensed by the local authority as fit for the purpose. One interesting footnote to this legislation is that it established, almost by accident, the legal basis for film censor-

Fig. 1.4 A nitrate release print (in this case from a government propaganda film made during the Second World War). The words 'nitrate film' can be seen in inverse on the left-hand edge, thus identifying the film as inflammable to anyone who is handling it.

ship in the UK which persists to this day. Though it was primarily intended to protect the public against nitrate fires, the 1909 act did not explicitly limit the grounds for refusing a licence to the sole criterion of health and safety. Councils immediately started imposing other, unrelated conditions on their cinema licences, including restrictions on the content of films that were shown. As a result of hundreds of different authorities making separate censorship decisions, the industry established the British Board of Film Censors (BBFC) in 1912 in an attempt at standardisation, which now effectively makes decisions on the authorities' behalf. Even now the BBFC's classification decisions carry no legal weight in themselves, but are advisory for the local authorities who issue individual licences to cinemas.

However, although these precautions and film-handling techniques gradually reduced and managed the risks posed by nitrate, they never entirely disappeared. Britain's worst film fire (defined as numbers killed) happened as late as 31 December 1929, when 69 people, almost all children, were killed and over 150 injured when a reel of nitrate caught fire during a children's matinee at the Glen Cinema, Paisley, Scotland.[30]

An obvious way of circumventing the need for the costly and restrictive health and safety requirements of nitrate would be to develop a non-inflammable base. As with panchromatic film and so many other moving image technologies, these were actually available decades before their widespread introduction. The reasons why nitrate remained in use for so long will also have a ring of familiarity, too. Early 'safety film', i.e. bases which do not have the uniquely dangerous combustion characteristics of nitrate, took two forms: those in which the nitrate base itself was modified to inhibit its combustibility, and bases in which acetic acid as distinct from nitric acid was used to dissolve the cellulose used to produce an inherently less combustible base material. Of the former category, W. C. Parkin in France developed a method of inhibiting the flammability of nitrate by adding a metallic sodium compound to the base during casting, which he patented in 1904. Various methods were invented along these lines of making nitrate more difficult to ignite, though if it did ignite it burnt just as fiercely as untreated stock. None of these techniques was ever reliable enough to convince any authorities to allow the treated stocks to be used in an unregulated way. The first mass-manufactured cellulose acetate base was marketed by Kodak in 1909, with Agfa following suit in November of that year.[31] It was intended for amateur use in the 28mm format, but there were significant difficulties in using the stock professionally. Early generations of acetate stock were far more fragile and dimensionally unstable than nitrate, because they retained a higher proportion of solvent in the base than nitrate. This tended to evaporate quite quickly in the weeks and months following manufacture, resulting in the base shrinking and becoming brittle. Acetate was more likely to tear and suffer perforation damage in the sprocket teeth of a camera or projector than nitrate, and the base was also a lot more prone to scratching. As with panchromatic emulsions, early acetate bases were also a lot more expensive to manufacture than nitrate and it took several decades for the film stock manufacturers to convince their customers that the additional costs of acetate offset the health and safety costs associated with nitrate. Brian Winston even goes as far as to suggest a conspiracy theory to explain the continued use of nitrate:

> It seems to me reasonable to suggest that the industry basically clung to an extremely dangerous substance long after an alternative was available as much as a protection against competition as an earnest of its commitment to top-quality images. After all, only professionals could handle real films; only professionals could be, quite literally, licensed as projectionists and cinema managers, creating thereby at the most potentially dispersed end of the industry, exhibition, a coherent, identifiable and controllable element. The business protection that this provided was worth the odd projection booth conflagration.[32]

Though elegant, this suggestion fails to account for why safety film was adopted by the industry following the introduction of an acetate base which offered comparable mechanical qualities to nitrate in 1948. On a more general level, the introduction of safety film followed the familiar industry pattern of separating the research and development and economic infrastructure phases of each major technological development.

During the period between 1909 and 1948 acetate bases were used mainly for applications in which nitrate fire precautions were not available. The largest by far was amateur film. George Eastman himself expressed the belief that nitrate should never be used in amateur technologies: on 4 June 1912 he wrote to Edison stating that he would only supply acetate base for use in the latter's 'Home Kinetoscope' projector, as, 'in our opinion, the furnishing of cellulose nitrate for such a purpose would be wholly indefensible and reprehensible'.[33] Thus when Kodak launched the first cameras, projectors and reversal stock in the 16mm format in 1923, the base was exclusively acetate. There were other, limited moving image uses for safety film before 1948. These included short cinema prints (such as advertisements and trailers) which did not need to withstand heavy-duty projection and which, unlike nitrate, could be sent through the normal post; prints for showing in temporary venues which did not have a separate projection room; and negative stocks for cinematography in situations where the risk of a nitrate fire was deemed unacceptable, e.g. medical photography and on board an aircraft. In Britain, prints of politically controversial films made on safety base were sometimes used in attempts to circumvent censorship (the 1909 act only applied to nitrate), and acetate bases were also used extensively for still photography. The mechanical properties of acetate gradually improved during the 1920s and 1930s, culminating with the introduction of cellulose acetate propionate in 1938, which was described as being 'midway between cellulose nitrate and the former acetate' in terms of its strength characteristics.[34] Even as late as World War Two, however, users of safety film still considered its fragility to be a major problem. By this stage 16mm had grown from a home-movie format to a professional medium that was used extensively for location actuality footage where the size and weight

Fig. 1.5 A 16mm triacetate release print. As with the nitrate example shown in 1.4, this element also contains an edge marking, only this time to indicate that it is not nitrate.

TECHNICAL INFORMATION

for Projectionists

Published by **G.B-KALEE, LTD.** **No. 6**

PROBLEMS of SAFETY STOCK

By **R. Howard Cricks.,** *F.B.K.S., F.R.P.S.*

Projectionists who show the Festival of Britain film, " The Magic Box," will find many of the sequences of technical interest. Among the earlier problems of William Friese-Greene in the invention of kinematography was to provide a material suitable for coating his emulsion upon, and for running through his camera. In one sequence we see him, amid noxious fumes, producing thin strips, 6 ins. wide, from large blocks of celluloid.

Celluloid is in many respects a very suitable substance for use as a film base; but several tragic fires in the early days of the motion picture drew attention to its dangerous inflammability, and for half a century efforts have been made to find a non-inflammable substitute having acceptable mechanical properties.

The raw material of celluloid is cotton, and recent research has shown why it is so suitable : nature has in fact been doing for millions of years what science has only recently succeeded in achieving—in building matter into long molecules (some large enough to be photographed in the electron microscope) which are tangled together and produce a material of high tensile strength.

THE FIRST PLASTIC

Celluloid was in fact the first plastic, and it is only within the last few years that scientists have succeeded in producing synthetic plastics having similar properties —and so far none of these synthetic plastics has properties equal to those produced by nature.

In the manufacture of celluloid —or cellulose nitrate, as it is called chemically—the cotton is reacted with nitric acid. Now this process is very similar to the process by which gun-cotton is produced, and the fact that it results in a highly inflammable material is not therefore to be wondered at.

Other acids may be used. For many years we have had so-called acetate film, made by substituting acetic acid for nitric; other acids that can be used are propionic and butyric. But none of the resultant materials had formerly characteristics equal to those produced by nitric acid; on the other hand, the results were comparatively non-inflammable.

CHARACTERISTICS OF BASE

Kodak safety base marketed since 1937 has been produced by a mixture of acetic and propionic acids. The Gevaert stock is a butyrate.

The characteristics of the base depend also upon the degree of chemical reaction permitted. While early safety base was known as a di-acetate, the new Kodak safety base is known as a tri-acetate.

Note that we describe the new base as ' safety ' rather than ' non-inflammable.' Safety base is in fact about as inflammable

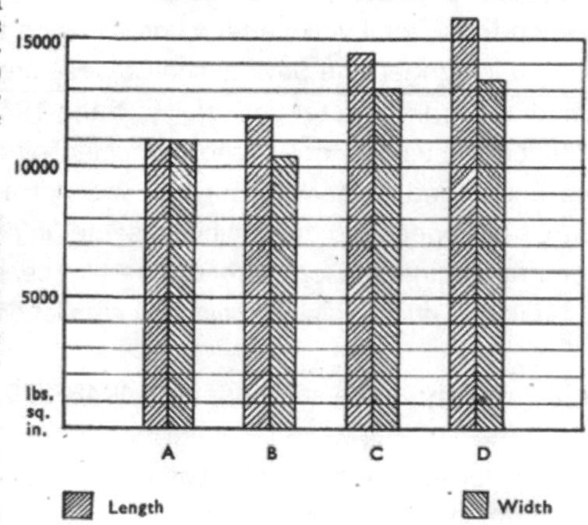

Fig. 1. *Tensile Strength of Kodak Film Stocks.* (**a**) *Acetate Base prior to* 1937. (**b**) *Acetate Propionate Base after* 1937. (**c**) *Tri-acetate Base.* (**d**) *Nitrate Base.*

Fig. 1.6 Technical information leaflet issued to projectionists in 1951. It warned cinema staff not to relax the health and safety precautions associated with nitrate while any of this highly inflammable film base remained in circulation.

of 35mm equipment was prohibitive, and there were frequent complaints about the unreliability of the exclusively acetate 16mm stock. Writing in April 1945, an armed forces cameraman argued that 'professionals have long been hampered by the shortcomings of this slow-burning base', noting shrinking, brittleness, jamming in camera mechanisms and extreme sensitivity to temperature and humidity extremes among the problems. 'There is only one possible solution', he believed, 'and that is the use of nitrate base.'[35]

The eventual catalyst for a wholesale conversion came when Kodak announced at the 1948 SMPE conference on 17 May 1948 that it had developed a new 'high acetyl' cellulose triacetate base which was significantly stronger than acetate propionate and almost as strong as nitrate. There is evidence to suggest that the new base was at least in part the result of work done on equipment and research findings captured from the Agfa laboratories in Germany and Czechoslovakia following the fall of the Nazi regime in 1945, because immediately before Kodak announced their two-year research programme which led to the launch of 'high acetyl' triacetate the US authorities had systematically blocked the British government's attempt to remove film base casting plant from Germany, apparently at the behest of Eastman Kodak.[36] The Nazi film industry certainly used acetate base 35mm stock far more extensively than was the case in Allied countries towards the end of the war, due mainly to the difficulties of transporting nitrate using a road and railway system under constant Allied attack, and also to facilitate the large number of screenings for armed forces in temporary locations that were ordered by the Propaganda Ministry to boost morale. Kodak started manufacturing the new acetate base in place of nitrate negative in the autumn of 1948 and ceased manufacturing nitrate in February 1950. Du Pont continued to produce nitrate for a further year and small quantities of nitrate remained in circulation for several years afterwards. Outside the West the production of nitrate went on a lot longer: the Soviets and Chinese are believed to have continued using it (including 16mm nitrate) until well into the 1960s, while Japan's main film stock manufacturer, Fuji, did not cease producing nitrate until 1958.[37] The relaxation of nitrate health and safety precautions was not as straightforward as some in the industry had hoped, and the exhibition sector in particular had to maintain fire safety equipment on cinemas long after nitrate film ceased being produced on account of back catalogue titles which remained in circulation.

Film technology in the safety period: 1948–2000

The main developments in film base technology during this period were gradual refinements of the triacetate base to improve mechanical strength and stability, and the introduction of polyester base stock. Polyester – polyethylene teraphalate – is an inorganic polymer, which, unlike all cellulose film bases, is not liable to dimensional change according to its varying moisture content. It is also far stronger then any other material used to produce film base and will not break, even under extreme pressure. The first polyester base was announced by Du Pont under the trade name Mylar in 1955,[38] though, as with acetate, it was not used on any significant scale as a base for moving image products for over three decades afterwards. Polyester offered significant advantages over nitrate and acetate: like acetate it was no more inflammable than paper, it had a far higher tensile strength and the surface was less prone to scratching. There were also two significant disadvantages. The strength of the new stock turned out to be as much of a curse as a blessing, since the fact that polyester is significantly thinner than acetate and prone to holding a static charge led to instances of it jamming in the mechanisms of cameras, printers and projectors. Whereas acetate film would simply break in these circumstances, the pressure on a jammed section of polyester can be transmitted to components in the film path and cause serious damage to equipment. Polyester also cannot be joined using cement (which works by breaking down the organic solvents used in nitrate and acetate), and can only be spliced using adhesive tape or ultrasonically. Heat-resistant adhesive tape, which is usually applied using a small machine that automatically cuts and perforates it during application, is used mainly on release prints in cinemas, where the momentary appearance of the tape splice on the screen is not considered a major issue. This is not considered suitable for use in production and post-produc-tion, where film is joined using an ultrasonic splicer, a device which softens the base by applying localised, intense heat and then fusing the two surfaces by means of ultrasonic energy. Ultrasonic splicers are a lot more expensive than tape or cement splicers, which is why they are usually only found in laboratories, archives or in the post-production departments of major studios.

Again, as with acetate, the amateur film market was used as a test-bed for the introduction of polyester. The earliest widespread use of polyester base stock was probably in a variant of the Super 8 home movie format (see chapter 2) which was known as Single 8 and marketed by Fuji from the late 1960s. Additional tensile strength was deemed necessary because the film was supplied in the form of self-loading cartridges (to make it easier for amateur filmmakers to load and unload film stock), which increased the mechanical stresses of film transport. Furthermore the perforations were extremely small, and thus needed extra strength and dimensional stability to withstand repeated engagement by the sprocket teeth of a projector. It was anticipated that, being an amateur medium, most consumers would be unlikely to want to edit their footage, whereas the added resistance of polyester to dirt and scratching would prove a major selling point for a medium that was likely to be subjected to repeated projection by untrained amateurs. Following these initial trials, polyester was also used in the Super 8 and 16mm prints of feature films shown on airliners, the rationale being similar (ability to withstand repeated handling by untrained operators). Polyester did not come into mainstream use for cinema exhibition until the 1990s. Apart from these applications its only widespread uses before that point were for X-ray plates and document microfilming. A survey of US cinema operators sponsored by Agfa-Gevaert in 1979 revealed widespread objection to its introduction as a release print medium, and that fear of equipment damage was 'the main source of resistance to polyester'.[39] During the mid- to late 1980s Agfa began an aggressive marketing campaign for its 35mm polyester release print stocks. In 1992 the National Association of Theatre Owners' (NATO) technology committee recommended that release prints be made on polyester film, in response to which Kodak announced on 9 May 1996 that it was investing $200 million in a new polyester film manufacturing plant at its New York headquarters.[40] This marked the start of the post-production and exhibition sector's conversion to polyester: in 1998 the new facility came on-stream and started producing the new Kodak 'Vision' print film (using the trade name 'Estar' to describe the base), type 2383. By 2000 almost all intermediate elements and release prints were being made on polyester. Camera stocks remain on triacetate, however, due to the potential for costly damage and lost production time (for example, if only one camera is available at a remote location) in the event of a polyester film jam.

The Eastman Kodak company has been cited extensively throughout this chapter. George Eastman was almost certainly the first to sell flexible, transparent film base sensitised with photographic emulsion for moving image use, and the company he founded has maintained the world's largest market share in this product throughout the history of its use covered by this book. By 1992 it was estimated that Kodak accounted for 75 per cent of film stock sales in the US domestic market.[41] Other significant film base manufacturers include the US chemical giant Du Pont, founded by the French scientist Eleuthère Iréneé du Pont in 1802 to manufacture gunpowder. In February 1925 it started producing nitrate film base at a factory in Parlin, New Jersey, and quickly established itself as the second largest supplier of stock in the US. In the 1950s Du Pont turned its attention to polyester film bases and became the market leader in this sector until Kodak's entry in the 1990s.

In the UK the firm of Ilford Ltd. (named after the east London suburb in which its main factory was located), was founded by Alfred Hugh Harman in 1879 and originally manufactured dry plates for still photography. Following the acquisition of a smaller company which operated a cellulose casting plant in 1895 it began manufacturing film, and sold its first length of 35mm for moving image use – 291 feet – to Birt Acres in the following year. Production-line manufacturing of 35mm negative and release print stock began in 1912 under the trade name 'Selo', but temporarily stopped during World War One because the operation was being run by two German expatriate engineers who were taken prisoners of war. Film manufacture at Ilford recommenced in 1920, but in 1923 the company abandoned its base-casting operation and thereafter imported 'raw' (i.e. unsensitised) cellulose nitrate base from the US. During the 1930s Ilford invested in the Dufaycolor additive process (see chapter three), which became one of the UK's leading colour film stocks for amateur use until 1951, and in 1943 launched the product for which it was best known in the post-war years: the HP3 moving image negative stock. This had a speed equivalent to EI250 in today's terms, which made it one of the fastest black-and-white stocks available and had a significant impact on newsreel and documentary production.[42] Ilford gradually concentrated more on the still photography market during the second half of the twentieth century, and in 2004 became the first major film manufacturer to go bust. The company went into receivership on 24 August, blaming the growth of digital imaging for an unsustainable decline in sales of film stock.

In Europe the pioneering firms of Pathé and Gaumont phased out their film stock manufacturing operations as other sources emerged during the 1900s and 10s, notably imported stock from Kodak in the US. The Gevaert Company in Belgium and Agfa in Germany (which had began production of nitrate in 1913) emerged during the 1920s as the major European suppliers. An Agfa factory at Wolfen in East Germany was taken over by the Soviets at the end of World War Two, and after two decades producing film stock under the name of VEB Filmfabrik AGFA, was renamed under East German ownership as OrWo (Original Wolfen) in 1964. OrWo became one of the main suppliers of black-and-white stock to the Soviet Union and Eastern Bloc countries until the collapse of communism. Due to the low cost of OrWo stock compared to Western equivalents, its output has traditionally sold well in India and elsewhere in Asia, and continues to do so under private ownership since 1998. In Japan, the Fuji company was founded in 1934 sensitising imported film base, but became the first Asian manufacturer of raw stock in 1936 and has been the main supplier to the Japanese market ever since. It started selling colour negative and print stock in the US in 1973, and since then Fuji's products have sold to a small but consistent niche market in the West.

Film technology in the digital period: 1992–2005 and beyond

Without doubt, film has proven to be the longest-lasting and most adaptable of any medium yet devised for storing and reproducing a series of photographs as a moving image sequence. The example of format obsolescence given at the start of this chapter describes the opposite extreme: an electronic medium consisting of a complex and unique combination of hardware and software, which became obsolete within 15 years of going on the market. Despite ever-increasing speculation that film will imminently be succeeded by computer-based alternatives it is still, at the start of the twenty-first century, the carrier used almost exclusively for projection in cinemas and for the production of feature films intended primarily for exhibition in cinemas. Furthermore it is still used extensively in television production. Digital processes have started to make inroads in the areas of intermediate duplication and special effects, but for the vast majority of theatrical features, initial cinematography and final output is still accomplished using a medium based on the three areas of technical knowledge which were converged by W. K. L. Dickson in 1889.

The key reason film has survived so long lies in the interface between the software and hardware used to create and reproduce film-based moving images. The essential functions of a camera, printer and projector remain unchanged from those of the 1890s. The optical precision of photographic lenses available has been developed and improved, as has the accuracy of film transport mechanisms and the mechanical stress exerted on the film in motion. The designs of shutters have improved to maximise the amount of light exposure during photography and projection, as has the quality and versatility of the film duplication process. But the essential mechanical properties and functions of film-based equipment have not; leaving aside the issues of nitrate inflammability (see above) and chemical decomposition (which will be discussed in chapter eight), unexposed film stock produced by the Lumières could be used in a modern 35mm studio camera and a finished Lumière film could easily be shown using the projector in a modern multiplex with only minor modifications. But that is not to say that the capabilities of film imaging have not changed, or as the advocates of digital imaging would argue, that we remain stuck in a time warp, saddled by the constraints of what is essentially a Victorian technology. The wheel is a stone-age technology, but it continues to be used extensively in the twenty-first century, because with minor modifications and additions to the basic principle (such as the addition of pneumatic tyres) it can easily be adapted for use in modern vehicles which are reliable, simple, cheap to maintain and easy to operate. The airship, on the other hand, was abandoned after barely thirty years as a form of public transport. This was because it was quickly discovered that airships could only fly in almost perfect weather, were very slow, depended on complex and unreliable control systems, needed almost as many highly-paid crew members as they could carry passengers and had an unfortunate habit of crashing and blowing up.

The same analogy can be used to compare film to the computer-based alternatives which are vying to replace it. The optical and mechanical devices that record on and reproduce from film are simple, effective, reliable and, to their users, a long-term capital investment. The research and development which led to improved image quality, versatility and economy of use are delivered through the medium itself – the film base and emulsion. A Lumière film and a modern Hollywood blockbuster can both be shown using the same projector, but that does not mean that the picture quality of the latter will be no better than the former. The polyester film base used in 2005 is still compatible with equipment made during the nitrate period, and even with such equipment will deliver a more stable picture, is less susceptible to visual defects such as dirt and scratching, is less likely to break and will withstand more intensive use than either acetate or nitrate. The emulsion in a 2005 release print will yield a finer grain structure, be more uniformly sensitised across the visible light spectrum and will enable much larger pictures to be projected from it, while its equivalent camera and duplicating stocks will be faster and able to withstand more printing across a greater number of generations than its predecessors. Successive generations of film technology have had inherent defects: the extreme flammability of nitrate, the fragility of acetate and the tensile strength of polyester risking equipment damage are but three examples. But the industry has consistently found that the cost of managing the use of film in order to minimise the impact of these defects has been more than offset by the long-term reliability and versatility of the medium, and improvements which address specific defects are introduced as and when it is economically viable to do so. This is not always at the point of invention, as the conversion from nitrate to acetate base illustrates.

This unique combination of the highest-quality, most reliable moving image medium yet invented and the fact that so much of the technological advance is delivered through the software rather than the hardware has also meant that film is, by a long way, the most expensive moving image medium yet invented. But the up-front costs of film stock and processing have proven to be small for those who require a high-quality, long-lasting carrier, in comparison to the equipment costs and rapid obsolescence of any of the alternatives. I have thus far been using the adjective 'digital' quite indiscriminately, and will seek to define its application to moving image technology more precisely in chapter eight. Since the launch by Sony of the first mass-marketed digital video format (i.e. a system which represents televisual images as digital data stored on magnetic tape), Digital Betacam, in 1992, speculation has been rife that 'digital' will signal the end of film. All digital imaging (and sound recording) uses computer technology to translate images which are visible to the naked eye into digital data and back again; therefore, the definition and quality of moving images produced digitally is essentially a function of the speed at which a computer can carry out this processing function relative to the volume of data it can store. The speed with which computers can crunch numbers and the volume of numbers they can store and manipulate is, and throughout the history of computer technology always has been, constantly expanding. Applying this principle to moving images, this means that a personal computer bought this year is capable of reproducing a sharper and more detailed moving image than one bought last year, just as a modern camera negative stock improves on the emulsion formulation it replaced. But a new generation of film stock can deliver improved performance when exposed in the same camera as its predecessor. A new PC is just that – a whole new computer, which requires the user to write off the capital investment in its predecessor. The guts of digital imaging lie in its hardware, whereas with film the technology is contained within the medium itself. £300 for a ten-minute roll of unexposed film stock (which can be expected to last for hundreds of years after processing when stored in appropriate conditions) might seem expensive compared to £30 for a one-hour broadcast standard videotape (which will probably be unplayable within two decades), but when the machine needed to record on and play the tape costs £40,000 and has a useful working life of three years, it becomes clear that the economies of scale with film are fundamentally different. A studio camera or cinema projector costs a similar amount and will last for decades, needing only minimal maintenance. Furthermore as film is an analogue, human-readable format there is little danger of it becoming obsolete as the Domesday Disc did. A producer making a substantial investment in originating a theatrical feature or major television programme on film can be reasonably confident that he will still have access to the footage in years to come. Like our airship, the hardware needed to read digital data from a magnetic or optical carrier is mechanically and electronically complex and intricate, is specific to an individual format, requires additional, specific hardware and software to drive it and most data storage formats become obsolete in less than a decade.

In 2000 computer technology was just about fast enough to enable the capture, manipulation and output of moving image content to a definition approaching that of lower-resolution film stocks. By 'just about', I mean that these functions represented the absolute limit of what the technology was capable of delivering, and that using it was hugely expensive because the commercial costs of this technology reflected the desire of the companies which produced it to recoup the enormous research and development costs. At the time of writing, digital video photography is used extensively for television, but in the cinema industry its use is confined mainly to the processing of special effects. It may well be that as processor speeds increase and the capital costs of film reduce, digital imaging technologies will become more cost effective overall than film. But even that cautious speculation also assumes that film technology will cease to improve (whereas in fact new generations of film stock have consistently outperformed their video and digital imaging equivalents in image quality, versatility, compatibility and longevity), and that sectors of the industry which currently operate on low capital equipment and high media costs (most notably cinema exhibition) will be able to find substantial additional funding to finance the conversion. It is likely, therefore, that film will continue to play a key role in the development and use of moving image technologies well into the twenty-first century.

chapter two | cinematography and film formats

Cinematography

> 'You know what wouldn't be bad? Ghosts. They come through the screen, fly over the heads of the audience ... ghosts that drip blood. I can see it all now: Bloodoscope ... Bleedorama!' – John Goodman in *Matinee* (1993, dir. Joe Dante)

Chapter one examined how the physical and chemical properties of the world's earliest and longest-lived moving image medium – photographic film – shaped the growth of an industry and culture which evolved around it. This chapter will expand on that overview to explore two specific areas of technology which are needed to produce moving image content on film. Cinematography, which can broadly be defined as the technology needed to expose photographic images (or 'frames') onto film which are intended to be reproduced as a moving image sequence, consists primarily of the cameras themselves and the peripheral technologies designed to facilitate their use (such as studio lighting). Film formats relate to the practice of standardisation, one which has proved to be a crucially important reason why film became the longest lived and most successful moving image medium. When Henry Ford famously declared that he would supply the Model 'T' 'in any colour, as long as it's black', he was pointing out that the private car could only be successful as a mass-produced consumer product if the cost of producing it was kept as low as possible relative to the disposable income of Ford's intended customers. The only way to achieve that was to develop a single design which could be manufactured many times over on a production line.

By the same logic, when a chief projectionist I once worked for expressed his complete disinterest in any of the cultural or artistic issues related to the films he was showing by declaring that 'as far as I'm concerned, it's this wide [holding up his right thumb and forefinger roughly 35 millimetres apart] and it goes through a projector', he unwittingly identified the key aspect of film technology which has made it so successful. The film he was referring to is 35 millimetres wide, has four evenly-spaced perforations alongside each frame and the dimensions and position of the frame and soundtrack(s) on the topography of the film's surface are defined according to published and universally accepted standards. That ensures that it will not only go through his projector, but almost *any* projector, anywhere in the world. But this has not always been the case. The film that we use today only became 'this wide' as a result of aggressive marketing, trade wars, political issues and other factors which had a greater influence on the history of media economics than many would readily admit.

When you consider that most sectors of the global film and television industries in general, and Hollywood in particular, depend for their economic survival on being able to sell their products in international markets, it becomes obvious that standardisation is not an obscure technical detail, but one of many instances in which technological issues have greatly influenced the historical development of moving image industry, economics and culture. Furthermore, they continue to do so.

This chapter, therefore, will address the historical development of two crucial areas of technology allied to film, but which in the main have evolved separately from the properties of the film base and emulsion itself, hence their omission from chapter one. It will offer a historical overview of the technology used for initially recording the moving image and for standardising the means of its distribution and exhibition, and suggest ways in which these processes of evolution impacted on the wider histories of moving image history and culture.

The moving image film camera

The film camera is essentially the combination of mechanical and optical technologies needed to expose still photographs, or 'frames' onto film in rapid succession in order that, after processing and printing, they can be reproduced as a moving image sequence. Professional motion picture cameras generally record the picture only, though more modern ones will include an electronic or mechanical system for interlocking the film movement with an external sound recording device. The camera, essentially, consists of five key components:

- THE FILM TRANSPORT MECHANISM, which contains two lightproof containers, or 'magazines' to house the unexposed film stock and to wind it up after exposure, plus the rollers and driven sprockets needed to feed the film through the gate.
- THE GATE, a metal plate containing an opening which corresponds to the dimensions of the exposed frame, against which the film is held stationary during exposure.
- THE INTERMITTENT MECHANISM, which advances the film through the gate between exposures. This usually consists of one or more pins which engage the perforations in the film stock, pull a predetermined length of film through the gate, retract and then return to the starting position.
- THE SHUTTER, usually a concentric metal disc with one or more protruding blades. The movement of the shutter is mechanically interlocked to that of the intermittent mechanism to ensure that a shutter blade is blocking the flow of light while the film is in motion.

• THE LENS, which consists of one or more ground and polished glass elements mounted in a cylindrical barrel. These refract, or alter the shape of, the light which passes through them in order to form the image to which emulsion of the film is exposed.

In the 110 years between the earliest known patent for a moving image camera using roll film (granted to William Friese-Greene and Mortimer Evans on 25 February 1890[1]) and the turn of the twenty-first century, all five of these components have undergone a continuous process of evolution and redesign. Raw materials, developments in manufacturing processes (especially for lenses) and the introduction of electronic and computer technology to control the mechanisms have also played a part in making the film camera of 2005 a vastly superior instrument to the ones used a century earlier. But, in a striking demonstration of the relative success of format standardisation, it would be possible to expose today's film stock in an 1890s camera and still produce a recognisable moving image.

Early camera technology

Although Friese-Greene's and Evans' patent was granted in 1890, there is no evidence that they ever got their camera to work. In general terms, there was a three to five year gap between the initial marketing of cellulose nitrate film by Kodak in 1889 and the beginnings of cinematography as we know it today. Whilst the inventors of the late nineteenth century realised that, in theory, film was the medium needed to record moving images, progress on the technology needed to make (and reproduce) the recording was slow. The engineers and scientists who had built either paper-roll cameras or glass-plate cameras with multiple lenses quickly realised that neither system was immediately adaptable to film. Possibly the earliest camera to be used in a systematic production schedule was Edison's Kinetograph, which was initially used to shoot films for showing in Kinetoscopes before the emergence of viable film projectors in 1895. It was large, heavy, could only be used in a fixed position and was driven by an electric motor. The intermittent mechanism was similar to that of a Maltese Cross design, which, ironically, would later become the standard form of intermittent mechanism used in most 35mm cinema projectors to this day (the different types of intermittent mechanism are covered in detail in chapter five).[2]

The other main camera design to emerge during the 1890s was the Lumière *Cinématographe*, in which a claw, mounted adjacent to the gate, was driven by two cams in order to produce the intermittent motion needed. The *Cinématographe* was in fact a camera, printer and projector in one unit, the change of function being accomplished by the addition of a light source and minor adjustments to the mechanism. Within the next two decades Maltese Cross mechanisms would become the intermittent unit of choice for projection, whilst claw mechanisms would become the standard for cameras, primarily because the latter produced a more accurate film movement but was less able to withstand the intense heat produced by the light sources used in cinema projection. Describing a Maltese Cross camera mechanism developed by the British pioneer R. W. Paul in 1898, Barry Salt observes that 'although it had the advantage of reversibility, it also had the disadvantage of poor registration, at least when compared to the Lumière mechanism'.[3] Throughout the late 1890s camera design and manufacture was essentially a small-scale cottage industry, with individual models being hand-built in small quantities for specific purposes. In the 1900s and 1910s their constituent technologies began to standardise, with the result that a small number of designs started to be mass-produced. Their features, capabilities and limitations would, when combined with the characteristics of the film exposed in them, prove a major influence on studio and location production technique throughout the silent period and beyond. Until the introduction of synchronised sound in the late 1920s, almost all cameras were driven by hand-cranking, with a handle that was directly linked by a system of gears to the sprockets, intermittent mechanism and take-up spindle. The only exceptions were cameras designed for amateur cinematography and some smaller models intended for newsreel use, which were driven by a clockwork motor that would enable takes lasting between ten and twenty seconds. Electric motors were virtually unheard of in cameras during this period because they were large, very heavy and required an external power supply. One of the few exceptions was the electrically-driven Edison Kinetograph, which restricted its use to the 'Black Maria' studio developed by Dickson and which was unable to be operated in remote locations.

By the mid-1910s, industry standards had been established for different categories of camera. The 'one-sprocket topical [newsreel] camera', for example:

> ... will usually have a first-class lens and should be capable of doing ordinary work, but will probably have a shutter with fixed aperture and will lack the inside focusing tube, film measurer and such-line accessories.[4]

This generation of camera, specifically designed for portability, was limited to a magazine capacity of 100 feet (just under 1¾ minutes continuous shooting at 16fps). The more advanced cameras designed for studio use during this period featured interchangeable lenses of different focal lengths, a larger magazine capacity (the largest capacities in general use were 400 feet – just over 8 minutes – rising to 1,000 feet – just under 17 minutes – by the late 1910s), a viewfinder and a footage counter. These models were not easily portable, however, and could only be used when mounted on a tripod. Among the models developed during the early period was the Pathé studio camera of 1903, which contained a claw-based intermittent mechanism similar to that of the Lumière *Cinématographe*, but which incorporated an idea patented by Woodville Latham in 1895: a loop of film between the continuously rotating sprockets and intermittent claw, in order to absorb the inertial shock of the film being pulled from and to the larger (and thus heavier) rolls of film needed for longer sequences of exposures, or 'takes'.[5] In 1905 the design was modified to incorporate external film magazines with a 400-foot capacity.[6] Another widely used camera was developed by James Williamson in Britain and manufactured from 1904. The magazines were enclosed, had a refined intermittent mechanism and featured a more accurate pull-down motion than its predecessors. In 1908 the French Debrie Parvo camera set the precedent for a series of small, portable designs intended specifically for newsreel and location use. Both 400-foot magazines and the mechanism were all built into a small box of around 21cm³, and other features included an adjustable shutter speed and a viewfinder which could be used during a shot. The Parvo could easily be operated while hand-held (i.e. without the need for any external support device, such as a tripod) in good lighting conditions.

Camera design and the rise of the studio system

The move towards standardised, mass-manufactured camera designs for studio use became firmly established in 1912 with the launch of the Bell and Howell model 2709 studio camera. This introduced a number of technical innovations which ensured that the 2709 would remain *the* standard studio camera used by Hollywood until the arrival of sound, almost two decades later. It also set an important industrial precedent, which was that of an external equipment manufacturing firm, with no direct financial links to the business of film production or exhibition, being responsible for introducing a key technical advance. This pattern – of equipment design and manufacture companies, separate from studios, distributors and cinemas, anticipating the film industry's needs and then developing technologies to meet them – would become the dominant means by which the global film industry adopted new technologies. Combined with technical and economic conflicts involved in the process of standardisation (covered in greater depth below), this pattern would establish the industrial/institutional model of technological development which held sway virtually throughout the twentieth century.

The idea of film technology being developed through a service industry distinct from the business of producing and exhibiting the creative content of the films themselves was the driving force behind the establishment of Bell and Howell in 1907. Of its founders, Donald J. Bell began working as a projectionist in Chicago in 1896. He soon recognised the difficulties and inefficiencies of operating and maintaining a large number of different projector designs, most of them the product of cottage industry design and manufactured in small quantities. Albert S. Howell was a recently-qualified mechanical engineer to whom Bell was introduced at the machine tool works where the latter was employed, and which fabricated spare projector parts individually and to order for Bell. The first mass-produced equipment marketed by Bell and Howell was a film perforating device in 1909; this rapidly became the industry standard, so much so that 'BH' shape perforations are still in common use for negative stock today. The 2709 camera, first sold in 1912, firmly established Bell and Howell on two counts: firstly, the camera introduced some technical features which marked a radical departure from any other 35mm studio camera commonly available at the time, and secondly it was manufactured on a production line basis, which increased efficiency and reduced operating costs through the economies of scale which apply to mass-production. Spare parts were readily available, and technical support was provided to studios in the form of service packages. The 2709 had a 1,000-foot magazine capacity, thus making it suitable for the longer takes needed by increasingly elaborate studio sequences. The gate mechanism and pin registration system on the 2709 incorporated two features which were novel for their time but which are a standard feature on virtually all 35mm cameras now. The first was the use of a pin registration system, in which a pressure plate, mechanically linked to the intermittent movement, held the film in place on fixed pins engaging the perforations during exposure, and then pulled it back to engage the intermittent claw to advance the next frame. When combined with an automatic perforator which ensured almost perfect accuracy in the spacing of the perforations in the raw film stock, this mechanism offered vastly improved vertical stability in the projected picture. The second was its all-metal body: as H. Mario Raimondo Souto notes, camera bodies before around 1920 were almost exclusively wooden, which required extensive maintenance and which were prone to light leakages.[7] Other features which had previously been available on a limited scale but not in a mass-produced piece of standard equipment included a variable shutter angle which could be adjusted during a shot (thus making it possible to produce fades in the camera and eliminating the need to introduce them in printing) and a rotating lens turret which enabled quick and simple changes to a lens of a different focal distance.

It is estimated that around 1,500 model 2709s were manufactured in total; not much in the context of today's consumer electronics, but a huge quantity for a specialised piece of industrial machinery.[8] Despite the initially high capital outlay of around $2,000 for the camera itself,[9] studios quickly realised that the economies of scale on which it was produced enabled highly cost-effective operation in a production environment. Production continued until 1958, and a small number of 2709s remained in use as rostrum cameras for animation until the mid-1990s, when digital mastering technology superseded this remaining niche in the market. There are very few other industries in which a standard piece of equipment remains in regular use for almost nine decades, and therefore the 2709's longevity is a powerful example of the relative mechanical simplicity of film-based moving image technology.

Although the 2709 was used for specialist processes over many years, its heyday as a mainstream studio camera was during the 1910s and 1920s. The conversion to sound in 1926–30 effectively rendered it obsolete for this purpose for two key reasons. Firstly and most importantly, the predominant studio practice during the silent period was to operate motors by hand cranking. Although an electirc motor was available for the 2709, the power supply needed made it more difficult to use the camera on location in this configuration, and further added to the noise it made (of which more below). Motor-driven cameras were essential for filming with synchronised sound, as accurate reproduction of the recording depended on a consistently accurate film transport speed, which was then matched in projection. Consistency of shooting speed was not considered a major issue during the silent period – indeed, variations were often deliberately introduced as an artistic effect. The 2709, like most other cameras built during this period, followed the convention of one complete revolution of the cranking handle passing eight frames (half a foot) of film through its mechanism. Steady cranking at two revolutions every second, therefore, would equal a shooting speed of 16 frames – or one foot – per second, which was considered the *de facto* norm until changes in projection technology during the 1920s precipitated a gradual increase (see chapter five for more on the differences in shooting and projection speeds). The second reason was the price to be paid for the 2709's extraordinarily precise pin-registration system compared with other cameras of its day: the mechanism was inherently very noisy, and would easily be audible on any sound recording made in close proximity to it. In the early days of sound some 2709s were fitted with electric motors and enclosed in a soundproof booth. This all but ruled out any form of camera movement (i.e. panning or tracking) during a shot, and it was clear that for use with synchronised sound, a fundamentally new design would be needed.

During World War One the nascent American film industry gradually relocated from the East Coast to the West Coast of the United States, driven by greater flexibility, cheaper land costs and the higher levels of light needed to shoot on location using the slow film stocks of the day. By the mid-1920s production activity had become largely concentrated in a hitherto obscure suburb of Los Angeles known as Hollywood, and had eclipsed that of war-battered France as the world's largest (in terms of monetary turnover) film industry. It was hardly surprising, therefore, that the more significant advances in camera technology during this period took place in the United States. These were provided chiefly by the Mitchell company, which sold its first camera in 1920. In its early versions, the functionality of the Mitchell mechanism did not vary significantly from that of the Bell and Howell. The main difference was a more accurate and easier to use viewfinder (but which as with all cameras before the advent of reflex viewfinders, could not be used during an actual shot). The intermittent mechanism used two simultaneously moving claws and cams, whilst a third cam operated a pressure plate which held the film in place during exposure. This was not quite as accurate as the pin-registration system used in the Bell and Howell, but the mechanism was a lot quieter and required less power to drive it (which meant that it could be adapted to be driven by an electric motor far more easily). The design was steadily refined throughout the 1920s, with a pin-registration system being added in 1928. This was a lot quieter. Another advantage of the Mitchell was that the shape of the camera body could more easily be encased in an outer shell, or 'blimp', which deadened the motor noise almost entirely for sound shooting, whereas the Bell and Howell was so noisy it had to be operated within a booth.

The Mitchell model BNC, introduced in 1934, was so quiet as to be almost undetectable by the studio microphones in routine use, and its introduction effectively rendered the Bell and Howell obsolete except for silent shooting (e.g. by second units on location) and specialist applications, most notably as a rostrum camera for stop-motion animation, for which the unsurpassed accuracy of its film registration system was used to great advantage. For studio use, the BNC enjoyed almost exclusive market domination until the mid-1960s. Mitchell also produced the first camera mechanism to be adapted for use with a widescreen format, when a 70mm version was developed for the short-lived Fox Grandeur process in 1929–30.

The only other significant studio camera design to emerge between the wars was the Technicolor beam-splitting camera, designed and produced by the Technicolor corporation for use in their three-strip colour cinematography process. As the camera was an integral part of that process and was not used for routine cinematography, its design and operation will be considered along with the other components of Technicolor in chapter three.

Camera design for actuality and documentary filming

Studio-shot features were not the only purpose for which moving images were originated on film, however, and a number of smaller, more portable models emerged during the 1920s. These were intended primarily for news and documentary filming, and incorporated some of the advances of their larger cousins. Three examples in particular are worth noting. The British Newman-Sinclair company had been around since the 1900s, and the wooden-bodied Newman-Sinclair Standard, first marketed in 1910, quickly became established as a favourite among news and actuality cameramen in the UK and Europe. The relatively large body size, simple claw-and-cam pulldown and fully enclosed magazines made for durability and easy maintenance, and among the Standard's notable users were Herbert Ponting and Frank Hurley, the cameramen who accompanied Captain Scott and Ernest Shackleton to the Antarctic respectively. In 1928 Newman-Sinclair produced the 'Autokine', which was used extensively among documentary filmmakers (mainly in Britain and Europe) for several decades. Oblong-shaped and roughly the same size as a Debrie Parvo, the Autokine featured a rugged, durable design, a pin-registration system, easily interchangeable lenses and – crucially – a double-springed clockwork motor which was capable of running the entire 200-foot magazine capacity in one take (Newman-Sinclair had experimented with electric motors in an earlier model of 1922 but found them to be too heavy, with insufficient battery capacity to film for any significant length of time). Without the need to hand-crank the mechanism, the Autokine was the first mainstream, professional 35mm camera which could reliably be hand-held. Probably the most extensive users of Newman-Sinclair Autokines were the British 'Documentary Movement' of the 1930s and 1940s, who valued their portability, versatility and ease of use in unorthodox locations. Well-known films shot (largely or entirely) with Autokines included *Man of Aran* (1934, dir. Robert Flaherty), *North Sea* (1938, dir. Harry Watt) and *Land of Promise* (1946, dir. Paul Rotha). As documentaries in those days hardly ever used synchronised sound (the soundtracks usually consisted of a post-synched commentary, effects and music), the noisy clockwork mechanism was not a problem.

Another camera designed for specialist actuality filming was the model developed by the American explorer and museum curator Carl Akeley in 1918. It was drum-shaped, with a cylindrical shutter occupying the entire surface of the inner chamber. A gyroscopic tripod head and a viewfinder assembly with an adjustable angle allowed easy use on rough terrain and in difficult locations, whilst the distribution of weight within the camera body and a twin viewfinder lens enabled a new generation of long and heavy telephoto lenses to be used with comparative ease. Akeleys were initially used by the military and by actuality cameramen, most famously by

Robert Flaherty in his documentary about Eskimo life, *Nanook of the North* (1921). But their unsurpassed versatility eventually led to their purchase by the Hollywood studios. From the mid-1920s onwards, Akeleys were increasingly used alongside 2709s for location work, so much so that the term 'Akeley shot' was used within the industry to describe a technically complex panning shot photographed in difficult or dangerous circumstances.[10] Akeley cameras were instrumental in the development of the action/adventure genre during the 1920s, being used on for the moving shots in spectacular epics such as *Ben Hur* (1925, dir. Fred Niblo) and *Wings* (1927, dir. William A. Wellman), as well as the stunt work for slapstick stars including Buster Keaton and Harold Lloyd. Aware of the niche market for small, portable 35mm camera mechanisms Bell and Howell launched its own, the Eyemo, in 1926. This design substituted the pin-registration system with a simple cam movement, thus reducing weight, and was able to be loaded and unloaded in daylight (essential, given that its film capacity was only 100 feet).

The other major technological advance in camera design to emerge before World War Two was the reflex viewfinder. Nowadays it is taken for granted that any camera – still or moving image, film, video or digital – will incorporate a means by which the photographer can accurately determine the composition of the image s/he is recording. Today these generally take one of two forms. The *rangefinder* system consists of an entirely separate optical system which is positioned close to the lens which actually receives the exposure to be recorded. While this is reasonably simple and cheap to produce, one serious drawback is the problem of parallax errors. This refers to the slight difference in composition caused by the distance between the optical centre of the viewfinder and that of the taking lens. This needs to be corrected – which is usually done by skewing the angle of the viewfinder lens slightly – in order for the former to give a reasonably accurate view of what the latter will record. However, slight errors can still be introduced, especially when filming very close subjects: the closer the subject is to the camera, the more greatly amplified the difference in angle between the rangefinder and taking lens will be.

The *reflex* viewfinder solves this problem by introducing a system of mirrors which enables light from the taking lens to be directed into a viewfinder as well as onto the film. The core of the system is a 'shutter mirror' which blocks the light path from lens to film while directing light into the viewfinder. During exposure this mirror is raised, obscuring the viewfinder and permitting exposure. Anyone who has a single-lens reflex still camera can see this working simply by removing the lens (having first ensured that no film is loaded), setting the shutter speed to 'B' and pressing the shutter. The mirror facing the lens will rotate upwards through 90°, enabling the operator to see directly into the film chamber. In a moving image film camera, the length of each exposure relative to the time taken for film movement is such that the repeated movements of the mirror will appear as a flicker through the viewfinder, but the operator is able to maintain accurate composition throughout a shot. Not only does this eliminate parallax errors, but it has the added advantage of enabling the photographer to see whether or not the subject is in focus.

Reflex viewfinders were not available in moving image cameras at all before 1937 (though their use in 35mm still cameras dates back to the late 1920s), and the viewfinding systems available varied from nothing at all to rangefinders which were initially of limited accuracy. Studio techniques evolved to deal with this issue: the distance between camera and subject would be measured, to ensure that focus was set accurately (hence the job title 'focus puller'), and by comparing the focal distance of the lens in use with the optical centre of the shot it was possible to determine, with reasonable accuracy, the outer limits of the frame. This procedure was not easy, especially with shots that involved camera movement, and therefore the development of more accurate viewfinding systems during the 1920s and 1930s had a significant impact on the composition and style of shots produced throughout the industry, from highly formal studio features to newsreel filming at sports events.

The first mass-produced 35mm camera to incorporate a reflex viewfinder originated in Germany in the mid-1930s. The firm of Arnold & Richter (Arri) was established in Munich in 1917, and produced its first moving image camera, the Kinarri, in 1925. August Arnold's Arriflex, launched at the Leipzig Fair in 1937, revolutionised the German film industry with its reflex viewfinder and an entirely new claw-and-cam mechanism which produced accuracy of registration approaching that of the Bell and Howell but without the need for a pin-registration system. In it, the claw assembly is moved by a triangular cam driven by a rotating shaft. This arrangement produces a very short period immediately before and after each exposure during which the claw remains fully inserted through the film perforations but entirely stationary, thus absorbing all vibrations produced by the film movement, a function which, in the 2709 and the Mitchell, is accomplished by pin registration. Arriflexes were used extensively by the Nazis both for location and studio work (such as filming on battlefields), and were greatly envied by Allied cameramen during World War Two because of these mechanical innovations. It was the first 35mm camera which offered a level of functionality approaching that of a full-scale studio camera, but which could also be described as truly portable. After the war Arriflexes found their way west, but initially only to a limited extent: Flaherty (again) used one to film *Louisiana Story* (1947, dir. Robert Flaherty), while the Warner Bros. cinematographer Sid Hickox put the reflex viewfinder to striking use in the half-hour opening sequence of the Humphrey Bogart thriller *Dark Passage* (1947, dir. Delmer Daves), which is shot subjectively, i.e. entirely from the hero's point of view. In spite of the obvious operational advantages of the Arriflex, the reflex viewfinder was not incorporated into a standard Hollywood studio camera until the Mitchell BNCR was produced in 1967.

The post-war evolution of 16mm cameras

From the 1950s to the 1980s the development of film camera technology was largely influenced by three factors: the desire to reduce the cost and increase the efficiency of studio practices, the introduction of widescreen and wide film formats and the growing use of film, and in particular 16mm film, as an origination medium for television.

No significant new 35mm camera mechanisms appeared during the 1950s. The evolution of 16mm from the system launched by Eastman Kodak specifically as an amateur format in 1923 to its growing professional use in film and television after the war will discussed more extensively under the heading of film formats below. It is noted here because developments in 16mm cameras during the late 1940s and 1950s made this transition possible and provided the technological basis for the expansion of 16mm into specialist cinema applications and television. A 16mm version of the Arriflex, the 16ST, appeared on the market from the early 1950s, incorporating a slightly modified mechanism from the 35mm version. Most notable was the use of pin registration, which, given that the much smaller size of the 16mm frame would serve to magnify any unsteadiness in registration many times more than with 35mm, was considered essential.

Other 16mm reflex models quickly appeared from manufacturers including Paillard Bolex in France and Canon in Japan, while the Bell and Howell Filmo – the 16mm ancestor of the Eyemo which had been manufactured since 1923 – enjoyed increased sales. Production continued until 1979, a longer run even than the 2709. By the late 1950s these cameras accounted for virtually all television output which was not broadcast live. In the 1960s these 16mm designs began to incorporate the means of synchronisation to an external magnetic sound recording device, such as the Arriflex BL, launched in 1965 and the last major 16mm professional camera mechanism, the Aaton, introduced in the mid-1970s. 16mm was also adopted enthusiastically by a new generation of documentary makers, notably the 'Free Cinema' group in Britain and the 'Direct Cinema' group in the US. Directors such as Lindsay Anderson, Peter Watkins, D. A. Pennebaker and Frederick Wiseman used the unprecedented portability of the new cameras together with related technologies such as faster film stock and magnetic tape sound recording to create a genre of actuality filming which simply would not have been possible with 35mm. It has been argued that this is the direct ancestor of today's 'fly on the wall' television documentaries. By the 1980s 16mm was effectively finished as a widely used origination medium, as its principle market in television news and documentary filming was being superseded by the ever more portable and reliable broadcast video technologies. As Brian Winston notes, 'the 16mm synch rig's heyday was to be a brief two decades',[11] from the late 1950s to the late 1970s, and during this period the

Fig. 2.1 The Arriflex 16MQ camera in use for actuality filming, circa 1960s. 16mm with magnetic sound recording was used extensively in the 1960s and 1970s before its replacement by videotape for television news and documentary production. Photo courtesy of BFI Stills, Posters and Designs.

most significant developments in camera design were driven by the growth of this format. As with Flaherty's use of an Akeley in the Arctic, the evolution of 16mm cameras marks a powerful example of an industry and culture being directly inspired by the possibilities offered by a particular technology.

Fig. 2.2 Two photographs taken with a zoom lens on a modern 35mm still camera. The top picture was taken at the widest setting, the bottom one at the closest. Without a zoom lens, the second picture could only have been taken either by physically moving the camera closer to the building, or by replacing the lens with one of greater focal length.

One constituent technology which was key to the growth of the 16mm camera was the so-called 'zoom' lens. These lenses, which enable the operator to variably enlarge or reduce the focal distance (and thus the area of coverage) of the lens are today (like viewfinders) provided as a standard feature on virtually every still, moving image, digital and video camera sold, except those for highly specialist professional applications. It is now almost impossible to enter a consumer electronics shop and buy any sort of camera without one. Before the mid-1950s, however, the only lenses commonly available were of a fixed focal distance.

In its simplest form, a photographic lens consists of a single piece of ground and polished glass, the curved surface of which 'refracts', or directs the angle of light passing through it to converge on a determined point (such as the film surface) behind it. The distance between the lens element and the surface on which the light is intended to converge determines the *focal length*, which is the physical area covered by the lens as illustrated here. In reality, single-element fixed lenses are no longer used at all except in disposable still cameras, with most designs incorporating additional, movable elements enabling the lens to be 'focused', i.e. the physical distance between the lens and the subject being photographed to be set. In a film camera, this setting, together with the film speed, level of ambient light, length of exposure and lens aperture (the size of the hole through which the final lens element directs light onto the film surface) are all variables which determine the depth of field in a shot, which is the proportion of the image behind and in front of the subject which is also in focus.

A 'prime' lens – one with a fixed focal length – does not enable the area it covers to be altered, except by physically moving the camera to which it is attached. For this reason lens turrets became a widespread feature on 35mm moving image cameras from the 1910s onwards. They were first introduced on the Bell and Howell 2709 and proved ideally suited for studio use. For example, the use of a lens turret would enable a cinematographer to film a medium long shot immediately followed by a medium close-up without having to move the camera and without the director or performers having to reconfigure the scene on stage. For location and actuality filming, however, the restrictions imposed by prime lenses were far more of an issue. While newsreel cameramen in the 1920s and 1930s were extremely skilful in selecting and unobtrusively changing between lenses in filming events such as football matches and parades, the introduction of a lens which permitted its focal distance to be infinitely varied during a shot was obviously far more preferable. The first zoom lenses appeared in 1925.[12] One of the earliest to be marketed for 35mm use was made by the Leicester-based Taylor-Hobson company, which had become one of the world's leading producer of lenses for cinematography and projection. The 'Varo', launched in 1932, had a variable focal length from 40mm to 120mm but there were two major drawbacks: it absorbed so much light that its use was difficult, even under intense studio lighting, and the focus setting could only be adjusted by taking the unit apart, thus further limiting the sort of zoom shot which was practically achievable. Furthermore the lens alone was almost as big and heavy as a 35mm studio camera.[13] Similar models appeared in Europe and the United States during the 1930s and early 1940s, all of which had similar drawbacks and were not used on any significant scale.

A modern zoom lens enables its focal distance to be adjusted by effectively incorporating two lens systems into a single unit. The distance from the zooming lens to the prime can be moved, while a mechanical system automatically adjusts the focus of the prime as this is happening. This, of course, requires the light to travel through a much higher number of glass surfaces, which has two implications: firstly, that a zoom lens absorbs more light, and secondly that a greater level of distortion is introduced into the recorded image. For these reasons zoom lenses were not (and still are not) used extensively in 35mm studio work; in a controlled environment such as a film studio the shooting requirements of a fixed prime can easily be met and they generally deliver a much higher image quality. Zoom lenses made their biggest impact in the 16mm market, where the versatility of a variable focal length for news, documentary and location shooting had a major impact. Interestingly the first 16mm zoom lenses which resembled the ones used today in operation (i.e. the zoom could be adjusted without the need to refocus) to sell in significant numbers both came from French companies: the Pan-Cinor, made by Berthiot, was a 16mm lens with a 17.5mm to 70mm zoom in 1956, while Angénieux launched a similar model two years later. By the early 1970s later variants of the Arri BL were being packaged with a more advanced 12mm–120mm Angénieux (first produced in 1963), a combination which, for news and documentary filming 'became ubiquitous'.[14]

Studio cameras after the 1960s: Widescreen and special formats

As has been stated above, there was no significant development in conventional 35mm studio camera design during the 1950s. Hollywood continued to use Mitchell BNCs and, to a lesser extent, 2709s and Arriflexes, the latter of which was used extensively in European studios. The Soviets and Chinese developed what were effectively clones of a number of Western cameras, most notably the Mitchell NC. Developments in studio camera technology which did take place in the post-war period can be placed into two categories: the emergence of camera mechanisms and lenses for use in widescreen and wide film systems, and the introduction of precision electronics and microprocessors to control and synchronise a camera's operation from the 1980s onwards.

Of all the widescreen and special format systems which emerged during the 1950s (which will be considered in greater depth below under the heading of film formats), the two which stuck required little or no modification to existing camera mechanisms. These were CinemaScope (and its numerous offshoots and variants), which optically compressed, or 'squeezed' a wide image into the conventional 35mm frame, and the Todd-AO system (and again, a great many variants thereof) which used film stock 65mm wide and with a slightly taller frame area to produce a much larger, uncompressed image. Stereoscopic film, or 3D, enjoyed a brief period of popularity during the early to mid-1950s. In order to photograph an approximation of the three dimensions as seen by the human eye, a pair of electrically interlocked cameras (usually Mitchell NCs) were mounted in a large casing which both positioned them for parallax and blimped them for sound. This assembly weighed almost as much as a small car and made any form of camera movement practically impossible. The resulting two 'left eye' and 'right eye' film strips would be shown by two projectors running in synchronisation onto the same screen. The viewer wore a pair of polarised glasses, which would enable the viewer's brain to 'translate' the two projected images into

Fig. 2.3 An anamorphic image as it would appear on the film (left) and in projection (right).

the impression of a three-dimensional image. Interestingly this technique has been revived in the 1990s with the addition of 3D to the large-format IMAX process, in which a pair of wide-format 65mm Arri cameras are used. The CinemaScope process used existing studio camera mechanisms without any significant modification, while the first generation of 65mm cameras were the same adapted Mitchell NCs which were briefly used in the Fox Grandeur process at the end of the 1920s (see below). From the point of view of cinematography, the key widescreen developments during the 1950s and 1960s were essentially in the lenses.

The essential technology in CinemaScope is the *anamorphic* lens, which through a system of barrel-shaped or prismatic elements compresses the image along its horizontal plane during exposure. In projection, the process operates in reverse and the picture is expanded. The earliest generation of anamorphic lenses, produced by Bausch and Lomb, could only be used in conjunction with 50mm primes. The two had to be focused separately, thus inhibiting camera movement. Henry Koster, who directed the first CinemaScope feature *The Robe* (1953), recalled that blocking and acting on set had to be configured as for a stage play, simply because of the need for a static camera and the fixed focal length of its lens. As the 1950s and 1960s progressed, lenses for widescreen cinematography became more versatile, could accommodate larger apertures and were produced in a wide range of focal lengths.

The Panavision company, founded by Robert Gottschalk in 1954, grew through the following decade and a half to become, along with Arri, the market leaders in 35mm studio camera technology. Panavision originally produced an alternative range of anamorphic lenses to the Bausch and Lomb anamorphs produced according to Henri Chrétien's original designs from the 1920s. The prime lens and anamorph could be combined into a single unit, making focusing easier and enabling quick and straightforward changes between lenses of different focal distances: by 1963 the lenses available ranged from 25mm to 360mm.[15] Panavision also introduced a new system for marketing its technology. As with the introduction by Bell and Howell of mass-produced, standard equipment design in the 1910s, this development would, by the end of the century, have been adopted in many different areas of the industry. Panavision only rented cameras and lenses to studios; it did not sell them outright.

There is one significant precedent for this, which was the three-strip Technicolor system in the 1930s and 1940s. A studio could not buy a Technicolor camera or film stock, and neither could the processing and printing be done at a lab of its choice. Instead, a producer had to enter into an all-inclusive contract with Technicolor. Its provisions covered the number of cameras available, where they could be used, the involvement of Technicolor personnel on set, the arrangements for laboratory work and the number of release prints of the finished film which the distributor undertook to purchase. Given the exclusivity and integrated nature of the process as a whole (the cameras by themselves were useless without access to the patented negative stock and laboratory techniques) together with the relatively small number of Technicolor films produced, this business model failed to make much impact on the industry as a whole; and in any case it was smashed in the early 1950s when single-strip tripack colour film, which could be used in any camera whatsoever, became available (see chapter three).

Panavision adopted a variant of this approach in which cameras and lenses were 'dry hired' to studios. Unlike Bell and Howell, which supported its equipment through sales of spare parts and maintenance contracts, Panavision cameras were supplied on a hire basis only. But unlike with Technicolor, no conditions were placed on how they could be used. This was accepted by studios because the inherent image quality produced by the Panavision lenses was far superior to anything the studios' own research departments or third-party suppliers were able to produce. Panavision's cameras were not fundamentally new designs, being based on Mitchell BNC mechanisms which the company acquired second-hand during the late 1950s and early 1960s, to which were fitted reflex viewfinders and more effective sound-proof blimps. The Panaflex camera, first introduced in 1963, was basically a hybrid of Arriflex optics and Mitchell mechanism. These designs were further refined through the 1970s and 1980s. The last fundamentally new design from Arri, the 535, incorporated computer technology to synchronise film to shutter movement and can also record data on the edge of the film needed to assist in cutting the negative and synchronising sound. A 65mm/70mm version, the 765, was also launched in 1989. Its first use on a widely-distributed feature film was to shoot *Little Buddha* (1993, dir. Bernardo Bertolucci), although the subsequent decline in the use of 65/70mm has prevented the 765 from entering mainstream production use. Arris, together with updated and modified Panaflexes, were the cameras used for the majority of 35mm filming at the end of the twentieth century.

This was the situation at the end of the 1990s, and it could well be that the two mechanisms which dominate 35mm studio cinematography – basically updated and heavily modified versions of the 1934 Mitchell BNC and the 1937 Arriflex – represent the final stage in the development of film camera fundamentals before the industry 'goes digital'. Not that this needs to happen soon: as we have seen, the most successful camera mechanisms have useful lifetimes measuring tens of decades, while the medium onto which they record – 35mm film – has been in constant use since the 1890s and continues to be used today.

The evolution and development of film camera technology has both shaped and been shaped by the industry and culture which has used it and its relationship with the other forms of technology needed to get an image from a studio set (or location) to a cinema or television screen. A number of economic models for the development and production of cameras and lenses have been seen. The earliest models were individually fabricated, usually by filmmakers themselves. Towards the end of the 1890s cameras began to be manufactured on a sale basis, though still in very small quantities. As the film industry rapidly expanded during the 1900s and 1910s the operation, performance and maintenance requirements expected of this technology made it apparent that mass-production would soon follow as the dominant economic model.

There then followed the 'Model T' approach of Bell and Howell in the 1910s and 1920s following their launch of the 2709. The 1920s and 1930s saw new cameras emerge, designed for specific uses and by more specialist sectors of what was rapidly becoming a large-scale global industry. These ranged from the Mitchell BNC and the Arriflex, which were intended primarily for studio use, to the Akeley and the Newman-Sinclair, designed for newsreel and actuality filming. The second half of the century saw the adaptation of pre-existing hardware to new formats (such as CinemaScope) and finally the marketing of integrated camera and lens packages by Panavision which, for studio use, established the supply of camera equipment very much as a service industry. The needs of the controlled production environment in a studio and those of documentary and news filming were addressed through separate forms of camera mechanism, and it was the growth of the latter which drove 16mm in the 1960s and 1970s. That such a relatively small number of mechanisms have accounted for the entire world's film production for over a century (extending the Model T analogy, imagine if only 10 designs of motor vehicle had accounted for 90 per cent of the world's road transport over a similar period) was made possible by a form of technology which did not in itself produce any hardware, but which rather have determined the way hardware has been designed and manufactured – film format standardisation.

Film Formats

> 'My fondest hope for the industry – standardisation.' – Donald J. Bell[16]

This is hardly a surprising comment, given that during the two decades before Bell made it, his company had risen to become one of the largest suppliers (probably second only to Eastman Kodak, which by the same token also had a vested interest in promoting standardisation) of moving image technology to the Western world on the back of that very strategy. If 100 per cent of the world's film studios use 35mm cameras, then a single mass-produced model could potentially take 100 per cent of the market share. If 30 per cent of those studios shoot on a different gauge, however, then the maximum potential number of sales is instantly reduced. There is a further need for technological standardisation within the industry, which relates to the comment by the projectionist at the start of his chapter, namely that film is 'this wide and goes through a projector'. Since the 1910s the world's dominant film industry – Hollywood – has depended for its success on being able to market its output globally, and not just in its country of origin (although the comparatively huge US domestic market gives Hollywood an inherent economic advantage compared to smaller countries).

Shortly after the conversion to sound the producer Samuel Goldwyn famously remarked that 'if the US spoke Spanish, Britain might still have a film industry'. He was of course referring to another type of standardisation: the spoken word. During the 1920s Hollywood was able to aggressively expand its overseas sales because American silent films could be made intelligible to non-English speakers by simply translating the intertitles, a process which cost virtually nothing in comparison with the overall production budget of a feature film. During the early years of sound, the technologies needed to achieve the same result, i.e. subtitling and dubbing, were difficult, limited, expensive and did not work very well. As a result Hollywood's international sales slumped, and about the only overseas market which did not fall victim to this was the UK. This was for the simple reason that as the two countries speak the same language (or at least, versions of it which are largely comprehensible to native speakers of the other), Hollywood talkies could continue to be marketed in Britain without any additional costs.

Even Hollywood's economic might could not standardise the English language to the extent that translated films are no longer necessary (so much so that British films featuring strong regional accents or dialects are routinely subtitled for release in the United States). But the principle articulated by Goldwyn applies to virtually every form of moving image technology, and indeed the fundamental principles of many other forms of technology are often dictated by pre-existing standards. For example, as we shall see in chapter three, the form of colour film which eventually dominated the industry throughout the second half of the twentieth century did so largely because it was 'backwards compatible', i.e. it could be used in pre-existing cameras and projectors. This was despite some major drawbacks compared to other colour systems which had been successfully used in the past. Standardisation was

The Bell & Howell Standard Cinematograph Camera

THE B & H Pioneer Standard Camera can be adapted for interchangeable regular and ultraspeed operation. It can be silenced for sound work; all sound recording systems can be used with this camera.

A special mechanism for color work by the Bi-Pack process, interchangeable with the regular mechanism, is available. This also can be silenced for sound work.

The Bell & Howell Standard Film Splicing Machine

THIS splicing machine makes neat and flexible splices of various standard widths. Splice is very quickly made and is stronger than the original film. Extra blades available for making different width splices on one machine. Special 16 and 35 mm. combination and other special splicers available. Safe, sure, and very easy to operate.

Bell & Howell

Bell & Howell Company, 1801-15 Larchmont Avenue, Chicago, Ill.
New York, 11 West 42nd St. . . . . . Hollywood, 716 North La Brea Ave.
London (B & H Co., Ltd.) 320 Regent St. . . . . *Established* 1907

Fig. 2.4 Advertisement for the Bell and Howell 2709 camera and standard perforating machine, stressing the economic benefits of standardised equipment (author's collection).

vital to television being able to work at all, as the mass-produced receivers in people's homes had to all be able to decode and display the same transmitted signal. The history of television (and of electronic imaging in general) is of products deliberately being designed to work in conjunction with some technologies but be incompatible with others in order to serve the manufacturer's economic interests. This was a coming together of technological innovation and business practices which had its origins right at the very beginning, in the early 1890s: the standardisation of film formats.

Establishment and use of the initial standards

To begin, let us consider the parameters which are understood to be defined by the term 'film format':

- Width of the film.
- Orientation (which way round the base and emulsion side of the film is positioned for correct use in a camera, printer or projector).
- Dimensions, shape and position of perforations on the film surface.
- Dimensions, position and orientation (i.e. whether vertical or horizontal) of each photographic image, or 'frame'.
- The characteristics of each frame which determine how it must be viewed or reproduced, e.g. its aspect ratio (proportion of width to height) or whether the image is anamorphic or stereoscopic.
- The 'pull-down', i.e. the length of film which needs to be advanced by a camera's or projector's intermittent movement in order to position consecutive frames accurately in the gate. This is normally expressed in numbers of perforations, because film is moved intermittently by means of sprocket teeth which physically 'pull' the film by its perforations.
- The film transport speed, expressed in the number of frames per second which are exposed and projected in order to accurately reproduce the impression of movement as recorded.
- Dimensions, position and type (e.g. optical, magnetic or digital) of any soundtrack recorded on the film (this will be considered further in chapter four).

All of these variables must be the same throughout the production, duplication and exhibition processes in order for film to be able to reproduce the impression of a photographic moving image in the same way it was recorded: a film shot at 18fps will appear too fast if it is projected at 24, and a film shot with an anamorphic lens will appear compressed if projected through a spherical one (as in the 'squeezed' image in fig. 2.3). The process of standardisation, therefore, ensures that equipment manufacturers and users, producers, distributors and exhibitors are all singing from the same hymn sheet in the way they handle film-based technology in the various stages of its progress from studio to cinema.

The process from which emerged the first amd most widespread standardised film format began when Edison's assistant W.K.L Dickson placed an order for celluloid (nitrate) roll film from Eastman Kodak in August 1889. By 1982, when the Kinetograph camera and Kinetoscope viewing device (see chapter 5 for more on

Professional Standard Film
Adopted by Soc. Mot. Pic. Engs July 1917

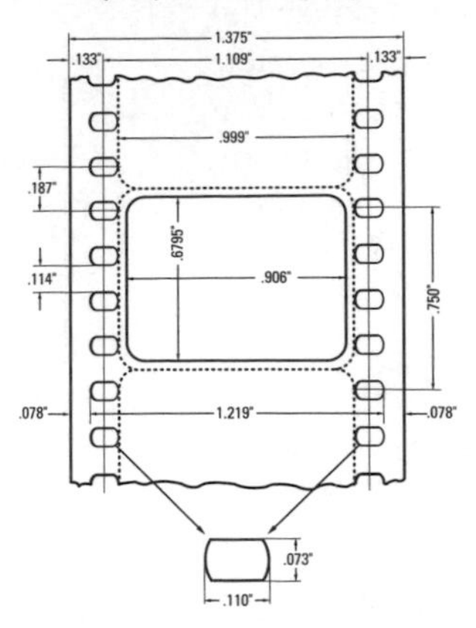

Frame Line Midway Between Perforations

Fig. 2.5 The film format specification for 35mm 'professional standard' as determined by Edison in 1889 and subsequently adopted by the newly-formed SMPE (above), and an example of release print in that format.

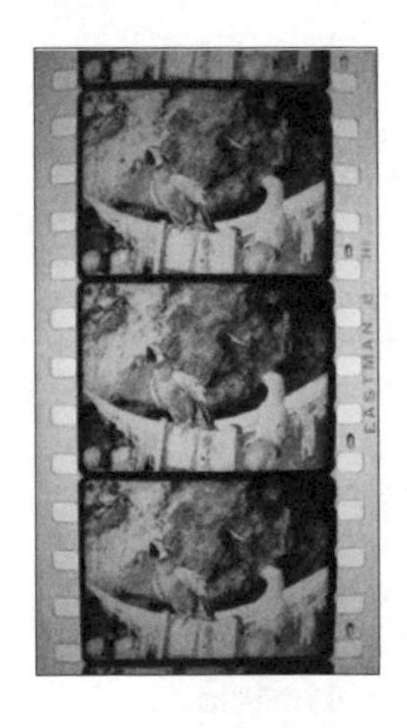

the latter) went on market, the basic format was there. With slight refinements and modifications, it is still in use today. It was 35mm wide, 'pulled down' a length of four perforations to advance each frame and had an aspect ratio of approximately 1:1.33 (vertical:horizontal). These choices were, according to John Belton, 'determined by a complex interplay of technological, economic and ideological factors',[17] but essentially they derive from two factors. Firstly the width and density of the film base needed could be manufactured using production lines designed for existing still photography stocks with only minor modifications, and its size and weight when handled in the lengths needed for moving image use were considered acceptable. Secondly, the ratio of 1:1.33 conformed to Dickson's aesthetic ideas of shot composition, again in still photography (of which he was a keen amateur practitioner). Paul Spehr has argued that this position oversimplifies the process of research and development which led to the initial standardisation of 35mm, suggesting instead that it was arrived at because of the mechanical requirements of early Dickson mechanisms, the limitations of the film casting plant in use at the time and the desire to avoid the use of technologies covered by existing patents.[18]

These two decisions set an extremely far-reaching precedent. Because the world's largest manufacturer of film was producing 35mm on an industrial scale, equipment manufacturers throughout the world followed suit and made cameras, printers and projectors which used it. By the mid-1900s the almost universal use of 35mm had made possible an international trade in film equipment, as European manufacturers increasingly sought to market their products in North America. The result was a sustained patent war which lasted for over a decade, with various parties trying to restrict or deregulate the rights to commercially exploit specific aspects of the technology in various different countries. Chief among them was the 35mm standard. Transatlantic imports of film – both raw stock and completed productions – from the UK and France (principally from the French Pathé and Lumière companies) had worried Edison and the main American film producers for some time, with the result that a consortium consisting of Edison, Eastman and a number of smaller players (notably the American Mutoscope and Biograph Company) formed the Motion Picture Patents Company (MPPC) with an alliance of producers at a meeting on 18 December 1908. This was in effect a 'closed shop' which used Edison's patent rights over the basic camera and projection technology to ensure that only Eastman Kodak film stock could be used in conjunction with it. Kodak undertook only to supply stock to members of the company, who paid a levy to the

MPPC for every foot they purchased. Only one non-American company, Pathé, was invited to join. It is believed that this was either an attempt by Edison and Eastman to 'divide and rule' the European market[19] or a reflection of the fact that Pathé was the only non-American company capable of supplying sufficient quantities of film to the US market to represent serious competition.

The MPPC eventually fell apart through a combination of legal action and market forces. Those smaller elements of the American film industry which were not a part of it produced or bought camera and projection technology which circumvented Edison's patents and used it in conjunction with imported stock. This undermined the MPPC's monopoly, as the technology regulated by its patents gradually lost market share to an emerging sector of competitors. A series of lawsuits culminating in a Supreme Court decision in 1917 effectively settled the issue: the practice of compulsorily tying the use of Eastman Kodak film to specific cameras and projectors was ruled to be a violation of antitrust legislation. Jeanne Thomas Allen suggests that 'the Edison/MPPC phase in the industrial development of film marks a transitional step from the entrepreneurial competition of the early years to consolidation and conservatism'.[20] It is important to bear in mind that the underlying basis for this development taking place was the existence of a *de facto*, almost universal film format: 35mm, 1:1.33, four-perf pulldown.

If Eastman's film had not been technically compatible with non-Edison equipment, there would have been no need to establish the MPPC in the first place, and even if it had existed there could have been no serious allegation that its activities were monopolistic, because producers would have been free to obtain film and equipment 'packages' from other sources. A more recent analogy would be the Apple computer company's policy of marketing computer hardware and the Mac OS operating system as a combined package. As no other operating system will run on Apple computers and Mac OS cannot be used on any non-Apple computers, it can hardly be considered a restrictive practice for the two to be sold together, because one is useless without the other. But if Microsoft were to form an alliance with a PC hardware manufacturer and then refused to allow its Windows operating system to be used with any other computers, that would be a completely different matter. Windows is compatible with personal computers produced by a wide range of manufacturers, and also on machines assembled by consumers from individually purchased components. Any systematic attempt to restrict this would thus be monopolistic, since, as with 35mm film, the technical standards implemented within the products allow for greater compatibility.

Although the MPPC did not last very long, the film format it unsuccessfully tried to control was rapidly becoming established as the industry standard. During the late 1900s automated perforating technology gradually standardised the size and pitch of 35mm perforations (see chapter one), and an international conference of film producers in Paris in 1909 adopted 35mm as the professional standard.[21] In July 1916 the Society of Motion Picture Engineers (SMPE, from 1950 the Society of Motion Picture *and Television* Engineers) was formed in New York, which by virtue of the American film industry's economic dominance would become the world's most prominent standard-setting body throughout the century. Its objectives were 'advancement in the theory and practice of motion picture engineering and the allied arts and sciences, *the standardisation of the mechanisms and practices employed therein* and the dissemination of scientific knowledge by publication.'[22] One of the first standards it set is the one reproduced in fig. 2.5, which enshrined 35mm as the 'professional standard film'.[23]

Film formats and the conversion to sound

The conversion process itself will be considered in detail in chapter four. This process, which in the West happened approximately between 1926 (the first commercial screenings of a feature film synchronised to a recorded soundtrack) and 1932 (the last cinemas without sound equipment either getting it or going out of business), had two significant impacts: the issue of formats and standardisation, which will be discussed here. The two main standardisation variables were the film transport speed and aspect ratio. The former had never been effectively standardised while the 'professional standard' of the latter was incompatible with the new sound technology and had to be changed.

As has been mentioned in the context of cameras, the speed of film through a mechanism is expressed in the number of individual images, or frames, exposed or projected per second (the acronym 'fps', which for some reason is usually given in lower case).[24] The majority of hand-cranked cameras followed a convention established by Lumière Cinématogràphe whereby one turn of the handle exposed eight frames. This was far too slow to give the illusion of continuous movement in projection, and therefore Lumière took to cranking at two turns, or sixteen frames, per second. This was also a convenient measurement because 16 frames are equal to exactly one linear foot of 35mm four-perf. With a three-blade shutter installed in the projector this was enough for the movement on-screen to appear seamless. Lower speeds resulted in visibly disjointed or jerky movement. Higher speeds give a more accurate rendition of movement, because less 'unrecorded' or 'lost' time elapses between one frame being exposed in the camera and the next. But higher speeds consume a greater amount of film stock and were generally considered unnecessary.

There was no standard shooting and projection speed during the silent period and no attempt was made to ensure that the two were identical in the case of individual films. Given that both cameras and projectors were generally hand-cranked in those days it would have been technically impossible to determine and enforce one anyway. Before the rise of the studio system after World War One, individual cinematographers would evolve characteristic shooting speeds and cinemas would usually determine the projection speed of a film in order to fit an allocated time slot for its screening.[25] During the 1920s, studios would advise cinemas of the speed they wished a film to be shown at, which often bore little relationship to the shooting speed. For example, comic effect could be heightened or action sequences made more spectacular by projecting a film at a slightly higher speed than it was shot at.

The result of the existing technology and industry practices of the late 1920s was that speed remained one variable not to have been fixed by the process of industrial development which led to the 35mm 'professional standard' of 1917. These industry practices notwithstanding, another major reason for this is that individual viewers tend not to be particularly sensitive to slight variations in the rendition speed of a moving image. As an SMPE member noted in 1927, 'the plain truth of the matter is that except for certain key actions, such as walking, eating and dancing, the projection speed can vary over wide limits without apparent falsity'.[26]

This was certainly not the case with an audio recording. The technological principles involved will be covered in greater detail in chapter four, though for the purposes of this discussion it is important to note that all analogue sound processes (i.e. all sound recording technology until the 1980s) depended on consistency and accuracy of the speed at which the recording medium is transported in both the recording and reproducing devices. Any variation will not only introduce differences in speed, but also in the frequency (pitch) of the sound being reproduced. Thus even a slight inconsistency of speed will sound obviously 'wrong', unlike its moving image counterpart which, thanks to the human brain, has a built-in margin for error. When moving images and sound began to be recorded and reproduced synchronously, therefore, there was, for the first time, a need for a standardised and consistent film transport speed to be implemented throughout all sectors of the film industry.

After a period of experimentation, the two competing sound systems which were launched both used a standard speed of 24fps, or 90 feet per minute. According to an engineer who worked on the Western Electric technology, the figure was arrived at as a 'reasonable compromise', in the middle of the range of projection speeds commonly used by exhibitors.[27] It is certainly true that the decision was largely exhibition-led. Cinema projection speeds gradually increased throughout the 1920s with the effect that by 1925, a Hollywood director remarked that 'as a matter of fact, most theatres show pictures at a speed of 85–90 feet per minute [22.6–24fps]', while the head of a cinema chain in Indiana also remarked that 24fps had become a *de facto* standard for projection.[28] The reasons for this will be covered in chapter five, but in relation to sound it would seem that the engineers involved in determining the sound speed decided to formalise what had become a standard by default rather than attempt to enforce a new one. The implementation of this standard required all cameras and projectors to be powered by electric motors, as neither

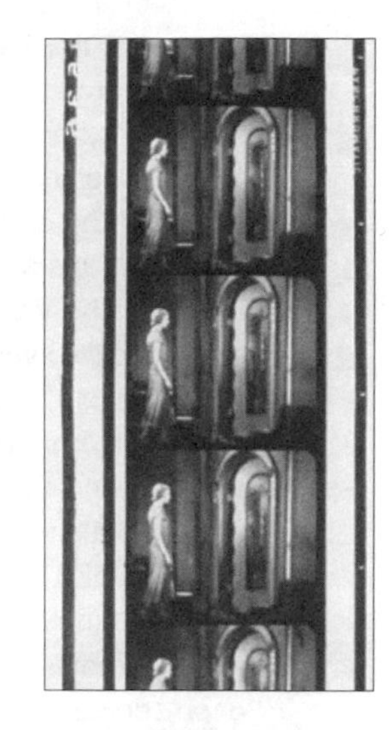

Fig. 2.6 The 'early sound' aspect ratio (above) and the 'Academy' ratio of 1932 (below). The frames above have been photographed with a 'full gate' silent camera and the optical soundtrack printed afterwards, which crops the image – note that the rounded corners of the camera aperture are visible on the right but not the left of the film reproduced above. The Academy ratio introduced a black bar ('matte') between each frame in order to restore the on-screen proportions of the old silent ratio.

clockwork nor hand-cranking could achieve the consistency of speed needed for sound reproduction.

The other format change required by the conversion to sound was to the 'standard professional' frame dimensions. The 1917 specification defined a frame which occupied the entire width of the film between the two rows of perforations, with a very small black matte of .071" separating each frame. This standard remained in use with the Warner Bros.' Vitaphone sound system, in which a film camera was mechanically synchronised to a disc (gramophone record) cutter, and the projector to a turntable. Thus the soundtrack did not affect the topography of the film in anyway, apart from requiring a constant speed of 24fps. The Fox Movietone, RCA Photophone and Western Electric systems, however, recorded optical sound: that is, the analogue waveform produced by the microphone was recorded photographically onto the film. The first sound films were either photographed as 'single system' negatives (meaning that both picture and sound were recorded simultaneously in the same camera and onto the same negative stock) using specially designed cameras, or on separate strips, using largely unmodified 2709 and Mitchell NC cameras for the picture, with the recording being made on a separate sound camera and synchronised during the editing process by means of a clapperboard.[29] For projection, both the picture and sound recordings had to be combined onto a single strip of film, with the result that the soundtrack obliterated a vertical strip of picture on the left of the frame (see fig. 2.6, upper illustration). The aspect ratio of the visible area of the frame remaining is thus changed considerably from the 1917 'professional standard', to a ratio of approximately 1:1.15.

This created problems both in production and in exhibition. In cinemas, projectionists discovered that either the optical soundtrack would be visible on the screen or the picture was too small to fit it. They soon started producing aperture plates (a piece of metal mounted between the light source of a projector and the film, with a hole corresponding to the dimensions of the printed frame) which masked off the soundtrack and a strip at the top and bottom of each frame, together with lenses which magnified the remaining picture area slightly more so that it fitted onto their existing screens. This practice was not approved of by the studios. Cinematographers found that if a sound film produced for disc release were subsequently printed with optical sound-on-film, an area along the left of the picture would be lost. And any sound-on-film in which shots were composed for the square ratio would risk actors' heads being cropped in projection. King Vidor, while filming *Hallelujah!* on location in 1928, was forced to cable his producer over the issue, asking him to 'please let me know if you are contemplating changing method of synchronising releases from records to Movietone so I can allow space outside of frame'.[30]

Possibly because tens of thousands of cinemas already existed with their stages and screens built for the 1:1.33 silent ratio, the early sound aperture did not last. One of the few people to defend it was Sergei Eisenstein, who preferred its aesthetic qualities,[31] but his was a lone voice: the industry consensus was that Dickson's ratio of 1:1.33 should be somehow restored. This was achieved by defining new frame dimensions which did not occupy the area in which the soundtrack was positioned and increasing the height of the black matte between each frame to the point at which the silent ratio was restored, albeit into a smaller frame size. Improvements in emulsion density in the previous three decades meant that there was no significant loss in the definition of a projected picture using the newer, smaller frame. This ratio was adopted by the Academy of Motion Picture Arts and Sciences in 1932 (hence it became known as the Academy ratio), with the SMPE implementing the standard specifications shortly afterwards.[32]

Widescreen film formats

The Academy ratio remained the predominant format for cinema exhibition for a further two decades, until the mid-1950s. In other areas of moving image technology it remains the standard, most importantly television (although the introduction of DVD and digital TV broadcasting in the very last years of the twentieth century began to stimulate sales of widescreen television receivers), but also in amateur film and, much later, the display monitors for personal computers. There are a number of factors why widescreen became established when it did, in the early 1950s. The emergence of television and the need for theatrical cinema exhibition to find a technological means of product differentiation from its new competitor (i.e. a selling point which television could not offer) is often cited as a reason, as are post-war demographic changes which adversely affected traditional cinema audiences.[33]

One question which is worth addressing is why several large-screen and widescreen systems were developed, used on an experimental basis and then abandoned during roughly the same period as the conversion to sound. One notable silent precursor was Magnascope, which used standard 35mm film and what was in effect a zoom lens on the projector.[34] During certain key sequences in a film the projectionist 'zoomed out' the lens (i.e. increased its focal distance) and opened motorised masking curtains around the screen, thus increasing the size of the projected image. Two notable productions which used this device were *Chang* (1927, dir. Merian C. Cooper & Ernest B. Schoedsack), a wildlife documentary in which shots of a stampeding herd of elephants were 'Magnascoped', and *Wings* (1928, dir. William Wellman), a World War One fighter-pilot melodrama in which the Magnascope lens was zoomed to project the aerial sequences, which were shot using Akeleys carried in the planes themselves. Another similar process which temporarily increased the screen size and/or ratio for dramatic effect was the French Polyvision system, which in anticipation of Cinerama 25 years later used three interlocked 35mm cameras and projectors to produce the image. The only significant feature film which used the system was *Napoléon* (1927, dir. Abel Gance), which even then was only shown in widescreen in a handful of venues.

In the following years attempts were made to introduce widescreen systems which were based on wide film gauges (i.e. film which was physically wider than 35mm). All stuck to the four-perf pulldown of standard 35mm, presumably so that the intermittent movements of existing cameras and projectors could be incorporated into the new equipment, although John Belton argues that the cinema architecture at the time precluded the use of frames higher than four perforations.[35] Paramount introduced Magnafilm, which was 56mm wide with an aspect ratio of approximately 1:2.0; Warner Bros.' 'Vitascope' used 65mm film with a similar ratio, and the RKO 'Natural Vision' format used 63.5mm film with a ratio of 1:1.85 (the same ratio as the most commonly used 35mm widescreen format today). All were used for a small number of features shown only in specially equipped city-centre venues in roughly the period between late 1929 and early 1931, before promptly disappearing without trace for a generation.

One widescreen format introduced during the 1920s is of particular interest, since its underlying mechanical technology was recycled into one of the more successful systems of the 1950s, and indeed was still in limited use at the time of writing. Used briefly between the first public screening on 29 September 1929 and the summer of 1930, Fox Grandeur used 70mm film in a process which had much in common with the 70mm Todd-AO format of the 1950s. The cameras were Mitchell NC mechanisms adapted for 70mm use and the projectors were based on existing Simplex machines, with the 70mm gate and film path components being supplied by Mitchell. Grandeur differed from the following generation of 70mm technology in that the pulldown was four-perf, the film speed was 20fps and the sound was optical. A number of shorts (including several releases of the Fox Movietone newsreel) and two features were shot in 70mm Grandeur: the musical revue *Happy Days* (1930, dir. Benjamin Stoloff) in which Marjorie White plays a showboat singer who makes it big on the New York stage, and *The Big Trail* (1930, dir. Raoul Walsh), a western charting the tribulations of an Oregon wagon train which was notable only for its use of widescreen and a hitherto unknown lead actor called John Wayne.[36] Both films were shot simultaneously using the 70mm Grandeur and standard Mitchell NC 35mm cameras.

But in the end, the first generation of widescreen technology disappeared almost as soon as it arrived. One possible reason for this could be the economic strategies of the studios for promoting its use. The 1929/30 introduction of widescreen was an attempt by the studios to link it technologically and institutionally to sound: the thinking was that while the industry was absorbing the huge capital outlay needed to convert to sound (e.g. the equipment needed in hundreds of thousands of cinemas across the world), widescreen could be introduced at the same time for comparatively little extra. The two were deliberately marketed as a package, as this quote from a Fox press release shows:

> The development of the Grandeur system has been an inevitable result of the revolution which Fox Movietone bought to the motion picture business. The new type of motion picture entertainment which came with sound demanded equal improvement in visual reproduction; Grandeur pictures do for vision what Movietone does for sound.[37]

The rest of the industry did not agree, and widescreen was effectively shelved for the next two decades. Another factor which mitigated against the introduction of widescreen in 1929/30 was again economic rather than technological: the Wall Street crash of 1929. Its timing could not have been better for sound and could not have been worse for widescreen. In the months leading up to their peak on 3 September 1929 US share prices rose by several hundred per cent, so much so that 'neither the World Series nor talking pictures could compete in fascination with the Stock Exchange ticker as it tapped out dreams – and realisations – of avarice'.[38] On 23 October – less than a month after the Grandeur premiere of *Happy Days* in New York – they crashed, marking the start of a recession which was to last until the middle of 1932. Compared to other US service industries Hollywood survived largely unscathed, but the new economic climate served to inhibit investment in technological research and development along with virtually everything else which was not essential to the core business of film production and exhibition. It could well be that if the Wall Street Crash had happened a year later, widescreen would have become an established technology by the mid-1930s. As it happened, the next significant technological change introduced by the American film industry – three-strip Technicolor – began to appear in the late 1930s, by which time the US economy was firmly on the way to recovery.

In this context sound just got in under the wire. By the autumn of 1929 Hollywood had committed to producing 'talkies', a large proportion of cinemas throughout the Western world had already installed the reproduction equipment and the rest were compelled to doing so whether they wanted to or not. Too much money had been spent and too many boats had been burnt for the industry to back out. Widescreen, however, was still in the research and development stage with the early experimental shows in first-run city centre cinemas taking place during the autumn and winter of 1929 at the moment of, or just after the crash. Sound had already passed the point of no return. This huge investment in technology had been made largely on the back of speculative investment by bankers, who now had no choice but to wait until it started delivering revenue. The crash ensured that however hard the studios tried to link the technologies of sound and widescreen, the money just was not available to enable an industry-wide roll out of the latter. But a further explanation for the failure of the 1929/30 widescreen experiment can be found in an issue that has recurred throughout this chapter: standardisation. This happened very quickly with sound: sound-on-disc was obsolete by the early 1930s and although various competing systems emerged for optically recording sound-on-film (which will be discussed further in chapter four), they were all reproducible using the same equipment in the cinema. This was not the case with the 1929/30 widescreen cinema processes: they used film gauges of differing widths, each requiring a totally different set of cameras and projectors, i.e. they violated the 35mm standard. Furthermore, other aspects of the format definitions (such as the speed and aspect ratio) were not standardised, thus adding to the cost and complexity of their use. Of these systems a contemporary technical manual remarked that 'just as there was no standard of film size, no rate of fps was established and the taking rate varied from 8 to 60 among the different systems, each of which was distinguished by some fantastic and polysyllabic name'.[39] Hollywood producers called for a widescreen standard to be established, but none ever was.[40]

The 1950s and widespread widescreen

Widescreen film formats returned – this time for good – in the early 1950s. An indication of the reasons why can be found in the opening scenes from *This is Cinerama* (1952, dir. Mike Todd *et al.*). This film, the very first of the new generation of widescreen productions, begins with a black-and-white, Academy ratio prologue in which the broadcaster and newsreel commentator Lowell Thomas presents a brief history of 'moving' images from Neanderthal cave paintings to the post-war growth of television. He emphasises the small, black-and-white images produced by the latter immediately before we see the curtains open out to reveal the full screen and thence the first wide, colour images.

Television has been blamed for many of the industrial, cultural and technological changes to film and cinema during the second half of the twentieth century, and it is certainly true that the 1950s saw the most intensive growth in the market saturation of this medium. But it would be oversimplifying the issue to say that widescreen technology was dusted off and put into full-scale use at the moment it was simply as a weapon with which the film industry could take on television. Wider forces for change and wider processes of change began to emerge during this period, of which this technology was an important factor but not the only one. Another example to sit alongside Lowell Thomas and television is a 1958 manual entitled *How to Make Good Home Movies*. The case-studies it explores include a country fair, a mother bathing her baby, 'Laura's Seventh Birthday' and 'Going to the Zoo'.[41] The many and various leisure activities cited in the book do not include going to the cinema. In Patricia Zimmerman's words, moving images had become 'a commodity for use within nuclear families' as distinct from a form of entertainment consumed in a communal setting.[42] The immediate post-war period saw a large proportion of the male population being released from the armed forces and starting families. This in turn led to a large expansion in house-building and the creation of new, suburban communities which were an appreciable distance from the town and city centres where cinemas tended to be located. Furthermore, a significant proportion of the cinemas in the UK and Europe had been damaged or destroyed by bombing. In the US a court decision in 1948 (the so-called 'Paramount case') found that the dominant Hollywood studios were operating monopolistic practices and ordered them to sell many of the cinemas they owned. A significant number closed as a result, and not only those which were studio-owned or controlled: independent theatre operators were also affected by these fundamental changes in the economic and regulatory framework of the exhibition sector. Film historians have characterised the late 1940s and early 1950s as marking the end of the 'Classical Hollywood' period, or in economic terms the dominance of the Hollywood 'system' in which studios, distribution infrastructure and cinemas were all owned and operated by the same companies. The film industry's core market was under threat from a number of economic, cultural and demographic changes; as with sound and the Great Depression, technology – this time in the form of a significant upgrade to the picture and sound technology – was the weapon Hollywood used to fight back on all these fronts.

Cinerama

Essentially, four significant widescreen systems came on the market during the early to mid-1950s. The first, Cinerama, was the invention of Fred Waller, a Hollywood research scientist between the wars, and Mike Todd, a Broadway theatrical producer. The process consisted of a cinema auditorium fitted with a deeply-curved screen intended to approximate the human field of vision, lit by three synchronised projectors positioned at reciprocating angles in boxes at the rear of the auditorium. Cinerama images were shot using three interlocked camera mechanisms mounted on a frame, using standard 35mm film stock but with a non-standard pulldown of six perforations, giving each image, or 'panel', an almost square ratio, and a non-standard speed of 26fps. There was no sound-on-film, with the six-channel magnetic soundtrack being reproduced from a separate device (known as a 'follower') which was electronically synchronised to the projectors. The first public screening of *This is Cinerama* took place in New York on 30 September 1952.

Cinerama followed the familiar pattern established by Vitaphone for sound and three-strip Technicolor in the case of colour: it established a market for widescreen and proved that the principle was an economically viable one, but the process itself was technically flawed and was soon pushed out by other systems which resolved the problems that quickly emerged. Cinerama camera assemblies were large, cumbersome, had a limited range of lenses and were very restrictive to use, especially in shooting fictional feature films in the way that Hollywood directors were used to. Cinerama required a specially designed auditorium which necessitated totally gutting and rebuilding an existing cinema to accommodate the projectors and screen. Cinerama exhibition was very labour intensive, needing as many as 17 projectionists to operate each screening. The joins between each panel were clearly visible, despite modifications to the projector apertures to try and lessen their impact. And the technology itself was just not very reliable. Four film elements running in separate mechanisms had to remain in constant synchronisation, and any show could be abruptly halted either by an equipment failure or a film break (and early triacetate film stock tended to break quite often). This happened so frequently that a number of short 'breakdown films' were produced. These featured Lowell Thomas explaining the intricacies of the technology and the various ways in which it could go wrong while the projectionists tried to resynchronise and restart the Cinerama projectors.

The other precedent set by Cinerama was to subtly change the exhibition context to meet the needs of widescreen technology. During the 'classical' period a key reason for Hollywood's economic success was that format standardisation and duplication technology allowed a single film to be copied many hundreds or even thousands of times at a relatively insignificant cost (insignificant in comparison with the production costs of a feature) and distributed all over the world. The studios were thus able to apply the 'pile 'em high and sell 'em cheap' principle, because the potential customer base was so large that even lavishly produced star vehicles could be 'sold' to consumers in the cinema at a very low unit cost. This was just not possible with Cinerama, given its technical and architectural demands. So Todd and Waller adopted the so-called 'road show' practice, in which a single film would be shown in a prestige city-centre venue for months at a time and then move on, rather than for two or three days at hundreds of venues across a country. Seats had to be booked in advance and the price was similar to that of a Broadway musical (i.e. many times more than a visit to a suburban fleapit). While this was essentially a practice borrowed from legitimate theatre, 'road-shows' had been used to a deliberately limited extent by Hollywood to market prestigious or highly-budgeted films from the teens onwards. Its significance in this instance was in showcasing form (i.e. the technology) over content.

Hollywood and 3D

In the end only seven films were produced using the three-projector Cinerama system, most of them travelogues and documentaries,[43] and less than thirty compatible auditoria were ever built.[44] Nevertheless, Hollywood took note of Cinerama's short-term economic success and looked for ways of capitalising on the public enthusiasm for new moving image technologies while also meeting the requirements of mass-distribution. Their first attempt did not last very long. Immediately before one of the major studios took on widescreen, independent producer and former radio impresario Arch Oboler produced and directed *Bwana Devil*, first shown in Los Angeles on 26 November 1952. Though unremarkable in terms of content – a cheap and nasty low-budget horror film in which a pride of man-eating lions pick off unsuspecting railway passengers – in form it marked the beginning of Hollywood's technological fight back. While Todd and Waller had changed the shape of the screen, Oboler tried to change the shape of the picture itself – with an image that appeared to be three-dimensional.

The technical processes he used were not new,[45] though Oboler's exploitation of them marked the first (and last) systematic attempt to mass-market 3D. This format uses the Cinerama method of multiple camera mechanisms physically positioned in relation to each other and electrically synchronised: in this case two cameras are mounted with their lenses roughly the same distance apart as human eyes, in order to expose a 'left eye' and 'right eye' image. The resulting films are then projected simultaneously through polarising filters while viewers wear spectacles which enable the human brain to perceive the two images as originating separately through their left and right eyes, thus reproducing the illusion of a three-dimensional image. Oboler's 3D system did not vary the standard aspect ratio or film format, which remained the 35mm four-perf Academy ratio. Therefore no architectural modifications to cinemas were needed and standard cameras and projectors could be used (thus reducing production and exhibition costs), though two of each were required.[46]

For a very short period during 1953 and 1954 two Hollywood studios attempted to introduce 3D as an alternative to Cinerama. Unlike Oboler's B-movie 'lion in your lap' Warner Bros. and MGM briefly attempted to produce first-run features in the format, for example the musical *Kiss Me Kate* (1953, dir. George Sidney) and the sus-

CinemaScope, with its revolutionary new lenses, achieves the illusion of depth without use of glasses. Its life-like, panoramic scope, plus stereophonic sound effect provided by strategically placed speakers permitting sound to originate from the part of the screen where the action takes place, combine to make the audience experience complete engulfment and participation in the action. Dotted lines show size of conventional screen as compared to new concave, all-purpose CinemaScope screen.

TWENTIETH Century-Fox's revolutionary CinemaScope has successfully passed a long series of exacting tests and has proved itself one of the greatest technological advancements since motion pictures found their voice 20 odd years ago.

Following President Spyros P. Skouras' and Production Chief Darryl F. Zanuck's momentous decision to go all out for the life-like curved screen process with stereophonic sound, the studio also announced that directors, cameramen and technicians had moved with speed, confidence and efficiency to make CinemaScope pictures available to cinemas this autumn.

After the audience-participation medium had been subjected to experiments to master the improved technique which it makes possible, it was put to work on the studio's biggest production for years, the $5,000,000 " The Robe," a Technicolor film of Lloyd C. Douglas' best-seller. At the same time several other films were scheduled for CinemaScope treatment while samplings of every conceivable action and locale were photographed to demonstrate the advantages of the new medium for every type of film — action, drama, musical or comedy.

Simple and inexpensive and not requiring glasses for viewers, CinemaScope achieves with one camera and one projector the

Fig. 2.7 Publicity brochure for CinemaScope produced by TCF for the British launch of the format in 1954 (author's collection).

pense thriller *Dial M for Murder* (1954, dir. Alfred Hitchcock). But within a short time the compatibility issues and technical shortcomings of 3D proved to be its undoing. As with Cinerama there was a lot to go wrong in the cinema (for example, a film break would halt the show), additional projectors were needed and the audience had to wear uncomfortable glasses for the system to work. Following Oboler's precedent 3D continued to be used in the occasional horror film for several decades afterwards, e.g. *Jaws 3D* (1983, dir. Joe Alves), the closing reel of *Freddy's Dead: The Final Nightmare* (1991, dir. Rachel Talalay), and more recently as an add-on to the IMAX special format process. But as a mainstream film format it never gained a foothold.

CinemaScope

During the brief 3D interlude of spring–summer 1953, Twentieth Century Fox (TCF) were working on an alternative widescreen system, to be known as CinemaScope. While production practices changed and a certain amount of auditorium rebuilding still had to be done (i.e. to accommodate the wider screen), this was not anything like as extensive or expensive as for Cinerama, because CinemaScope was able to produce a wide image using existing cameras and projectors with only minor modifications. CinemaScope used standard 35mm, four-perf film with frame dimensions of 18.16mm by 23.16mm, giving an aspect ratio of approximately 1:2.55, and four stripes of magnetic oxide (one on each side of either row of perforations) to carry the stereo soundtrack. Smaller perforations, known as 'Fox holes' were used to maximise the available surface area, and the picture was horizontally compressed using anamorphic lenses on the camera and projector. TCF originally licensed the lens design from the French scientist Henri Chrétien, whose 'Hypergonar' lenses were used in France during the brief widescreen boom of the late 1920s. A succession of American companies – initially Bausch and Lomb, and later Panavision – subsequently developed anamorphic lens technology to improve the speed, focal distance and optical characteristics, and within a few years it had become a core part of American film production, later becoming established in Europe and across the world. On 16 September 1953 the first CinemaScope feature, *The Robe*, opened in New York, almost exactly a year after Cinerama had done likewise.[47]

Of the widescreen systems introduced during the 1950s, CinemaScope was by far the most successful, if measured by the criterion of market saturation. This is because its inventors had followed the principle which had characterised all the key developments in moving image technology up to that point: it maximised the on-screen effect, minimised the investment needed in technology and infrastructure and was entirely backwards compatible. CinemaScope consisted of equipment which was added to existing cameras and projectors and which did not affect their ability to shoot or show any of the pre-existing 35mm formats. Interestingly, the one aspect of CinemaScope to be rejected by the industry was TCF's attempt to package the process with stereo sound. Just as a generation earlier Fox had tried unsuccessfully to sell Grandeur as an integrated package with Movietone, on this occasion there was sustained resistance from smaller, independent exhibitors to being compelled to buy

Paramount feature productions photographed
in
VISTAVISION

White Christmas ✓
Three Ring Circus ✓
Strategic Air Command ✓
Far Horizons ✓
Hell's Island
We're No Angels
Run for Cover ✓
To Catch a Thief
Lucy Gallant
The 7 Little Foys

The Trouble with Harry
You're Never Too Young
The Girl Rush
The Desperate Hours
The Court Jester
The Vagabond King
The Rose Tattoo
Artists and Models
Anything Goes
The Ten Commandments

These short subjects also photographed
in
VISTAVISION

VistaVision Visits Norway ✓
VistaVision Visits Mexico
VistaVision Visits the Sun Trails
VistaVision Visits Hawaii

"Strategic Air Command" is presented with horizontal double-frame projection—This is a sample of the release print.

Color by Technicolor

Fig. 2.8 Promotional literature for Vistavision, including frames from a release print of *Strategic Air Command* (1955, dir. Anthony Mann).

both CinemaScope and magnetic stereo at once. In 1956 TCF followed the lead of other studios (most notably Warner Bros.) which had started producing anamorphic prints with mono optical sound and announced that it would do likewise. In the new variant, the width of a frame was reduced to the 'scope' ratio we have today of 1:2.35 in order to accommodate standard perforations. Fox holes had proven susceptible to tearing and straining in repeated projection, and had significantly shortened the useful life of each print, although four-track mag prints continued to be produced on a limited scale until the late 1970s. The MPPC episode in 1909–17 had shown that attempts to make the purchase of one technology (film stock) conditional on another (cameras and projectors from approved manufacturers) in a way which defied market forces would not work. Hollywood found this out again a decade later when they tried to introduce widescreen on the back of sound, and a third time when TCF tried to introduce stereophonic sound as an integrated package with CinemaScope.

VistaVision

The following spring, on 27 April 1954, Paramount unveiled the Bing Crosby musical *White Christmas* (1953, dir. Michael Curtiz) at New York's Radio City Music Hall, and with it yet another widescreen system intended to make technology succeed where predicable genre cinema was rapidly failing. VistaVision also used 35mm film, but instead positioned the wide frame horizontally. With frame dimensions of 24mm x 36mm (the same size as the frame exposed in a 35mm still camera), this offered a vastly higher definition image to CinemaScope, without introducing the slight distortion caused by even the highest quality anamorphic lenses. As with CinemaScope, only one camera and projector was needed, but unlike with CinemaScope, these could not be used with any other format.

VistaVision lasted as long as it did – until *One Eyed Jacks* (1961, dir. Marlon Brando) as a primary production format – largely because VistaVision originals could be optically printed onto standard 35mm (spherical or anamorphic) with the release prints retaining a significant increase in picture quality when compared to films originated in a lower resolution format, hence the slogan 'Motion Picture High Fidelity'. With a native aspect ratio of approximately 1:1.5 VistaVision did not claim to offer a wider image than any of its competitors but did produce a sharper and denser picture, even after printing to a different format for exhibition. The only other studio to use VistaVision was the Rank Organisation in the UK, which licensed it from Paramount in 1955. According to Geoffrey Macnab this was purely and simply to combat the spread of television,[48] and support for that argument can be found in one of Rank's first VistaVision productions, possibly one of the most atypical examples of a big-budget widescreen premiere. In *Simon and Laura* (1955, dir. Muriel Box), the ostensibly happy couple of a television soap opera are in real life on the verge of divorce, and finding it increasingly difficult to separate their idyllic fictional relationship on the small screen with their mutual hatred of each other on the larger VistaVision one. To warn of the perils of live broadcasting, the film's *dénouement* shows the transmission of the series' Christmas special episode, in which the two stars conduct an unscripted, spontaneous and almighty row as the producer looks on in helpless horror, contemplating imminent unemployment. Referring to Paramount's somewhat more saccharine subject matter, one critic suggested that the film be more appropriately titled *Black and White Christmas*.

One of the main casualties of VistaVision was the Rank-owned Kalee company, which by the mid-1950s was one of the world's largest producers of cinema projectors. With Rank's decision to adopt VistaVision Kalee invested heavily in the research and development for a VistaVision machine, which reached the prototype stage just as Paramount was abandoning the format. VistaVision had never been used for exhibition on any significant scale in the US, and projectors had never been mass-produced. Rank discovered that the vast majority of cinema projection rooms simply did not have the physical space to accommodate two sets of projectors (the VistaVision machines could not show standard 35mm as well), and in the end only four were built, two of which were used briefly in Rank's flagship cinema, the Odeon Leicester Square, for premiere runs of Rank and Paramount VistaVision productions. It is thought that this failed investment was a major reason for Kalee's subsequent closure, and thereafter Rank cinemas imported their projectors from the Italian firm of Cinemeccanica. VistaVision continued to be used on a very limited scale for back-projection plates in studios until superseded by digital technology in the late 1990s.

Fig. 2.9 Todd-AO 70mm film threaded in the magnetic sound pickup of a modern projector (picture courtesy of City Screen, York).

Todd-AO and 70mm

The last of the 1950s widescreen formats was conceived as a modification of Cinerama which offered similar picture and sound characteristics but which could be delivered using much cheaper and more reliable technology. It was also the final system to appear, the first screening of *Oklahoma!* (1955, dir. Fred Zinneman) taking place on 10 October 1955. The core technology was essentially a resurrection of Fox Grandeur, consisting of a 65mm camera negative and 70mm print stock, the additional width being to accommodate four stripes of magnetic oxide which held six separate channels of soundtrack. The differences were that the pulldown was increased to five perforations and the film transport speed was initially 30fps. As with Cinerama, Todd-AO was projected onto a curved screen (with a horizontal angle of 128°), using optically compensated lenses to produce the curve from a single strip of film.

This format was the product of Mike Todd, who had left Cinerama in 1952 to work on a more technically viable alternative, and Brian O'Brien of the American Optical ('AO') company, who was primarily responsible for the lens technology. CinemaScope and the Todd-AO 70mm format are the two 1950s widescreen systems which still survive in some form today, because as with CinemaScope, the Todd-AO projector was compatible with the pre-existing standard. By replacing a few components it could also be used to show 35mm (unlike the Grandeur projector of 1929). In the end only two films were produced in the original curved-screen, 30fps version of Todd-AO: *Oklahoma!* and *Around the World in Eighty Days* (1956, dir. Michael An-

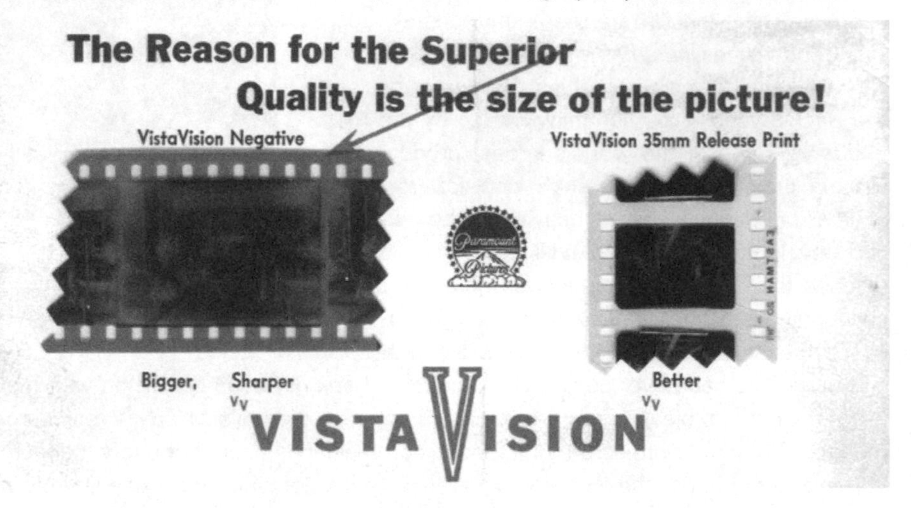

Fig. 2.10 When it proved impossible to market VistaVision horizontal projectors for cinema exhibition, Paramount promoted it as an origination medium which would yield higher quality release prints for conventional 35mm, non-anamorphic widescreen release (author's collection).

derson). Subsequent variations to the format enhanced its backwards-compatibility. The film speed was reduced to 24fps, thus enabling productions to be more easily released on 35mm anamorphic prints, while the 'bug eye' lenses were replaced with spherical ones so that 70mm could be shown on a flat screen (in 1962 the director David Lean demanded that the curved screen at a London West End cinema be replaced with a flat one lest it distort the mirage sequences in *Lawrence of Arabia*). Furthermore Cinerama abandoned their three-strip process in 1963 and replaced it with a system based on 70mm film with anamorphic compression. 70mm production declined throughout the 1960s and 1970s, while optically-enlarged prints from 35mm originals (using wet-gate printer technology developed by Panavision) were made on a regular basis, primarily to utilise its six-channel sound capability. Before the launch of digital sound-on-film in 1992 there was no other way of presenting six-channel stereo besides 70mm, hence the reason blow-ups were made on a regular basis, even if in most cases the increase in picture quality was marginal.

So, of the four key widescreen technologies launched during the 1950s, one (Cinerama) lasted only a decade before effectively disappearing without trace, another (VistaVision) lasted only a short time as a production and exhibition format but remains in limited use for special-effects production and special venue exhibition, a third (70mm) retains a niche market for high-quality imaging, while 35mm anamorphic formats (i.e. CinemaScope and its derivatives) remain a standard format for which virtually every cinema is equipped and which accounts for a third to a half of feature films produced in the Western world today. It is interesting to note that the two formats which survived are those which sought to modify and improve on existing norms rather than abandon them altogether, which is yet another example of standardisation as the moderating influence between the aspirations of the technologists and the economic realities of the film industry. And in the absence of a dedicated widescreen format, producers have since the 1950s been extending the technique used to create the Academy ratio of placing a black matte between each frame on standard 35mm in order to produce wider ratios but keep the four-perf pulldown. Following the initial success of CinemaScope, Paramount began to use a 1:1.66 matte on standard 35mm for reduction prints of its VistaVision features; other studios settled on 1:1.75 and 1:1.85.

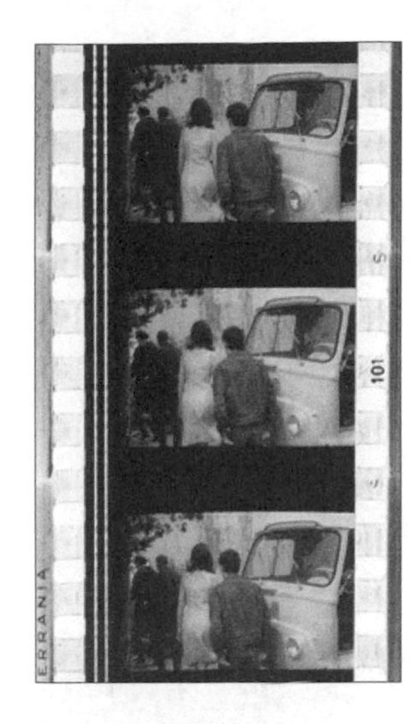

Fig. 2.11 Widescreen pictures from standard film. Originally devised by Paramount as a means of showing VistaVision without the need for a dedicated projector, this format is now used to release virtually all theatrical features which are not shot anamorphically. The film above is 'hard matted' to a ratio of 1:1.66; below is film (which also has a stereo optical soundtrack) shot in a camera with an Academy mask, but intended to be projected with a lens and aperture plate which magnifies the area of the frame between the black lines to give a ratio of 1:1.85.

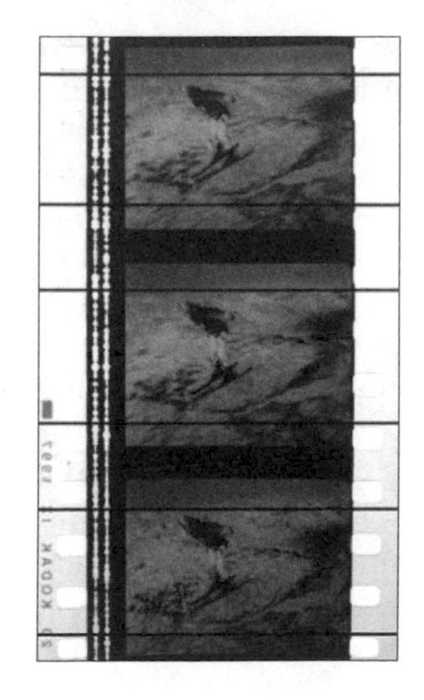

Thereafter format confusion reigned. In the absence of any firm standard for 35mm spherical widescreen most cinemas equipped themselves for one or other of the ratios in use, the decision being made more on the basis of architectural conditions inside the auditorium and the cost of lenses than with any regard to the ratio in which films were intended to be shown. By the late 1970s Hollywood had more or less settled on a 'flat' widescreen ratio of 1:1.85, which was also adopted in the Far East, while Europe and Asia tended to use 1:1.66. This remains the case today, although 1:1.66 is increasingly falling out of use (even in France, where 1:1.66 had become firmly established as the ratio of choice among the *Nouvelle Vague* directors such as Jean-Luc Godard and Claude Chabrol), as 1:1.85 more closely approximates the proportions of the 16:9 widescreen television screen.

By the end of the twentieth century 1:1.85 had effectively replaced Academy as the standard 'flat' ratio for cinema exhibition, although unlike with Academy, no universally accepted standard exists which says so. Even by the late 1990s a projectionist with a choice of widescreen lenses available would usually have to determine which one to use by holding a section of print against a light source and making a guess based on the apparent composition of the frame. The British cinematographer Walter Lassally observed that this state of affairs has led to a policy of 'shoot to protect' among production designers and cameramen, in which the frame area between the 1:1.85 boundaries and the edge of the Academy matte is kept clear of any action (for display on a TV screen or in case this entire area is masked in projection) but included in the frame composition (lest the film be projected with an Academy mask), as in the lower example in fig. 2.11.[49]

The aspect ratios in use in 2005, therefore, have their origins in a spate of research and development undertaken by the Hollywood studios during the late 1920s. This did not initially lead to industry-wide implementation, largely because the formats which resulted were not in any way compatible with pre-existing 35mm production and exhibition technology and because the prevailing economic conditions impeded their introduction. Hollywood tried again in the 1950s, and of the spate of formats developed this time two eventually stuck: CinemaScope because it offered a substantially wider image simply by attaching a new lens to existing equipment and 70mm because the projectors could also show standard 35mm. Furthermore the economic conditions this time were such that significant changes to the technology used in moving image production and exhibition were likely to address a problem rather than cause one.

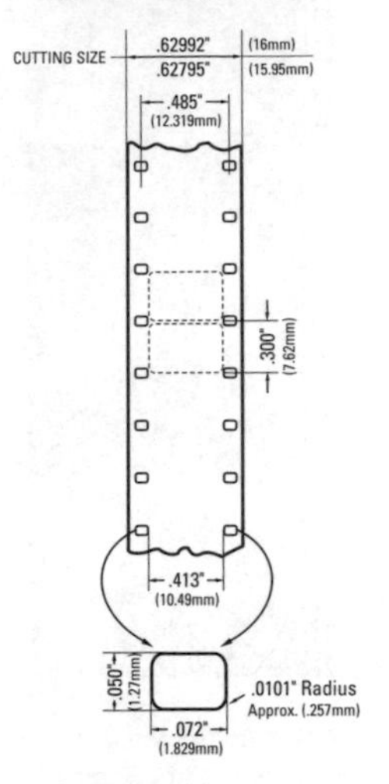

Fig. 2.12 SMPE specification for 16mm film.

Sub-35mm film formats

'In 1918 I proposed at a meeting of the SMPE held at Rochester, NY, the creation of a new and separate standard for motion pictures used *outside* the theatre, or on what is usually called the non-theatrical field.'[50]

It is easy to see what prompted this remark, made by Albert F. Victor of the Victor Animatograph Corporation at a lecture in 1944. Between 1898 and the launch of the first mass-marketed sub-35mm gauge – the Cine-Kodak 16mm system in 1923 – at least 41 camera and projector systems were designed for amateur use, 30 of which used nitrate![51] During the early years of the twentieth century, amateur cinematography was clearly not a hobby for the faint-hearted. Although there were efforts to promote equipment aimed specifically for the 'home movie' sector, the reality was that non-professional filmmaking before 1923 was sporadic and limited, and most film not intended for revenue-earning exhibition was shot by people with some connection to the film industry (which gave them access to equipment and film stock) and on 35mm.

Between the appearance of the first acetate stocks in 1909 and the 16mm gauge in 1923 the use of sub-35mm gauges for cameras and projectors intended for non-professional users was attempted by several manufacturers, without commercial success. As early as 1898, in fact, the British pioneer Birt Acres had laterally split 35mm film in two, producing a 17.5mm strip for use in his 'Birtac' camera. The two major drawbacks with this and all the other pre-1909 systems were that the stock was nitrate and that reversal film was not available for moving image use at this stage (the technique was first developed in 1899 – see chapter one), meaning that amateur cinematographers had to go to the trouble and expense of shooting a negative and then having a print made. The first gauge intended purely for amateur use which took advantage of Eastman's earliest acetate base was sold by Pathé Frères in Paris from 1912.[52] Again using a negative-positive system, the Pathéscope system featured a film gauge of 28mm. Pathéscope camera negatives (and, in the case of commercial titles sold for home exhibition on the format, printing intermediates) were nitrate in order to take advantage of its durability, but projection prints were acetate to minimise the risk of fire. Nitrate elements had four perforations on both sides of each frame. Acetate elements, however, had four perforations per frame on one side, which were engaged by the transport sprockets in the camera and projector mechanisms, but only one on the other. This was adjacent the frame line and was engaged by the projector's intermittent movement. This asymmetrical arrangement of perforations on the print stock ensured that it was physically impossible to thread nitrate stock in a Pathéscope projector (and thus ignite it) and equally impossible to thread prints the wrong way round, thus making life easier for the non-professional operators. It was this system – only with acetate negatives as well – which Albert F. Victor called for the SMPE to standardise along with 'professional standard' 35mm in 1918, and which it duly did.

Meanwhile Eastman Kodak had been working during the 1910s on adapting the chemistry of reversal processing to an emulsion for moving image use. By the end of the decade the company believed that the combination of an acetate base and reversal processing (i.e. the customer got a projection positive straight out of the camera) had made amateur film ready for the mass-market. Safety was a big selling point, and George Eastman did not believe that the compromise offered by the Pathéscope system was sufficient in the hands of untrained amateurs. In June 1923 Eastman Kodak launched the Cine-Kodak system ('system' meaning a package of camera, projector, sale and processing of film stock), using a gauge of 16mm with a single perforation on either side of each frame. Eastman was reflecting widely held concerns about the use of nitrate in private homes, as this quote from a 1929 manual on home cinematography demonstrates:

Fig. 2.13 The 9.5mm Pathéscope system was originally marketed in 1923 as a way of viewing commercially produced films at home, but after Eastman Kodak launched their 16mm format the same year, cameras were soon made available.

> It [safety film] renders the home projection of films safer than much that ordinarily goes on in the home – cigarette smoking, for instance – and incidentally, disposes of any questions that conceivably might, and not improbably would, arise in relation to tenancy and fire insurance where an accumulation of films was known to exist or frequent exhibitions took place.[53]

As with the 28mm Pathéscope perforation arrangements, the choice of 16mm was made deliberately on health and safety grounds. 16mm *could not* be produced by slitting 35mm in half, and as 35mm was the only film gauge for which nitrate stock was routinely manufactured, there was no danger – as there could be with 17.5mm

– of 16mm nitrate ever getting into circulation, either by accident or design. Eastman made it a selling point that 16mm was and would always be a safety-only gauge. When 16mm entered systematic professional use during World War Two this became a contentious issue, as we shall see below. The width of 16mm was determined by calculating the dimensions of a frame needed to produce a six-foot by nine-foot image in projection (deemed to be the largest image a tungsten-lit projector would ever be capable of) and adding space for perforations.[54]

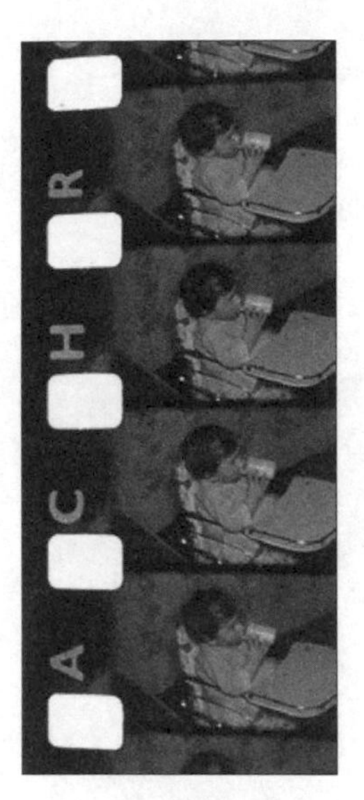

Fig. 2.14 Standard 8mm (1932, above) and super 8 (1965, below) film gauges.

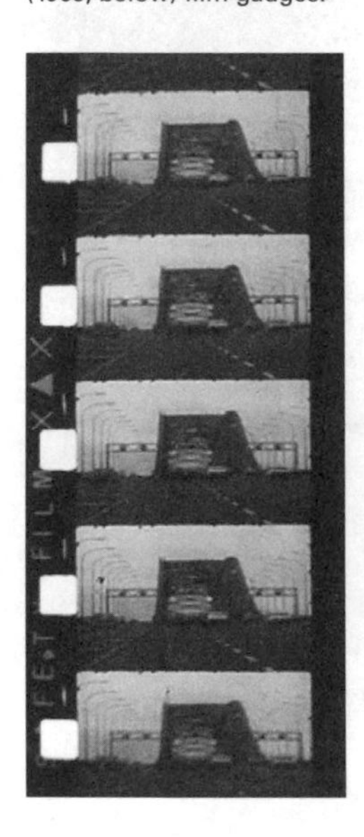

Within a few years 16mm dominated and had substantially enlarged the market for amateur cinematography, with cameras and projectors also being manufactured and sold by Bell and Howell and Victor Animatograph. In the autumn of 1922 Pathé abandoned 28mm and produced an alternative, much cheaper gauge: 9.5mm. Uniquely, the perforations were not situated along the edges of the film stock but in the centre on the frame line. Initially the format was used only as a home entertainment medium, for viewing reduction prints of theatrical features and shorts on home projectors, but the following year Pathé introduced the first 9.5mm camera. This format achieved a significant proportion of the market share in Europe, where cameras and film stock continued to be sold until the early 1960s, but was never used on a significant scale in the US. 9.5mm enthusiasts pointed out that the frame dimensions (and thus the picture quality) was almost as large as 16mm, but the cost of stock and processing was around half that of the Kodak system. To reduce costs even further, Pathé developed a reversal system which enabled home processing kits for 9.5mm to be used, but the temperature control and timing requirements of reversal development proved to be too precise for most amateurs to achieve. By the mid-1920s virtually all processing was being handled by the company's own laboratories.

The next significant sub-35mm format to enter the marketplace was 8mm, in 1932. 'Latest Eastman Achievement Cuts Cost of Movie Making Nearly Two Thirds' declared a Kodak advertisement from July of that year.[55] The principle on which this format had been created showed that the deliberate incompatibility in the 35mm to 16mm reduction process was an exception (implemented for health and safety reasons) which broke a rule that was now being observed again: reduce research and development costs as much as possible by recycling as many characteristics as possible in an existing format in creating a new one. 8mm film did not even slit 16mm down the middle to start with. The raw film stock supplied for use in the camera was standard 16mm only with additional perforations added.

The first 8mm film was supplied in lengths of 25 feet of specially perforated (but otherwise standard) 16mm reversal stock. It was exposed in two passes, then slit and joined at the lab after processing. Some later camera designs (most notably models from Bell and Howell) were made to take pre-split raw stock, which had the advantage of a greater footage capacity and a wider range of laboratories which could handle the new stock.

A further refinement of the 8mm system was the 'Super 8' format, launched by Kodak in June 1965. This essentially reduced the size and increased the pitch of each perforation by almost half and used the extra space to increase the frame dimensions, thus increasing the image definition available in projection and making space for an optical or magnetic soundtrack. As with 9.5mm, the original purpose of Super 8 was as much a pre-VCR format for viewing 'pre-recorded' (i.e. reduction prints made from 16mm internegatives) material in the home as an amateur-filmmaking medium. Standard 8 had proven a very resilient gauge for amateur use, mainly because the large and closely-spaced perforations could withstand a certain amount of mishandling. But the format could only carry a very low quality magnetic soundtrack (during the 1950s and 1960s a number of devices were marketed to synchronise 8mm projectors with early consumer tape recorders, though none was easy to use or particularly reliable)[56] and therefore the designers of the new 8mm variant were trying to adapt the gauge in order both to increase the picture quality and accommodate a soundtrack. Initial problems caused by the smaller perforations tending to strain and tear in projection were largely overcome when in 1967 Fuji marketed a variation of the system known as Single 8, which used polyester stock as distinct from triacetate. This marked the first widespread use of polyester film base, and was followed subsequently by its use in 16mm prints for airline use in the 1970s and eventually 35mm cinema release prints in the early 1990s. Super 8 was the last film format marketed for amateur use before consumer moving image technology started to become video-based in the late 1970s.

Fig. 2.15 16mm with optical soundtrack

From the late 1930s 16mm started to be used in a wider range of applications than simply for home movie making. In 1932 Bell and Howell launched the Filmosound projector, which enabled optical sound to be added to 16mm. Research during the 1920s had established that, unlike 35mm, 16mm film did not need to be perforated on both sides in order to produce an acceptable level of horizontal stability in camera and projector gates. The unperforated side could be used as a guided edge which, as long as it maintained physical contact with the gate assembly, could accommodate a soundtrack.

Thereafter 16mm started to be used for moving image applications other than home-movie making, but which for a number of reasons could not be accommodated by mainstream cinema exhibition. These included educational films for schools, promotional and training films produced by industry and government information and propaganda films. As 16mm was an exclusively safety format none of the health and safety restrictions associated with nitrate applied to 16mm. Films could therefore be shown in venues other than licensed and regulated cinemas, including schools, factories, church halls, community centres and political organisations. From the 1930s onwards 16mm was also used by groups of enthusiasts known as film societies to show reduction prints of commercially produced features; this often included films which had been banned from 35mm theatrical release for political reasons, and the increasingly professional role of the format attracted increasing opposition both from the mainstream film industry and governments around the world. As we have seen in chapter one, it has been argued that the reason why nitrate remained in use as long as it did was a deliberately maintained restrictive practice.

The use of 16mm was given a further boost during World War Two. The need to transport release prints and equipment for showing to troops in remote locations and for raw camera stock to be used in news reporting made a format which was smaller, lighter and a lot less dangerous than 35mm very attractive. Even with the slightly improved mechanical tolerances of acetate propionate, cameramen frequently found that the stock would jam and tear, as this writer commented:

> ... but the 16mm medium is no longer confined to amateur use, and has not been for a long time. Being used by professionals for professional purposes in a professional manner, professionals have long been hampered by the shortcomings of this slow-burning base ... There is only one solution to this problem, and that is the use of nitrate.[57]

16mm nitrate was never manufactured in the West, although it is believed to have been used extensively in the USSR and China as late as the 1960s.[58] Although there are a number of powerful factors to undermine the argument that 35mm nitrate was maintained as an industry standard in order to prevent the use of 16mm from expanding, such an argument does point to the increasingly complex issues which emerged as 16mm became the first (and, as it turned out, the only) sub-35mm format to be used on any significant scale professionally.

The introduction of cellulose triacetate safety film from 1948 onwards removed any remaining issues of durability and tensile strength, and with them the last barrier preventing the 16mm format from entering widespread industrial use. During the immediate post-war period to the mid-1980s, the most extensive use of 16mm was as the *de facto* recording and reproduction medium for television. The relationship between television and film will be considered in greater depth in chapter six, though for the purposes of this discussion it should be noted that video recording technology did not exist at all before 1956. The earliest videotape recorders were about the size and weight of a small car and almost required a PhD in electronics to operate. Video technology could not be considered portable in any meaningful sense until the advent of 'ENG' (electronic news gathering) equipment in the late 1970s, and therefore film – specifically 16mm film – was an important part of television technology during the three decades between the introduction of mass-broadcasting and this role being superseded by videotape. Meanwhile, the introduction in 1955 by Kodak of two high-speed, low-grain 16mm black-and-white reversal stocks (types 7276 and 7278) enhanced the suitability of 16mm for this application.

The format's versatility was further extended by the introduction of the Super 16 format in 1971, in which the frame area is extended to cover the strip of film normally occupied by the soundtrack (the sound is recorded and synchronised from separate magnetic tape). This produces an aspect ratio of approximately 1:1.75 (known as '16:9' in the television industry) without any loss of definition relative to standard 16mm. Super 16 was initially used by feature filmmakers on a limited budget or who needed smaller and more portable cameras than is possible with 35mm, for subsequent enlargement to 35mm intermediates and release prints. It also started to be used by television following the introduction of HDTV and subsequent widescreen television in the 1990s. By 2000 16mm was almost obsolete, and was only used on any significant scale for archival preservation. In the early 1990s, portable video projectors became available at similar prices to the 16mm machines of a decade earlier, and rapidly superseded 16mm for non-theatrical exhibition. In cinemas, the growth of multiplex exhibition in the 1980s (more on that in chapter five) increased the demand for 35mm release prints, causing labs to equip themselves for 35mm mass production. 16mm release printing remained a labour-intensive, manual operation, so much so that by the early 1990s a typical 16mm release print of a mainstream studio feature actually cost more to produce than its 35mm equivalent. Therefore the film societies and part-time venues re-equipped for 35mm. Increasingly high-definition videotape formats, which could withstand copying through multiple generations (essential for editing) without any visible loss of image quality in transmission, largely superseded 16mm for television production. By 2000, the only remaining uses of 16mm and Super 16 as a production medium were for low-budget theatrical features and prestigious, high-budget television drama and nature documentary filming, applications for which even the image quality of modern digital videotape was not considered sufficient, and by students.

Large film formats

This topic has been covered to a certain extent in the section on widescreen above, in that Cinerama, VistaVision and 70mm were all technologies developed to produce a larger (in terms of absolute size as distinct from aspect ratio) image than could be obtained from a single strip of 35mm. However, I have counted them among conventional cinema processes because they were first and foremost new technologies for making conventionally acted, directed and produced narrative feature films for showing in cinemas. In the event Cinerama failed because the auditoria had, in effect, to be specially built. There is another group of large film formats which were not primarily designed for conventional cinema auditoria, but specifically for special entertainment venues such as fairgrounds, theme parks, museums and major exhibitions.

The first and longest lasting was Imax, first demonstrated at an international trade fair in Japan in 1970.[59] Developed by the filmmakers Graeme Ferguson and Roman Kroitor and the engineer Robert Kerr, it was in principle a 70mm variant of VistaVision: 70mm film with the same perforation size and pitch as Todd-AO but which moved horizontally, each frame occupying the length of 15 perforations. This frame was so large (three times as big as normal 70mm) that it was possible to project it onto a screen 24 metres tall – the equivalent height of an eight storey building. While the initial generation of Wilcam-Imax cameras used the pin-registered cam-and-claw mechanism similar to that of the 1970s generation of Panaflexes, a conventional, mechanical intermittent mechanism was found to be impossible in projection because of the speed of film transport and the intense heat produced by a lamp needed to illuminate a screen that big. Instead, the projectors used the so-called 'rolling loop' method in which both film and picture gate rotate in synchronisation. Other variants of Imax soon followed, including Omnimax (curved screen) in 1973, Imax-3D in 1986, which simply applied the same technique used for *Bwana Devil* and the 1950s generation of 3D films (two synchronised projectors with the viewer wearing polarised glasses) to Imax, and Imax-HD in 1992, which enabled viewers to sit closer to the screen without experiencing visual distortion.

The output of Imax films has until very recently been confined to the genres of documentary, travelogue and music video. Notable public successes have included *The Dream is Alive* (1985, dir. Graeme Ferguson), about the US space programme, a music film *At the Max* (1991, dir. Noel Archambault *et al.*) featuring the Rolling Stones and *Fires of Kuwait* (1992, dir. David Douglas), dealing with the reconstruction in the Middle East following the 1991 Gulf War, which won the best documentary Academy Award. One key limitation of the system was that until the late 1990s the capacity of an Imax projector limited a film's running time to approximately 45 minutes (though *At the Max*, running 89 minutes, was shown in two halves with an interval). This was eventually overcome and the 82-minute animated feature *Fantasia 2000* (2000, dir. Hendel Butoy *et al.*) was shown in the small number of converted Imax cinemas which could accommodate the larger reels, as well as being released conventionally on 35mm. Unlike any other special format/special building system, Imax has survived successfully for over three decades because it only ever aimed for a niche market. Unlike Cinerama, the system's promoters never tried to persuade every suburban multiplex to equip themselves with the format, instead opening small numbers of purpose-designed venues in prestigious city-centre locations. From 2002 digital film processing technology has been used to produce Imax enlargements of feature films originally shot in conventional formats: at the time of writing *Apollo 13* (1995, dir. Ron Howard), *Star Wars: Episode 2 – Attack of the Clones* (2002, dir. George Lucas) and *The Matrix Reloaded* (2003, dir. Andy & Larry Wachowski) have been given a limited roadshow release in this format. However, and despite the ability to convert ordinary films into this large-screen format, Robert Carr and R.M. Haynes correctly point out that, 'regardless of the extremely high definition or clarity of Imax, it will never be a general film process due to its logistical and financial implications'.[61] The UK market vindicated this prediction in January 2004, when one of the country's nine IMAX screens was forced to close due to falling revenue. 'For the purists it will remain unbeatable', commented the commercial director of another venue, 'but that doesn't point to a growth market, which is a worry.'[61]

Other special large film formats which have been used since the 1970s include Showscan, which used conventional 70mm film only running at a speed of 60fps in order to heighten the clarity of movement and totally eliminate visible flicker, and a number of hybrid 70mm systems with a vertical, 8-perforation pulldown for showing reduction prints of Imax originals in venues such as fairgrounds and theme parks.

Conclusion

The decision which marked the start of film being used in a systematic way to originate, duplicate and display moving images was the definition of a format and a standard. Before the early 1890s the *principles* of film-based moving image technology – a flexible, transparent base, coated with a photosensitive emulsion which was exposed and projected intermittently between movements of a fixed distance – had been established. But in order for them to work efficiently and economically in practice the variables inherent in these principles had to be fixed. How wide would the film be, what were the dimensions of each frame, what length of film would be pulled down by each movement and how many times per second would these movements take place? Unless those variables are the same in both the camera and projector, the illusion of a moving image cannot be reproduced effectively.

In the early 1890s, therefore, W. K. L. Dickson determined a standard which he and his employers hoped would be adopted throughout the nascent industry, not least because the latter owned the patent rights. It was the same logic that would later motivate Donald J. Bell's 'fondest hope' for equipment standardisation. Standard 35mm proved to be ideally suited for conventional cinematography and auditorium projection, so much so that (with a few minor variations), this format is still being used for this purpose in 2005. But as the range of applications for film expanded to encompass systems designed specifically for amateur use, widescreen, film as an recording medium for television and film for special venues, new formats were evolved to address these purposes. They stood or fell on their compatibility with (and in the case of 16mm, its deliberate incompatibility with) pre-existing formats and the ways in which the technologies were commercially exploited. The MPPC episode and TCF's attempt to sell CinemaScope and stereo sound as a package demonstrated (as would the Betamax consumer video format in the 1970s) that an increasingly sectorised industry would not put up with the use of one form of technology being made artificially contingent on another. Formats which were designed and marketed in a way which were most compatible with existing technologies introduced the smallest additional cost element possible and functioned in an essentially open marketplace. Therefore the Academy ratio, Super 8, CinemaScope and 5-perf 70mm all survived, while 3D, the 1920s widescreen systems, 9.5mm, Cinerama and VistaVision were only used for a short time and on a limited scale, and were never really commercially viable. About the only significant exception to this rule is the triumph of 16mm over 17.5mm, and this was artificially engineered for a specific political reason: the danger of allowing nitrate film to be used in an unregulated environment. A convincing demonstration of this phenomenon can be found in the growing importance of standard-setting bodies, most importantly the SMPE/SMPTE, and to a lesser extent the ISO and DIN.

The next two chapters will consider how two additional technologies – colour and sound – were developed and adapted for compatibility with established film and equipment formats in order to meet the demands of the commercial climates in which they emerged.

chapter three | **colour**

> 'I wonder if he'd like to have me bring my magic lantern over some evening?' Alexandra turned her face toward him. 'Oh, Carl! Have you got it?' 'Yes. It's back there in the straw. Didn't you notice the box I was carrying? I tried it all morning in the drug-store cellar, and it worked ever so well, makes fine big pictures.' 'What are they about?' 'Oh, hunting pictures in Germany, and Robinson Crusoe and funny pictures about cannibals. I'm going to paint some slides for it on glass, out of the Hans Andersen book.' Alexandra seemed actually cheered. There is often a good deal left of the child in people who have to grow up too soon. 'Do bring it over, Carl. I can hardly wait to see it, and I'm sure it will please Father. Are the pictures colored? Then I know he'll like them.' – Willa Cather, *O Pioneers!*

A processed photographic image in its most basic form consists of an emulsion formed of pure metallic silver of varying density. This is often referred to – incorrectly – as 'black-and-white'. The correct term for such an image is 'monochrome' (from the Greek: 'mono' – one, 'chrome' – colour). The metallic silver appears to the human eye as a single colour, but its shade, or intensity, varies according to the volume of the chemical present on a given surface area of the film. In an analogue photographic image the number of shades is theoretically infinite, so perhaps a more accurate term for such an image would be 'black *to* white'. In the digitally processed image opposite, there are 256 shades of the printing ink used to produce this book between the darkest and the lightest. A genuinely black-and-white image is comprised of only two shades: either you see the colour of this paper with ink on it (black) or without (white). The photographic and moving image research scientists of the nineteenth century had, by 1889, arrived at a combination of technologies which enabled the recording and reproduction of (seemingly) moving images using one colour. It was not long before they started work on adding the rest of them.

Fig. 3.1 A genuinely black-and-white (top) and monochrome (bottom) photographic image (showing a reel of 35mm monochrome release print stock).

From 1889 until the introduction of optical sound in the early 1930s, moving image film reproducing more than one colour did so in one of two ways, which I will classify as artificial colour and photographic colour. Artificial colour refers to processes which introduced coloured dye to the film independently of the recorded monochrome image, while photographic colour systems attempt to record a greater range of the visible colour spectrum than is possible with silver halide alone at the point of photography, and then to reproduce that recording accurately in projection.

Artificial colour (1889–c.1930)

The earliest form of artificial colour was the method Carl intended using for his lantern slides – hand colouring. Carl's slides were not, of course, photographic: they were images produced by making a small oil or watercolour painting on a transparent base (i.e. glass) which was then displayed by projection. Needless to say hand colouring of 35mm moving image film was a far more difficult and time-consuming process. Not only were the frames a lot smaller than in opaque-base paintings or even lantern slides, but the volume of images was far greater: approximately 960 per minute of screen time. During the nineteenth century the pioneers of still photography had identified a number of chemical compounds which could be used to dye the metallic silver emulsion, such as hydrogen sulphide for brown or copper ferrocyanide for red. They were applied to nitrate film by being mixed with alcohol and then placed on the emulsion surface with a very small brush. The dye was absorbed by the gelatine layer which binds the emulsion to the base, and after the alcohol had evaporated the gelatine was dyed. Colours were applied as the film passed a bench-mounted machine similar to the gate assembly in a camera or projector which was lit from below and in which each frame was held stationary while being painted.

Brian Coe cites the earliest instance of a hand-coloured film being publicly shown in the UK as that of an 'Eastern dance' by R. W. Paul at the Alhambra Theatre (later the Odeon Leicester Square) in London on 8 April 1896,[1] and it seems that by the turn of the century the technique was in regular, though limited, use. The longest surviving films to be hand coloured were made by the fantasy director Georges Méliès; by 1899 he was selling 'some' release prints of fictional narrative films up to 10 minutes in length with hand colouring.[2] Hand-coloured prints of two of his best-known productions, the science fiction drama *Le Voyage dans la lune* (*The Journey to the Moon*, 1902) and the sci-fi spoof *Le Voyage à travers l'impossible* (*The Voyage Across the Impossible*, 1904), in which explorers from the 'Institute for Incoherent Geography' travel to the sun by train, survive in the UK's National Film and Television Archive and have been preserved on modern colour negative stock. They reveal an astonishing accuracy of colour registration between the different parts of the image, given that the original element would have been coloured by individually painting onto almost 10,000 separate images, each of which was about the size of a postage stamp. By the middle of the 1900s it became apparent that hand colouring was just too labour intensive to be economically viable for significant numbers of release prints, and ways were introduced of automating the process.

But that was not quite the end of this technology. In the 1930s the avant-garde animator Len Lye revived it for some sequences in a series of short films commissioned to advertise the Post Office: *A Colour Box* (1935), *Trade Tattoo* (1936) and *Rainbow Dance* (1936). *A Colour Box* was an experimental film intended to highlight and distort the effect of the 'induction of continuous movement phenomenon' as described in chapter one. No camera at all was used to create the film, which was made entirely by painting directly on to strips of unsensitised, raw film stock. The abstract painted shapes in some sections disregarded the positioning of frame lines, and so a continuous, flowing shape on the film itself appears in projection as disjointed, four-perf sections. These scenes were intercut with repeated drawings occupying four-perf sections, thus creating the familiar image of movement. Lye explored these ideas further in *Trade Tattoo* and *Rainbow Dance*, superimposing hand painting on live photography, just as Méliès and the early colourists had done. These films were duplicated and distributed using photographic colour systems (Dufaycolor, Technicolor and Gasparcolor respectively) which had become available during the intervening three decades and which are discussed below. Lye's ideas were adopted on a limited scale by other experimental animators, most famously by the Canadian Norman McLaren, whose short *Begone Dull Care* (1949) set hand-painted animation to jazz music and in doing so launched the career of the pianist Oscar Peterson, and by the American avant-garde pioneer Stan Brakhage.

Reverting to the mid-1900s, Méliès and his colleagues quickly discovered that hand-coloured films were enormously popular with audiences. But as the nascent film industry grew to require ever larger quantities of release prints for each title, the labour-intensivity of this process quickly made it impossible in its original form: Barry Salt suggests that a hand-coloured release print would cost an exhibitor 'three or four times' a monochrome one (this being before the days of distributors when prints were purchased outright by exhibitors).[3] In 1905 the Pathé company devised a means of mechanising it. Branded as Pathécolor this system originally involved cutting stencils from a release print by hand, but in 1907 was modified by the introduction of a device which involved back-projecting each frame onto a screen approximately the size of a modern portable television set. The colourist traced the outline of the area of the frame to receive the colour dye using a pointer, which was linked by a pantograph mechanism to a stylus which cut away the corresponding area on a frame of raw film stock. The process was repeated for each frame and the resulting reel of cut film became a stencil. Separate stencils were produced for each colour to be used in dyeing the release print. To dye the film itself, a machine was used which laid a stencil on top of a release print and applied a dye to the surface. The dye passed through those areas of the stencil which had been cut away, thus dyeing the print itself in the areas selected by the colourist. The print was then rewound and the process repeated with the next stencil and colour dye. The dyeing process could be repeated for as many prints as were required, effectively meaning that the hand colouring only had to be done once. As Roderick T. Ryan has noted, the principle behind this system was very similar to the dye transfer process developed by Technicolor (see below), the crucial difference being that the Technicolor dye process was intended to reproduce colour which had been recorded photographically rather than added later.[4] Nevertheless, stencil colouring remained a labour-intensive process requiring highly skilled colourists: a manual from 1915 estimated that a typical Pathécolor colourist was capable of cutting three feet of stencil (or 48 frames) per hour.[5]

Pathécolor and variants thereof were used on a limited scale until the mid-1930s. In 1916 Max Handschiegl devised a means of mechanising the application of each colour dye and increasing the accuracy of registration, and his system was used by a number of Hollywood studios for high-budget features during the late 1910s and 1920s. However, in the last analysis, stencil colouring was also too time-consuming and expensive to be a viable means of adding colour to feaure-length narrative films. In the early to mid-1910s feature films of several reels duration started to be shown, notably *Cabiria* (1914, dir. Giovanni Pastrone) and *The Birth of a Nation* (1915, dir. D. W. Griffith), the first European and American features to significantly exceed 10,000 feet in length respectively. At the rate of three feet per colour per hour, these features would literally have required several man-years to prepare a full set of Pathécolor stencils (assuming the average of three to six colours). A very small number of prestige, highly-budgeted feature-length films were coloured in this way, for which artificial, individual colouring systems were used during the 1920s. But this approach was clearly no longer viable for the mainstream, given the industrial-scale production of feature-length release prints which was now taking place. By the end of the 1910s there were still no photographic colour systems available which matched even the aesthetic impression of reality achieved by artificial, individual colouring systems at the time. Another means of artificial colouring was needed which was quicker and cheaper. The 1910s and 1920s, therefore, became the decade of tinting and toning.

In terms of the image on the screen, it would be fair to say that tinting and toning was actually a step backwards from Pathécolor or Handschiegl. Unlike these systems a maximum of two colours could be applied to each frame, and furthermore there was no way of applying the dye to selected areas of the film – the whole surface had to be dyed uniformly. *Tinting* is the application of a layer of dye that is absorbed by the gelatine 'subbing layer' which binds the emulsion to the base, resulting in the light or clear areas of the picture taking on the colour of the tinting dye. Examples include Amaranth for red, Quinoline for yellow and Naphthol for blue. *Toning* is the application of a dye that reacts with the metallic silver which forms the emulsion, resulting in dark or opaque areas of the picture taking on the colour of the toning dye. Examples include Prussian Blue, Chrysoidine for brown, Safranine for red and even Uranium Sulphate for brown! Put in crude terms, toning colours the black and tinting the white. A combination of two colours could be used for the tint and tone (such as Prussian Blue and Croceine Scarlet for a red tint with blue tone), and in some rare cases stencil colouring was also applied to a print after tinting and/or toning.

Because toning dyed the developed silver image, it could only be done to a release print after processing. Tinting, however, was carried out in one of two ways. Tints could also be applied after processing in the same way as tones, which was done in a laboratory by simply winding reels of processed release print through a set of rollers immersed in a bath containing the dye. Because the gelatine layer which receives the tinting dye does not undergo chemical change during processing, release print stock could also be tinted *before* exposure. The first reels of pre-tinted stock were sold by the Belgian Gevaert company in 1912, and within a few years they were marketed by all the major film manufacturers. According to Ryan, by the early 1920s pre-tinted film accounted for 80–90 per cent of commercial release printing.[6] In 1921 the range of pre-tinted stocks sold by Eastman Kodak included red, green, blue, light amber, dark amber, pink, yellow and orange.[7] In order to colour different scenes using different dyes, the negative being used to strike the release print (which, until the use of intermediate elements became widespread in the late 1920s, was usually the cut camera negative) would be made up into reels of sections, not necessarily in their order of appearance in the finished film, but with each reel of negative to be either tinted and/or toned with the same dye or printed onto the same colour of pre-tinted stock. Each section was identified with a reference number identifying its place in the finished film. After processing, the sections of coloured positive were cut together in order of their reference numbers and intertitles added to form each complete print. By today's standards this method was also very labour-intensive, though nowhere near as much as with stencil colouring.

The practice of tinting and toning came to an abrupt and almost total end with the introduction of sound in 1926–32. This was because the optical, or photographic recording and reproduction of a soundtrack depended on a consistent amount of light illuminating the photoelectric cell which 'read' the soundtrack in a projector for accurate modulation. As the presence of a uniformly applied colour dye which changed from scene to scene varied the density of the emulsion, it affected the signal level being fed to a cinema's amplifiers, and thus the volume of the sound being played in the auditorium. To address this issue Eastman Kodak introduced a range of stock which was pre-tinted in the picture area but not the soundtrack, known as 'Sonochrome': however, it was considerably more expensive, the dyes had a habit of leaking and it was only ever purchased in small quantities. A few sound feaures were nevertheless released (either in their entirety or with selected sequences) on Sonochrome print stock, or in a combination of Sonochrome and two-colour Technicolor scenes, *Hell's Angels* (US 1930, dir. Howard Hughes) being a notable example. Pre-tinted stock continued to be sold during the 1930s for amateur use in the (silent) 8mm and 16mm formats, but would never again be used on any significant scale for release printing for cinema projection. For a few years the majority of cinema release printing reverted to monochrome for the first time since the 1900s. In the early 1930s cinema audiences may have gained sound, but in doing so they (temporarily at least) lost colour. As the managing director of British International Pictures noted on his return from a trip to the US in the autumn of 1932, 'there is practically no colour in America now. Colour has virtually collapsed and there is no hope of it becoming a commercial possibility again.'[8] He was right in that due to the incompatibilities between the colour technology in widespread use during the 1920s and sound-on-film, the latter had pushed the former out of the market. But by this stage, however, truly photographic colour systems were almost at the point of being mass-marketed. By the end of the decade they were a firmly established technology, and colour as a 'commercial possibility' was back; this time, for good.

The theory of colour photography

Colour photography – including its use in moving image technology – differs fundamentally from the artificial colour systems described above. With hand colouring, stencil colouring, tinting or toning, colour dyes are added to a photographic image according to the perception or personal taste of the individual doing the colouring. No colour information is recorded at the point of photography. True colour photography consists of two processes: making a recording of the colour perceived by the naked eye at the point of photography (i.e. more than just the 'monochrome' of a silver halide latent image), and then reproducing that information in projection.

Colour photography as we know it today has its origins in an experiment which the scientist James Clerk Maxwell demonstrated in a lecture to the Royal Institution in 1861. This showed that every shade of colour which can be perceived by the naked eye (the 'visible colour spectrum') can be reproduced with light or dyes on a solid base by mixing varying proportions of what he termed the three *primary* colours: red, green and blue. For his experiment Maxwell photographed a piece of tartan ribbon three times with a red, green and blue filter in front of the lens. Filtration is a crucial technique in colour photography, which involves the use of a semi-transparent solid that will allow some areas of the visible light spectrum to pass through it but not others. For example, a piece of glass coated with a blue dye will only allow blue light to pass. A more sophisticated form of refractive device known as a *prism* has triangular surfaces facing a parallel axis. When light is shone through one surface it will project through the reciprocating one at a specific point in the visible spectrum, depending on the angle of projection.

In the lecture theatre Maxwell projected his photographs using three magic lanterns (i.e. slide projectors) fitted with the same filters he had attached in succession to the camera. The resulting image on the screen showed the tartan pattern in roughly its original colours. The colour reproduction was not entirely accurate as the emulsion used to form the original photographs was not only mono*chrome* (i.e. the processed image only displayed one colour) but mono*chromatic*, meaning that the unexposed silver halides were only sensitive to blue light. But Maxwell's experiment worked well enough to prove the principle.

The process of mixing the primary colours can be achieved in two ways, known as additive and subtractive. *Additive* colour involves adding to black the red, green and blue light needed to produce a given shade. For example, by shining a red light onto a screen and then a blue one, the screen will become purple. By projecting equal proportions of red, green and blue the screen will be white. This is the way in which colour is produced in television. *Subtractive* colour involves the use of filters to take away from white the proportions of red, green and blue light needed to produce a given shade. Because the three primary colours have reciprocating negatives (thus enabling duplication in a negative-positive system), all film colour today is subtractive. These three colours are cyan, magenta and yellow, and are produced by removing (i.e. filtering out) their reciprocating primary colour from white light. They are sometimes described as the *secondary* or *complementary* colours, but here will be termed the *subtractive negatives*. For example, the subtractive negative of red is the colour you would see if all the red area of the colour spectrum was filtered out of a white light source, i.e. cyan. By the same token the subtractive negative of green is magenta, and of blue is yellow.

Additive colour systems (1899–1952)

The earliest forms of photographic colour used in moving images were all additive, and worked either by a refinement of Maxwell's system – filtering the light used to expose and project the photographic image – or by incorporating colour dyes within the film itself. By the 1880s ways had been discovered of extending the photosensivity of the silver halide solution used in monochrome photography to include green, thus producing *orthochromatic* film.[9] As noted in chapter one, *panchromatic* film, which is uniformly sensitive to all three primary colours,[10] was not introduced on any significant scale for monochrome moving image cinematography until the 1920s, though it was used in additive colour systems a lot earlier.

The first generation of additive systems were restricted to two colours, usually red and green, due to the mechanical limitations of the cameras and projectors used. Colour systems which incorporated filters or dyes into the film itself did not become reliable or commonplace until the mid-1920s; all the systems developed before then used an arrangement of colour light filters in the camera and projector, some being more successful than others. The earliest known systematic research and development in this area was undertaken by the so-called 'Brighton School' of film pioneers based in the seaside town in southern England. As Luke McKernan has argued, it embraced 'a wider group, all active in the Brighton and Hove area in the early 1900s' who achieved the first demonstrable results in the process which ultimately led to the truly photographic systems for recording and reproducing colour we know today.[11]

The first step was an attempt at a full, three-clour system by the photographer Edward Turner and the entrepreneur Frederick Marshall Lee in 1899. This involved fitting an additional shutter disc to a conventional camera, which consisted of three filtered 'blades' for the primary colours. Thus, red green and blue records would be registered on three successive frames of exposed film. In projection, the frames were projected their reciprocating filters. The Lee and Turner system had a fundamental flaw. Because the camera used the 'successive frame' method but the projector displayed all three colour images simultaneously, parallax errors (see chapter 2) would have been visible as a 'fringing' effect, blurring the image on the screen. Interestingly, an account of a demonstration of Lee and Turner system written by George Albert Smith, who a few years later would develop the rather more successful Kinemacolor process, does not seem to recognise this fundamental mechanical flaw: that the projected image would inevitably be distorted because the three colour records were not exposed simultaneously. He suggested that: '...the difficulty is due to the fact that cinematograph pictures are too small to begin with', and that it was the scale of magnification in projection which caused the fuzzy picture.[12] What had worked for Maxwell did not for Lee and Turner because Maxwell's three-colour experiment was with a still photograph of a stationary subject. So it did not matter how much time elapsed between the exposure of the three colour records, and those records could be shown using three separate projectors focused on the same screen without any need for synchronisation. As film-based moving image technology depends on rapid successive exposure and projection, this obviously creates problems when each individual picture occupies more than one frame of film.

However that was by no means the end of mechanical, additive colour. George Albert Smith, in collaboration with another figure associated with the Brighton film industry, Charles Urban, continued to develop Lee and Turner's idea, and specifically to work on the problem of registration. Smith realised that the time gap between exposure of the colour records was a problem, though it would seem from his account of the Lee and Turner experiments that it did not occur to him to try and devise a means of simultaneous exposure. Instead he reduced the number of colour records from three to two, doubled the film speed, to 32fps,[13] and developed a panchromatic emulsion – a key element in the advance his system represented relative to Lee and Turner's. The result was Kinemacolor, first demonstrated publicly in London on 1 May 1908.[14] The camera used was a standard Moy and Bastie with a filter wheel mounted behind the existing shutter, timed to hold alternating filters in front of the lens while the shutter blades were open. Different filter combinations were used in the camera to suit the scene being shot (red and cyan being the most usual combination), while in projection the filters were red and green or blue/green. The combination of the slow film emulsions in use at the time, the limited exposure time available at a rate of 32fps and light absorption by the filters meant that to all intents and purposes, only exteriors in bright sunlight could be shot.

As Nicola Mazzanti discovered when printing preservation masters of Kinemacolor original release prints, the issue of 'time parallax' (or 'motion fringing' as Barry Salt termed it) is still there: Smith's improvements vastly reduced the visible flicker and colour smearing, but could not entirely get rid of them.[15] There was a gap of one 32nd of a second between the exposure of the two colour records, meaning that a fast-moving subject could move during the time taken by the film pulldown, resulting in a visible visible lack of registration in projection. Nevertheless, contemporary reviews agreed that, even with this limitation and the fact that Kinemacolor could only reproduce two of the three primary colours, the impression of colour reproduction on the screen seemed surprisingly life-like. During the early 1910s Kinemacolor was used with notable commercial success on a number of short subjects and finally for the World War One propaganda film *Britain Prepared* (1916, dir. Charles Urban), before patent litigation by Urban and Smith's commercial rivals forced them out of business.

At around this time there were several other two-colour systems based on successive frame exposure through mechanically operated filters, but eventually the researchers began to realise that truly accurate colour registration could only be obtained if the colour records were exposed and projected simultaneously. A number of two-colour systems emerged during the late 1910s and early 1920s which managed to achieve this, for example Colcin in the UK and the Busch process in Germany. Out of the research being undertaken all over Europe into mechanical additive colour systems eventually came the first successfully demonstrated three-colour additive system: Léon Gaumont's Chronochrome, developed as a rival system to Pathécolor (with the added advantage of being truly photographic) and first publicly demonstrated in Paris on 15 November 1912. The Chronochrome camera had three separate filtered lenses to simultaneously expose the red, green and blue records onto successive frames of panchromatic stock, and an enlarged aperture and triple lens assembly in the projector to reproduce them, again simultaneously. Coe notes that 'at the cost of introducing some mechanical complication, Chronochrome achieved good quality by using three almost full-size component images'.[16] In doing so it also established the principle (of using three, simultaneously created colour records) which would be incorporated in the world's first successfully mass-marketed three-colour subtractive system, Technicolor (see below).

The development of mechanical, additive colour systems during the 1900s and 1910s, first using successive frame exposure and projection, then later simultaneous, proved that Maxwell's findings could be successfully applied to moving images as well as still ones. However, these systems never made it into mass production and the mainstream cinema industry, largely because of the issues of technical standardisation discussed in chapter two in relation to film formats. To borrow Coe's phrase, the cost of introducing some mechanical complication was not one which the rapidly globalising film industry was willing to pay. All of these technologies, from Kinemacolor to Chronochrome, required specially designed cameras and projectors which could only be used in conjunction with the colour system for which they were marketed. Some, for example Kinemacolor, also required special film stock. As with Cinerama, VistaVision, Vitaphone, CDS and countless other examples, these colour processes all fell by the wayside due to their cost, complexity, unreliability and incompatibility with pre-existing technologies. By the end of the 1910s they probably accounted for 2–3 per cent of the total film footage shown in the world's cinemas at most, while the vast majority of audiences were seeing tinted and/or toned prints, and would continue to do so for a further decade. What was needed was a means of photographing and reproducing colour in a way that was compatible with the 35mm, four-perf standard as it had been enshrined by the SMPE in 1917. To achieve this, successive or simultaneous multiple-frame systems were out of the question and the use of moving filters attached to the camera and/or projector was at best undesirable. A way was needed of incorporating all three colour records on a single frame of 35mm, and ideally they had to be represented in the form of colour dyes on the release print itself, meaning that colour projection would be possible without any modification whatsoever to the equipment already installed in cinemas.

No full, three-colour system achieving the latter was available which supported the production of multiple release prints in significant quantities until the introduction of three-strip Technicolor in 1935 (see below). There were, however, two notable transitional stages between the mechanical additive systems of the 1900s and 1910s and the dye-based colour processes which emerged during the 1930s and which are still with us today. They also represented the final stage in the development of additive colour for film before this approach was largely abandoned and the subtractive method took its place.

These technologies were both attempts to record and reproduce all three primary colour records from a single frame of film, thus eliminating the 'mechanical complication' which ultimately killed off Kinemacolor and Chronochrome. The first was *lenticular* colour (sometimes referred to as *mosaic* or *dry-screen* colour). The earliest description of this process is variously attributed to Robert Berthon or Gabriel Lippmann in 1908 or 1909 respectively. It involved embossing a series of vertical indentations or lenticules into the film base, and exposing the film through a colour filter with three 'bands', for red, green and blue. They were so focused that the indentations would act as miniature 'lenses', recording an image through each separate filter onto the appropriate section of the panchromatic emulsion. After developing to a reversal positive the film was projected through a reciprocating filter. This time the indentations acted as projection lenses, directing light through the appropriate filter band in front of the projector's objective lens, thus producing a full-colour picture on the screen. This technique was first successfully demonstrated by Berthon in Paris on 17 December 1923. The patent rights were subsequently bought by Eastman Kodak, and after further development it was marketed as the first colour system intended specifically for amateur use, known as Kodacolor, in 1928.[17]

Kodacolor achieved moderate success as a home-movie medium between 1928 and its withdrawal from the market in 1938 following the launch of Kodachrome (see below). It, and the small number of rival lenticular systems developed around the same time, represented a move in the direction the industry required: all three-colour records were on a single strip of film and furthermore did not require the mechanical complication of a successive frame system. But there were crucial drawbacks. As with the mechanical systems, the amount of light needed for exposure was very high. In this case, not only was the film exposed through a light-absorbing filter fitted to the camera, but as the lenticular embossing was on the base of the film, it had to be exposed with the base side facing the lens, meaning that the base absorbed yet more light before it struck the photosensitive emulsion. As with Kinemacolor, Kodacolor could only be used successfully in bright sunlight. Furthermore, each different coating of panchromatic emulsion varied slightly in its relative sensitivity to the three primary colours, which without any compensation would upset the overall colour balance in projection. A special paper stencil had to be supplied with each roll of Kodacolor which the customer fitted in front of the filter to ensure that exposure was compatible with the colour balance of each batch of film. Failure to use it frequently resulted in poor colour rendition on the screen.[18] And finally, despite intensive research by Eastman Kodak in the early 1930s, lenticular colour could not be made to work in a negative-positive process, only reversal. This prevented its use within the professional film industry, for which large numbers of release prints were needed.

The Kodacolor lenticular process was briefly revived when, in November 1951, the 35mm Eastman Embossed Print Film, type 5306, was first demonstrated. This was the result of collaborative research between Eastman Kodak and Twentieth Century Fox's head of research, Earl Sponable. The idea was to produce a way of striking large numbers of single-strip colour release prints which was cheaper and more flexible than the Technicolor dye transfer process which at the time represented the only colour system capable of yielding multiple prints of consistent quality (see below). Kodacolor mk. II involved optically printing colour separation internegatives produced using the Technicolor beam-splitting prism, through the lenticular filter bands onto 5306. Although TCF decided in 1953 to use it as a release printing format, they quickly abandoned the idea because by this stage tripack stocks, which did not need a filter attachment on the projector and did not absorb as much light as a lenticular print (see below), were rapidly being adopted industry-wide. This rendered the need for a lenticular release printing system obsolete.

One other additive colour process in use during the 1920s and 1930s deserves a mention here, because it represented another small step on the road to a fully stand-alone, single-strip colour system which could also be used to produce large numbers of prints. In 1908 the Paris photographer Louis Dufay started marketing 'Diopticolor' reversal glass plates for still photography, and over the following two decades worked on making his process suitable for cinematography. The process itself consisted of a film base onto which was printed a matrix of embossed lines, coated with alternating patterns of red, green and blue dye; termed the 'réseau' (network) by Dufay. Underneath the réseau was a layer of conventional panchromatic emulsion. In exposure the réseau acted as a filter, registering a latent image or not depending on whether light of the appropriate colour passed through it. After processing the silver halides were washed away in the exposed areas, allowing light from the projector to pass through the coloured réseau and illuminate the screen. In unexposed areas the halides were fixed to a metallic silver, preventing light from passing through the colour dye and showing on the screen as opaque.[19] Because the colour filters were in the form of dyes built into the film itself, no modification was needed to a camera or projector designed for use with conventional monochrome stock in order to make it compatible with Dufaycolor. The issue of mechanical complication had thus been overcome, but two problems remained. Like Kodacolor, Dufay film had to be exposed through the base, and again, this was a reversal system and the production of multiple prints was difficult, expensive and involved considerable loss of image quality. Dufaycolor was therefore used predominately by amateurs, and from 1935 was sold in the Kodak 8mm and 16mm formats. It was also available in the Pathé 9.5mm gauge, and was thus the only colour film available to European home-movie makers using this format in the 1930s.

The British film manufacturer Ilford had entered into an agreement in 1932 to market and promote Dufaycolor, initially in small gauge reversal form; but it also undertook to invest in research and development aimed at adapting it for 35mm negative-positive use in the professional film industry. There were a number of complications. Because the Dufay system was additive – the three colours had to be evenly balanced so that together and in equal intensity they would form white – the colour temperature of tungsten (incandescent) or carbon arc studio lighting frequently caused problems. And although a negative-positive variant of Dufay was eventually developed, the production of mass print runs was complex, expensive and had a high wastage rate. The end result was that for professional cinema film use Dufaycolor represented a substantial extra cost per print: around 3½ pence per foot, which Simon Brown's research suggests was around six times the cost of monochrome.[20]

Dufay had undoubtedly cracked the nut of photographically recording all three primary colours and reproducing them additively without the 'mechanical complication' of specially modified equipment which had dogged all the mechanical systems from Lee and Turner to Kodacolor. But as with all non-mechanical additive systems, the fatal flaw was that it was a reversal process. Thus the production of multiple prints was difficult, expensive and the result was of poor quality. Unlike Technicolor, Dufaycolor did not cost a lot more *to shoot* than conventional monochrome: that extra 3½ pence per foot was for release printing. Another problem was that the dense Dufaycolor réseau absorbed a lot of light in projection. The carbon arc projector lamphouses installed in the 'picture palace' cinemas built during the 1920s and 1930s often struggled to produce an acceptable level of illumination on the screen due to the long throws which were necessary. Dufaycolor exacerbated this, yielding an even dimmer picture. The only Dufaycolor films printed in any great quantity for cinema release were a newsreel of King George V's silver jubilee in 1935, a short sequence in the musical revue *Radio Parade of 1935* (1935, dir. Arthur Woods) and the only full-length Dufaycolor feature: a 'simple minded, naïve' (to quote Halliwell's Film Guide) propaganda film made to drum up recruitment for the Royal Navy shortly before the outbreak of World War Two, *Sons of the Sea* (1939, dir. Maurice Elvey). It was shot mainly at Dartmouth Naval College and took advantage of the intense natural light found in south-west England during the height of summer. Dufaycolor made a brief revival as an amateur medium between the end of the war and Dufay-Chromex (the company to which Ilford had sold out in 1937) going into liquidation in 1952, and for a small number of short travelogues and documentaries. And that was pretty much that for additive colour in commercial moving image technology.

Subtractive colour I: Technicolor

The inventor of the Technicolor process, Herbert Kalmus, argued that the successful marketing of colour film technology depended on addressing two issues in relation to standard industrial practice:

> 'How far will it [the film industry] permit departure from standard equipment and materials, and how will it attempt to divide the additional requisites of recording and reproducing colour between the emulsion maker, the photographic and laboratory procedure and the exhibitor's projection machine?'[21]

The inventors and promoters of additive colour discovered the hard way that the answer to Kalmus' first question was 'as little as possible', and that the answer to the second was to concentrate non-standard equipment and procedures in areas of the process where its economic impact was lowest, i.e. at the production end, where the effect of the economies of scale invoked by mass-production (Henry Ford's 'any colour, as long as it's black') was minimised. All the additive colour processes developed up to and including Dufaycolor were ultimately a commercial failure because they fell foul of one or both of these criteria. The successive frame method was fundamentally flawed, because it could not record or reproduce more than one colour record simulataneously (though Kinemacolor managed to keep the visible impact of this down to a minimum). Both the successive frame and the simultaneous mechanical systems required modification to both the cameras and the projectors in cinemas. Only the film processing chemistry was largely unaltered, and even then the use of panchromatic stock was not common and required special procedures (e.g. handling in total darkness). And while lenticular systems could be used in almost standard and Dufaycolor in totally standard cameras and projectors, the manufacture of lenticular raw stock was non-standard and expensive. Furthermore, being essentially reversal systems, the mass duplication of both was very difficult and very expensive, meaning that these processes were effectively useless as mass media.

Just as W. K. L. Dickson realised in 1889 that cinema film technology had to be standardised along a 'one size fits all' philosophy in order to make it commercially viable, Kalmus realised that colour had to be made to fit that economic model, too. Furthermore he realised that exhibition was the sector of the industry where standardisation was most important and where introducing any standard which violated that technology would seriously jeopardise its chances of success. This meant that whatever happened in the studio or laboratory, his colour film had to be showable via the hundreds of thousands of 35mm projectors operating worldwide, without any faffing about with mechanical filter wheels, multiple strips of film, special lenses, non-standard pulldowns or anything that would increase costs or reduce reliability at the exhibition end. The result was a combination of two technologies which together comprised the original Technicolor system and which were developed in the two decades between the formation of the Technicolor company in 1915 and the release of the first three-colour Technicolor feature, *Becky Sharp* (1935, dir. Rouben Mamoulian). The Technicolor camera combined the principle of mechanical, additive colour (i.e. optically filtering light to record the three primary colours separately) with the technique upon which tripack colour coupler processes would later depend: exposing each colour record individually, simultaneously and initially as a negative (i.e. not reversal) in order to permit mass-duplication. The final version departed from all previous systems in that the three colour records were exposed onto three separate elements of monochrome stock running in synchronisation: it was termed the 'three-strip' Technicolor camera, in order to distinguish it from earlier 'two colour' (red and green only) versions used on a limited scale for shorts and a small number of Hollywood features during the 1920s.[22]

Though Barry Salt suggests that the camera was 'to a certain degree modelled on the Mitchell',[23] it was essentially a design unique to Technicolor. Cinematographers appreciated several refinements which were not to be found in any studio camera formerly in use, notably a viewfinder which virtually eliminated parallax errors and focusing that could be remotely controlled.[24] The mechanism consisted essentially of a magazine which held three 1,000-foot rolls of 35mm stock and two gates mounted at right angles to each other. Between them was a 45° beam-splitting prism, mounted behind a single prime lens. Directly behind the prism was a gate holding a single strip of panchromatic stock, which recorded the green image. At a 90° angle to it was a second gate, through which passed two strips of film in contact with each other, emulsion to emulsion. The strip nearest the prism was sensitised to blue light only, and had a red/orange dye which blocked green and blue light, with a layer of panchromatic stock behind that which therefore recorded the red image. However, these stocks were all monochrome negative stocks, i.e. not reversal, so unlike any previous photographic colour system which had used successive frames or multiple strips of monochrome film to record separate additive colour records, the colours actually recorded on the Technicolor films were the subtractive negatives of the primary colours, i.e. cyan, yellow and magenta.

The three camera negatives were then developed in the same way as normal monochrome stock, and cut to produce three edited negatives holding the three colour records for the final film. Release prints were made using an extraordinarily complex method of physically transferring three organic dyes onto the surface of the release print stock. Each strip of negative was printed to produce a *matrix* element, in which a gelatine layer varied in thickness in proportion to the silver density on the original negative. These matrices were then immersed in a dye of the subtractive negative colour to the element used to print the matrix (e.g. the red matrix is immersed in a cyan dye). The gelatine absorbed a quantity of the dye in proportion to its thickness, and thus to the density on its source negative. It is for this reason that the Technicolor printing process is known as the *imbibition* process (from the Latin verb *bibere* – to drink or absorb).

The raw release print stock carried a black-and-white emulsion to receive the soundtrack and a uniform gelatine coating to absorb the dyes. Each dyed matrix was placed in contact with the print stock in three separate passes, using the 'pin belt' mechanism developed by Technicolor to ensure that the two elements remained in precise registration. While in contact with each other, dye passed from the matrix element to the print in proportion to its density in the matrix (hence the reason imbibition prints are also sometimes referred to as 'dye transfer' prints). When all three matrices had been printed, the result was a full, three-colour image on the print stock. The remaining dye was then washed out of the matrices, which were then redyed and used to make the next print. Because this was not a photochemical process, Technicolor printing could be done in full daylight. However, there is anecdotal evidence to suggest that there were initial problems with ensuring the accuracy of registration (alignment) between the three colour dyes, resulting in blurred edges and colour fringing on the prints. A distributor's print manager who worked in London during World War Two told me that due to wartime film stock shortages, Technicolor prints which ordinarily would have been junked at the quality control stage had to be put into circulation. Such copies became known as 'north of Watford' prints,[25] the implication being that the good ones were reserved for prestigious central London cinemas while the slightly misregistered prints were sent to the provinces.

These teething troubles apart, the original Technicolor process was, in terms of its technical performance, a phenomenal success. It represented the culmination of two decades of research and development by Kalmus and his staff, and was truly the first three-colour system which enabled large quantities of high-quality release prints to be made according to the same economies of scale as monochrome, and which did not require any modification to equipment or practices in cinemas. The extent to which it was accepted by the public as the first genuinely mass-producible system for recording and reproducing full, three-colour moving images is summed up by the 'glorious Technicolor' marketing campaign: the phrase has subsequently entered the English language as a colloquial expression denoting high quality, glossy production values (not to mention the phrase 'Technicolor yawn', meaning to vomit). But Technicolor was phenomenally expensive at the production end. The historian Rachael Low estimates that immediately following its launch, the use of three-strip Technicolor on an average studio feature added between £20,000 and £25,000 to the production budget (around 20–30 per cent).[26] The increased production costs included hiring the cameras themselves, three times the negative stock cost of a monochrome production, more complicated and expensive studio lighting (even by 1939 the three-strip negative stock only had a speed equivalent to EI40, requiring far more intense studio lighting than a monochrome production) and of course the production of matrices and prints. As Panavision was to do two decades later, the three-strip cameras were never sold outright to studios. Furthermore (and unlike Panavision equipment), they were not even available on a 'dry hire' basis.

Fig. 3.2 The three-strip Technicolor camera mounted in a soundproof blimp with operator Jeff Seaholme and 'colour consultant' at Pinewood Studios, near London, circa late 1940s. Picture courtesy of BFI Stills, Posters and Designs.

Technicolor did a lot more than just supply cameras and lab services. They marketed a complete package which impacted on virtually every stage of the production and distribution process. In Technicolor UK's standard contract with a production company from 1943, the provisions included: all raw camera stock to be supplied exclusively by Technicolor, all cameras to be hired exclusively from Technicolor, all photography to be undertaken under the supervision of a technician supplied by Technicolor (who could override the director of photography's creative decisions), all the lab stages from processing of the camera negatives to the production of release prints to be done exclusively by Technicolor (with a minimum order of 100 prints). In addition, clause 15 of the contract stated that 'the Producer shall use the services of and consult with a Colour Consultant to be supplied by Technicolor'.[27] There have been many documented instances of these consultants exerting considerable influence on set and costume design and much else besides. For example, when directing his first Technicolor production, *Rope* (1948), Alfred Hitchcock noted that following differences of opinion between the director of photography (Joseph Valentine) and Technicolor's appointed consultant (William V. Skall), the former became 'sick' and the latter shot most of the film.[28]

The late 1930s and 1940s were the heyday of the integrated beam-splitting camera and imbibition printing Technicolor process. After the war the cameras were largely superseded by dye coupler stocks (see below). The last Hollywood feature to be shot using the three-strip camera was *Foxfire* (1955, dir. Joseph Pevney). In the UK (which operated the only IB printing facility outside the US), it was last used on *The Ladykillers* (1955, dir. Alexander Mackendrick). Imbibition printing from coupler negatives (see below) continued into the 1970s, with the matrices being made from separation intermediates derived from dye coupler camera negatives. More recently, an updated version of the imbibition printing system was introduced at Technicolor's Hollywood plant on a limited scale in 1998. It was subsequently used to produce small print runs of major Hollywood features and high-profile rereleases, most notably Robert Harris' 2000 restoration of *Rear Window* (1953, dir. Alfred Hitchcock). The new generation of imbibition prints were clearly sharper and more saturated then even the most modern generation of coupler prints. But sadly the updated dye transfer process was abandoned after only a couple of years. Despite the higher image quality the industry was not willing to absorb the increased cost, with the result that Technicolor was not able to find enough of a market to make the system commercially viable.

The combination of technologies and services which constituted the original Technicolor 'package', therefore, marked a significant step forward in that it proved that the technique of recording and reproducing colour subtractively successfully enabled the production of full, three-colour cinema release prints on the scale needed to support a mainstream release and (eventually) of consistent quality. But with echoes of additive systems like Dufaycolor which almost made it into mainstream use but not quite, there was one aspect of Technicolor that did not quite comply with established industry standardisations, and thereby ensured that the 'glorious Technicolor' package would usually be restricted to high-budget, 'A' movie features (a bit like 65mm/70mm would a generation later). This was because it was just that – a package. Studios could not simply buy the hardware outright and then go away and do what they liked with it: they had to allow Technicolor a considerable degree of involvement and decision-making power on a film produced and released using their system. The technological success together with the associated restrictive practices of Technicolor resulted in the American film industry starting to look for the form of colour film technology which eventually superseded it, the one which is primarily with us today: 'tripack' or dye-coupler film stock.

Subtractive colour II: dye-coupler stocks

What this managed to achieve which Technicolor could not was to provide a single roll of unexposed film that could be used in the same camera, without any modifications whatsoever, as monochrome film. Thereafter it could be duplicated through intermediate stages to yield mass-produced release prints, just as monochrome could. In fact the only differences as far as production was concerned were in the studio lighting and processing chemistry. It is in the latter that the guts of this technology lies, as we shall see shortly.

It is somewhat ironic that, though this technology would eventually crack the nut of making full, three-colour photography usable in mass-produced moving images, its first significant commercial application was in a reversal system marketed primarily to 35mm still photographers and amateur filmmakers.[29] Kodachrome, as it was eventually marketed, was the brainchild of two professional musicians and amateur photographic chemists, Leopold Mannes and Leopold Godowsky. Since the early 1920s they had been attempting to create a form of photographic emulsion which is now known as the 'dye coupler' or 'chromogenic' process. In very simple terms, the-single strip film emulsion contains three layers which are sensitised to the primary colours. When it is developed, a chemical reaction converts each layer of photosensitive emulsion into a visible dye of the corresponding colour. The technique was initially described by the German chemist Rudolf Fischer in the early 1910s, but its first successful commercial implementation was with the launch of Kodachrome, first in 16mm movie form in 1935, and for 35mm still cameras a year later. The fact that Kodachrome was a reversal stock which could not be adapted to a negative-positive process meant that for moving images, its use was largely confined to the amateur domain. However, its comparatively slow speed and very fine grain ensured that it rapidly gained a commercial foothold for still photography, especially in the glossy magazine market: as the photo editor of *National Geographic* remarked, 'we knew the millennium was here for magazine colour reproduction. It [Kodachrome] had the possibility of almost infinite enlargement.'[30] At the time of writing (March 2005), 35mm Kodachrome stock is still being produced for still photography in the EI64 and EI200 variants, and for moving image use in Super 8 EI40 stock. The very fine grain EI25 emulsion was discontinued in 2003 due to rapidly falling sales.

The film industry did use Kodachrome on a very limited scale in conjunction with the Technicolor dye transfer printing system, as a cheaper and more versatile substitute for the three-strip camera. 'Technicolor monopack', as this hybrid was termed, was first marketed in the US in the autumn of 1942. 35mm Kodachrome reversal film was exposed in a conventional studio camera, the only difference in production technique being the need for more studio light (or bright natural sunlight on location) relative to black-and-white: by the early 1940s monochrome film emulsions were available with speeds of up to an equivalent of EI200, while Kodachrome had speeds of EI10 for the daylight-balanced stock and EI16 for the artificial light variant respectively. After exposure and processing, Technicolor produced three black-and-white separation negatives from the Kodachrome original, which were then dye-transfer printed in exactly the same way as if they had been exposed in the three-strip camera. Technicolor monopack was used for a number of productions between 1942 and the end of the decade, when it was superseded by negative-positive coupler stocks (see below). These were mainly features in which the three-strip camera was unsuitable, either for reasons of portability or lighting requirements. Arguably the best known feature to include Monopack footage was the war propaganda film *Western Approaches* (1944, dir. Pat Jackson), much of which was filmed in a lifeboat in the North Atlantic using natural light only.[31]

From the film industry's point of view, its economic requirements dictated that for colour to achieve mass-market saturation, a form of coupler technology which would work in a negative-positive process was needed. This eventually materialised in the West in the late 1940s, almost certainly as the result of Allied forces having helped themselves to the infrastructure which remained of the Nazi film industry following the end of World War Two.

Since the mid-1930s the Nazis had been anxious to develop a colour film system to rival Technicolor. Hitler's propaganda minister, Joseph Goebbels, was a well-known admirer of Hollywood genre cinema, and specifically its potential to communicate political or ideological messages under cover of 'entertainment'. He understood that if audiences realised that they were being fed propaganda, they would reject its message. This was demonstrated by a string of high-profile box-office flops commissioned by the Nazi government shortly after it came to power in 1933. Goebbels had noted the commercial success of Technicolor in high-budget, prestige features – in particular he praised the 'magnificent artistic achievement' of *Snow White and the Seven Dwarfs* (1937, dir. David Hand)[32] – and was determined that the Third Reich should have something similar. But the form of technology they used to deliver it was dramatically different from that of Technicolor; it would also prove to be cheaper and far more versatile.

The first colour film produced by the German Agfa company was a reversal stock for still photography, 'Agfacolor Neu'. It was launched in 1936, the year after Kodachrome went on sale in the US.[33] However, there was a crucial difference in the way its chemistry worked, one which would enable its conversion to a negative-positive process shortly afterwards. With Kodachrome, the coupler elements (i.e. the chemicals which converted the exposed film into three visible dyes) themselves were not present in the emulsion as exposed, but introduced during processing. This made both the chemistry of the film emulsion (five layers, each one to three microns thick) and the processing of it enormously complex. With the first generation of Kodachrome, this consisted of 28 separate procedures, 'all of which had to be carried out with the utmost precision.'[34] The Agfa system incorporated the coupler elements into the film emulsion during manufacture, which were 'activated' during processing, i.e. the developing chemical simply induced the change from an emulsion to a dye. This reduced the number of processing steps to four, and any existing black-and-white lab could easily adapt its equipment to process the new stock.

In 1939 Agfa launched a negative-positive version of the system, which was used to produce a number of feature films throughout the remaining life of the Nazi regime. Two notable examples were *Münchhausen* (1943, dir. Josef von Baky) and *Kolberg* (1945, dir. Veit Harlan).

As I have argued elsewhere, the Allied plundering of captured Nazi film technology infrastructure may well have hastened the global film industry's conversion from nitrate to safety stocks,[35] not to mention the Nazis' role in developing magnetic sound technology (see chapter four). With colour there is even less speculation implicit in such an argument. As the Agfa plant which manufactured and processed the bulk of its colour stock was in Prague, the equipment and chemicals remaining there were quickly removed by the Russians after the end of the war, and some years later a cloned version of the system emerged, dubbed 'Sovcolor'. The first major Soviet feature to be produced after the war, *Ivan Groznyi* (*Ivan the Terrible*, 1946, dir. Sergei Eisenstein), induced scenes that were shot and printed on leftover Agfacolor stock abandoned by the Nazis. Meanwhile the Americans sent federal investigators to interrogate Agfa scientists being held by the Allies as prisoners of war. With them went representatives from Agfa's US subsidiary (which by that stage was wholly American-owned), the General Aniline and Film Company of Santa Monica, California (Ansco). Ansco lost no time in obtaining both the physical and intellectual property associated with Agfacolor, and in 1948 the first Anscocolor stocks – initially reversal – went on sale to the industry.[36]

Although a two-strip only (red and blue) system known as Cinecolor had been used on a limited scale in the US for B-movies and documentaries during the mid-1940s, the introduction by Ansco of a tripack stock which eliminated most of the complexity and expense associated with Technicolor marked the start of a conversion process. Within two decades, black-and-white film stock would have virtually disappeared from Hollywood studios and mainstream cinemas. The launch of Anscocolor was quickly followed by a number of other manufacturers starting to produce tripack stocks including Gevaert in Belgium and Ferrania in Italy. In Japan, Fuji non-substantive reversal coupler stock was used to shoot *Carmen Comes Home* in 1949 and in October 1955 the company launched a negative-positive process suitable for moving image use.[37] The most significant development in this period was the launch of Eastmancolor in 1950. In this stock the Kodak company introduced a number of refinements, most notably the use of coloured couplers. The couplers are coloured in two of the layers to provide masking, which improves the clour reproduction of the duplicate elements by correcting for dye deficiencies in the negative stock. The coloured couplers are what give modern colour pre-print elements (including 35mm still negatives) their characteristic orange tint, even though this is not visible on the finished print.

Eastmancolor and the range of masked coupler emulsions which followed were a phenomenal success. It was a technology which fulfilled the industry's needs for the mass-rollout of colour. Although early versions of Eastmancolor were significantly slower than their monochrome equivalents (EI16 at first, increasing to EI24 in 1953 and EI50 with the launch of type 5250 in 1959),the additional cost (and difficulty) of using this process relative to black-and-white and where a large run of release prints was needed was negligible compared to those of any other colour process which had ever been available. Although Eastman Kodak was (and remains to this day) the market leader in masked colour emulsions, the distribution of patents and the commercial realities of the day ensured that the restrictive practices which had applied to earlier colour systems – most notably Technicolor – no longer applied. Eastmancolor stock could be purchased by a studio off the shelf, shot by its own cinematographers using its own cameras, be processed in a lab of its choosing (as a matter of policy Eastman Kodak did not own, operate or franchise processing labs for motion picture film – they just sold the raw stock) and the cut negative used to produce multiple release prints using the same sequence of intermediate elements as black-and-white. The fact that masked coupler emulsions were also being rolled out across the still photography sector (both amateur and professional) helped also to tip the economies of scale in this technology's favour.

This form of colour film technology has effectively remained the industry standard for the last five decades, although the colour saturation, grain and definition available from tripack colour emulsions has continued to evolve and improve throughout that time. As F. P. Gloyns puts it:

> The story of the laboratories from those days up to the present is a record of gradual improvement of technique leading to improved consistency and quality rather than of any fundamentally new innovation. In principle, the products of 1950 are those which we use today, but they have all been vastly improved in detail.[38]

Another issue was the need for quality control across the mass print runs which, given the quantities involved, had never been a significant issue with any previous system. The use of chemical analysis to maintain, or 'replenish' the developer solutions used in film processing to a consistent strength became routine in the 1960s (it was desirable to do this with black-and-white developers too, but many labs did not). The Bell and Howell model 'C' printer introduced a system of dichroic mirrors (not unlike the beam-splitting prisms in the three-strip Technicolor camera), and 'light valves', which allowed precise control over the colour temperature of the light used for exposure in the printer. In the 1980s the huge increase in the volume of release printing necessitated by the advent of multiplex cinemas lead to the development of high-volume, high-speed printing and processing which today is largely computer-controlled. This will be discussed in greater depth in chapter five. With the benefit of hindsight, we now know that the early generations of masked coupler tripack emulsions had one very serious flaw: their chemistry was highly volatile, making them susceptible to serious colour dye fading over time. It was a flaw which went undetected until the 1960s, and is now one of the main problems which archivists and restoration experts have been grappling with in recent years. Colour dye fading and the techniques which have been developed to reverse it will be covered in chapter seven, but are beyond the scope of this discussion.

By the mid-1970s, black-and-white had become an exception which proved the rule. The conversion process was accelerated as American network television increasingly moved to full colour broadcasting in the late 1960s, with the result that broadcasters were increasingly reluctant to license black-and-white films for transmission from the Hollywood studios. *Who's Afraid of Virginia Woolf?* (US 1966, dir. Mike Nichols) was probably the last major first-run studio feature in black-and-white to be licensed by the US networks.[39] Thereafter, films as diverse as *Alice in den Städten* (*Alice in the Cities*, 1992, dir. Wim Wenders), *Manhattan* (1976, dir. Woody Allen) and *The Elephant Man* (1984, dir. David Lynch) used black-and-white in order to make an artistic statement, just as colour features in the 1930s and 1940s – such as *Becky Sharp*, *Gone With the Wind* (1939, dir. Victor Fleming) and *Münchhausen* – had used this technology to make a commercial or political statement.

Conclusion

The story of colour film processes, and specifically the reasons for different forms of this technology being developed, appearing and disappearing as and when they did, lies as much in the cultural and economic domains as it does in the purely technological. Since Maxwell's experiments in 1861 had shown that photographically recording and reproducing colour variations as perceived by the naked eye was theoretically possible, a large number of scientists and engineers tried to apply the principles involved to both still and moving image photography. As with television, there is no 'great man' theory or individual process or technique which can explain the push towards full colour as a standard: Eastmancolor would not have become that standard if the earlier, flawed systems had not demonstrated the existence of a market which that product eventually serviced.

This had, of course, been what generations of scientists and engineers had been working to achieve from Frederick Lee and Edward Turner onwards. Most of them thusfar were ultimately thwarted: the results of their research were either technically flawed, incompatible with the economic realities of the film industry, or both. Non-photographic colour was always perceived to be 'second best' to colour information recorded at the moment of photographic exposure. In any case, even state-of-the-art non photographic colour (Pathécolor, Handschiegl), which attempted to retrospectively 'add' scene-specific colour information to different areas of the frame, proved to be so labour-intensive as to be uneconomic for entire features. While tinting and toning was able to be applied to large print runs at minimal cost, it proved incompatible with sound-on-film, which in the late 1920s was a more economically attractive proposition. By that stage full 'three strip' photographic colour was on the verge of becoming a technical and economic reality.

The very first generation of truly photographic colour systems used the successive frame method. It was a combination of technologies which, as a package, quite simply did not work. The combination of panchromatic emulsions with filtered lenses enabled multiple colour records to be recorded and reproduced, but the successive frame method could not enable these records to be exposed or projected simultaneously. The visible colour 'fringing' in Lee and Turner's system was so bad as to preclude its use for three-colour reproduction altogether. Kinemacolor used a number of devices to minimise the shortcomings of successive frame, notably a reduction in the number of colour records from three to two, an abnormally high film transport speed (32fps) and only photographing static or slow-moving subjects. While this mitigated the complete failure experienced by Lee and Turner, Kinemacolor also proved that a means of photographing the three colour records simultaneously was needed for any further progress to take place. While the short-lived Gaumont Chronochrome process proved that this could be done mechanically and additively, by the early 1920s it, like tinting and toning, was about to be eclipsed by the next stage of development.

The early single-film systems of the late 1920s and early 1930s cover the transition from additive to subtractive colour. In different ways they focused the industry's mind on the idea of a colour film process which would meet its own economic needs, i.e. the Henry Ford model of reliability and mass production according to increasing economies of scale. Lenticular Kodacolor, Dufaycolor and Kodachrome all succeeded in enabling the full visible colour spectrum to be exposed and projected from a single strip of film. But, being reversal processes, none were capable of mass-duplication without significant extra cost and loss of image quality. This was where Technicolor stepped in, applying the principle of subtractive colour recording to enable large-scale printing, albeit using the horrendously complex dye-transfer process. But cost still remained a problem, one which restricted Technicolor's use to big-budget feature films in which colour was used as an explicit selling point.

One interesting side-effect of this is that most colour cinematography which did take place before the Eastmancolor revolution of the 1950s was by amateurs, for whom the inability to produce large numbers of prints was not an issue. The resulting wealth of colour 'home movie' material made during the 1930s and 1940s has come to the attention of television documentary producers in recent years, helped by the public sector film archive movement (which has always believed that amateur footage is as culturally valuable and worth preserving as much as commercially made films) and the fact that many of these systems (and in particular post-1938 Kodachrome) seem virtually immune to the dye fading which affected early generations of Eastmancolor. Television series including *The Third Reich in Colour* (Spiegel TV, 1999) and *The British Empire in Colour* (TWI, 2002) have done much to increase awareness both of a hitherto ignored aspect of our moving image heritage and (although to a lesser extent) the early colour film technologies themselves.

Ironically the origins of the colour process which would ultimately dominate moving image film technology throughout the latter half of the twentieth century have their origins not in the economic domain, but in the political. In the mid-1930s Technicolor had come to be seen as a symbol of Hollywood's global domination of world film culture, in response to which Goebbels decided that the Nazis had to have an alternative. There is certainly no evidence to suggest that Agfa sought to develop dye coupler emulsions into a viable technology in order to offer a cheaper and more flexible alternative to Technicolor, even though this was the eventual result. In 1950 Eastman Kodak released the first masked coupler emulsions to the global market, and the rest is really a side issue to the history recounted in this chapter.

If a more convincing demonstration were needed of the economic prerogatives which drive and have always driven technological evolution in the global film industry, it can be found in the fact that the use of black-and-white is now more expensive than colour, despite the chemistry being so much simpler. Demand for stock and processing has decreased to the point at which the former is only manufactured by Eastman Kodak to special order, and the latter is offered only by specialist labs which primarily serve the archive market.

An ironic illustration of the extent to which colour made the transition from an embryonic series of research and development projects to an industry norm, both in film and television (for more on the latter see chapter six), was the thankfully short-lived phenomenon of 'colourisation'. This emerged in the 1980s and involved a variety of techniques: they ranged from what were in effect electronic versions of hand colouring and stencil colouring to computerised attempts to automate the process of adding colour information selectively to black-and-white originals. It was driven by the television industry, which perceived that black-and-white footage was in some way inferior and would be rejected by consumers. Adding colour would therefore be making classic movies 'better'. This assumption reached its zenith (or nadir, depending on your point of view) with the establishment of a company called American Film Technologies (AFT) in 1988, the idea being to offer a colourisation service to film studios and broadcasters which were sitting on the rights to large collections of black-and-white archive material. AFT started by colourising *Meet John Doe* (1941, dir. Frank Capra), *The Scarlet Pimpernel* (1934, dir. Harold Young) and *They Made Me a Criminal* (1939, dir. Busby Berkeley) for television broadcast. A trade paper report made their reasoning brutally clear in predicting that 'if these succeed in the ratings, hundreds of currently unsyndicatable black-and-white series could rise from the dead to funnel streams of new revenues into the coffers of their distributors'.[40] A sarcastic response to the colourising phenonenon is offered in the film *Gremlins 2: The New Batch* (1990, dir. Joe Dante) in which a thinly-disguised caricature of the cable TV impresario Donald Trump offers his subscribers '*Casablanca* ... now restored in full colour and with a happy ending!'

Towards the late 1980s the colourisation bandwagon gathered steam, with the Hollywood-based Color Systems Technology signing a deal with the TV magnate Ted Turner to colourise much of the MGM archive, the rights to which he had bought in March 1986. This led to a classic 'industry vs. artistry' debate: as with panning and scanning (see chapter six) Turner and his supporters regarded the issue purely as a business decision, while those for whom film was an object of cultural integrity condemned what they saw as the 'vulgarisation' of historically important footage.[41] There were even calls for legislation to ban the practice, though, unsurprisingly, none ever materialised.[42]

By the mid-1990s colourising had all but disappeared, one suspects because it never found much of a market in the first place. However, it should be born in mind that – technically, at least – the process of adding colour information to a photographic image which had never captured it in the first place was nothing new. In the commercial film industry it had been widespread – in the form of tinting, toning and the various forms of selective non-photographic colour – in the three decades before the transition to sound. But, when revived in the late 1980s, it had failed because more effective ways of recording and reproducing colour photographically had become an industry standard, the technical quality of which consumers expected as a bare minimum and which was cost-effective to deliver as part of the production process.

The case-study in the next chapter follows a very similar pattern. The idea of synchronised sound had existed throughout the development of moving image technology. Indeed Thomas Edison was only interested in developing the latter in the first place because he thought it would add commercial value to his audio technologies. Yet the mass-integration of picture with sound did not take place until the late 1920s, just as the mass-integration of colour with moving image photography did not take place until the early 1950s. Again, with sound, we will see that a comparatively rapid series of events precipitated the commercial rollout of technologies which has been in active development for several decades previously; really the last instance of this happening before we move into the 'post-film' era of technologies centred around television, video recording and digital imaging.

chapter four | **sound**

> 'Ladies and Gentlemen – isn't this a marvellous invention?'[1]

> 'Unlike a portrait, the reproduction of a dead voice gives one an uneasy feeling.'[2]

> 'A cinema patron and his wife have been driven from one of our few remaining places of public worship by an overdose of decibels ... It ranged from rock-pop to the factory floor, the wilful infliction of punk rubbish on our eardrums to the enforced hammerings and screechings of workshops.'[3]

Despite this customer's somewhat less than enthusiastic reception of the newly released Dolby stereo process, audio recording has been always been associated with moving image technology in some shape or form throughout the period of the latter's existence. Recordings have been produced for playing in synchronisation with the picture for all but the first three decades of commercial film (and even then some activity in this area took place), and for almost all publicly broadcast television. Synchronised sound is also an integral component of all video recording technology, of digital offline and of Internet-based moving image content. The evolution of these audio technologies will be considered alongside the video technologies with which they are associated in subsequent chapters. This discussion, therefore, is concerned primarily with the evolution of audio recording, manipulation and reproduction technologies specifically associated with film.

The overwhelming majority of historical research related to film sound technology tends to focus on the 'key moment' of the conversion process, which in North America and Europe took place roughly between 1926 and 1932. It is not difficult to understand why. At the outset of this process, hardly any feature-length films were produced and shown commercially with a synchronised audio recording. By the close of it, 'silent' films, as Scott Eyman elegantly puts it, 'belonged to the permanent, irremediable past'.[4] Three generations of writers and historians, therefore, have (correctly) identified this period as being one of great industrial and cultural change. So has the Hollywood film industry itself, mythologising the conversion through productions such as *Hollywood Cavalcade* (1939, dir. Irving Cummings) and *Singin' in the Rain* (1952, dir. Stanley Donen & Gene Kelly), which celebrate the arrival of sound, and *Sunset Boulevard* (1950, dir. Billy Wilder), which condemns it. But many writers have fallen into the trap of reading huge, agenda-setting *ideological* developments into this process as well, when the real significance of the conversion as a 'key moment' lies in the convergence of a complex framework of technological, cultural and economic factors which enabled the mass rollout of synchronised sound to take place when it did. For example, Laura Mulvey suggests that Warner Bros.' collaboration with AT&T during the mid-1920s, which resulted in the Vitaphone process, enabled the 'negotiation of the technology into a showbusiness reality'.[5] But however valid her arguments are as to Warner Bros. having made the business case for sound *per se*, her essay completely ignores the fundamental technological flaws in Vitaphone which made it essentially incompatible with the film industry's industrial practices (and which were well known a decade earlier), the fact that it only remained a 'showbusiness reality' for about four years and that, in all likelihood, silent cinema would have returned if other, more suitable methods of sound recording and reproduction had not been introduced as the result of Warner Bros. making that business case.

The 1927–30 'conversion', therefore, gave rise to what is probably the most extreme example of the 'key moment' model of technological historiography, one which in this chapter I shall try to set in the context of developments before and since. In doing so, I will argue that the long-term impact of audio technology on film production and consumption is better understood through a more balanced approach to its ongoing development and implementation (or, in some cases, lack of implementation) throughout the late nineteenth and twentieth centuries. From this standpoint it is possible to extrapolate a clearer picture of the ways in which audio technology had affected the production and exhibition of films.

The origins of audio recording and their links to moving image technology

Given that the following four decades would be spent trying to add synchronised, recorded sound to moving pictures, it is more than a little ironic that in all likelihood, the earliest successful moving image technology was invented for the sole purpose of adding illustration to sound recordings! One of Thomas Edison's patent applications relating to the Kinetograph stated that 'I am experimenting upon an instrument which does for the eye what the Phonograph does for the ear',[6] and his assistant W. K. L. Dickson claims to have successfully demonstrated a Phonograph mechanically synchronised to a projected film on 6 October 1889.[7]

Audio recording came first by a decade or so. On 7 December 1877 Edison demonstrated his 'Phonograph' to the public at the offices of the *Scientific American* magazine. The device, though crude, demonstrated the basic technical principles through which sound would be recorded until the mid-1920s, and would be reproduced domestically well into the 1950s. It consisted of a steel cylinder coated with tin foil, onto which a recording was made acoustically. As the cylinder was rotated at a (more or less) constant speed by hand-cranking the shaft, the operator spoke into a diaphragm. This was mechanically linked to a stylus which inscribed a linear indentation (groove) into the constantly rotating and laterally moving (thus preventing the groove from being overwritten in the cylinder's following rotation) foil surface. As the noise level increased, the greater air pressure forced the needle deeper into the foil, while at lower volumes the indentation was lighter. Edison discovered that when, after recording, a similar needle was passed over the groove at the same constant speed but at much lower pressure, its vibrations could be reproduced as changes in air pressure created through a similar diaphragm, thereby playing back the recorded sound.

With the apocryphal words 'Mary had a little lamb',[8] Edison had successfully demonstrated the basic technical principles of analogue audio recording, and shown that they could be made to work in practice. As with film, the issue was then to identify how this technology could be commercially exploited. A decade later, his (or, more accurately, Dickson's) Kinetograph/Kinetoscope system for recording and reproducing film-based moving images would ultimately be eclipsed by minor variations to the technology introduced by Armat and Latham in the US and Lumière, Paul and Prestwich (among others) in Europe, in that projection before a large audience was what made the sums add up. The Kinetoscope's fatal flaw, in other words, was that it was an expensive piece of capital equipment which could only show films to one person at a time.

As with Edison's movie equipment, the Phonograph represented an impressive scientific feat, but one which, in the first instance, did not seem to have any obvious commercial application. Edison's original intention was to market the Phonograph as a device for consumers to create and replay their own recordings, similar to George Eastman's simple and mass-produced stills cameras. As David Morton explains:

> Edison's original phonograph was merely a clever parrot, or better yet, an aural mirror. The phonograph's lacklustre sales soon made it clear that few Americans would be satisfied with simply recording themselves. Instead, buyers rewarded those who used the phonograph to create a system for mass-produced entertainment, purchasing millions of records (or later tapes or compact discs) for their personal enjoyment.[9]

The promoters of sound recording technology, it seems, took two to three decades to work out what the nascent movie industry (represented by the Lumière brothers) discovered in one fell swoop on the evening of 28 December 1895: that selling a single recording to multiple customers at the same time was the route to commercial success. Not that it is too difficult to find a justification for Edison's line of thought. The previous chapter discussed a number of colour film processes which produced impressive results on the screen, but which ultimately fell by the wayside because it was too difficult and/or expensive to produce multiple copies from a single original. Edison's Phonograph suffered from the same flaw. As a rapidly emerging market for the hardware and software needed to reproduce pre-recorded sound in the home emerged (which itself had a number of Victorian precedents, such as pianola rolls),

Edison's cylinder technology found itself unable to economically supply it because, as with reversal colour film processes, that was not what it was either intended or designed for.

Initially, every cylinder had to be individually recorded. It was not until 1912 – long after the film industry had perfected the means of producing release prints in the quantities needed to satisfy the exhibition market as it then existed – that Edison had developed a mass-duplication system for cylinders (the 'Blue Amberol' process), based on electroplating, which offered an acceptable quality of reproduction compared to the original. By this stage, Edison had pretty much been eclipsed in the audio market by a variant of the technology more suited to customer demands, just as he had been by the systematic emergence of film projection in Europe.

Acoustic recording using discs, as distinct from cylinders, as the carrier, emerged not long after Edison's initial experiments. In New York the German émigré Emile Berliner launched the first commercially marketed form of this technology, which established a rapidly growing market share during the 1890s. Although the basic technical principle of mechanical, acoustic recording used in disc technology was identical to that of Edison cylinders, Berliner managed to adapt it to facilitate the mass-production of release copies from a single original recording far more efficiently than was possible with Edison cylinders. The master recordings consisted of a rigid disc coated with wax, into which a stylus inscribed the groove in a spiral formation, starting at the outer edge and moving inwards. Variations in input volume registered as lateral movements in the wax rather than the depth of the indentation. Both of these changes meant that the wax original could be electroplated, a metal 'stamper' produced and release copies pressed on a highly durable medium far more cheaply and efficiently than with cylinders; and the records sold in shops offered a much higher sound quality than their cylinder equivalents.[10] By the mid-1900s, sales of pre-recorded music on discs had overtaken those of cylinders, although production of the latter struggled on into the early 1920s.

I have cited Edison as a major case study around the origins of industrial practice in audio recording because, by the mid-1900s, his experiences offer a clear demonstration of how the film and audio industries were heading in the direction of two separate economic models. Both were, by this stage, reliant on the means of producing multiple copies from a single original. But films were increasingly being consumed by mass audiences in a formal, auditorium setting; whereas most copies of audio recordings were sold outright to private individuals for playback in a domestic setting (the impact of the latter on film industry economics will be discussed in relation to home video in chapter six). Those economic models were greatly influenced by the strengths and weaknesses of the technologies themselves: characteristics which, with both moving images and audio, Edison's designs found themselves hopelessly fighting against. Despite a last-ditch rearguard action in the form of the MPPC 'adventure' of 1908–17 (see chapter two), which attempted to package Edison's essentially obsolete moving image technologies with the commercially successful Eastman film stock, he ended up paying the price of the pioneers on both fronts. By the mid-1920s the Edison Company had disappeared, virtually without trace, from both the film and the recorded sound industries.

The reconvergence of those two economic models into a single mass-medium consisting of edited films shown in synchronisation with a recorded soundtrack took place gradually over the following two decades. It was driven largely by developments in audio technology which made it increasingly compatible with the existing infrastructure of film production, duplication/distribution and exhibition. The 1926–32 'key moment' certainly marked a period of intense activity in the rollout of synchronised sound. But to overemphasise its overall importance would be to risk falling into the trap of assuming that synchronised film and sound did not exist on a commercial basis before then, and/or that research and development in the area suddenly stopped once the last cinema had been wired up. The cultural, commercial and technological developments in film sound which happened during the intervening period (very roughly, 1905–26), place the traditionally understood conversion stage into a somewhat more organic and progressive context. This stage can usefully be considered in two separate categories: live and recorded film sound.

Film sound before the conversion: live performance

It has become a recurrent cliché among cinema historians that 'there was no such thing as a silent film': rather, there was a period during which the dominant cultural and economic mode of practice dictated that the film and the sound which went with it were produced and supplied separately. Although some early film exhibition, notably of newsreels and other non-fiction subjects, did take place in total silence, this would eventually become the exception which proved a soon to be established rule.[11] The 'sound of silents', therefore, was generally arranged by the exhibitor, and tended to consist of one or a combination of four forms of live performance: a spoken lecture or commentary; a live musical performance, ranging from a single instrumentalist to a live ensemble or orchestra; sound effects corresponding to, and generated in approximate synchronisation with, the action in the film; and, more rarely (in the European and North American film industries, at least), a theatrical performance by actors which took place near the screen and in view of the audience.

It is likely that lectures were the dominant mode of performance during the first decade of industrial-scale film production and exhibition. The reason for this is twofold. Firstly, this was already a well-established practice in 'magic lantern' slide exhibition, where sets of slides and printed lecture texts were supplied by the producing company; the latter were usually read verbatim. Secondly (and as we shall see in the following chapter), cinema exhibition was initially an itinerant business. Fairgrounds, music halls and other 'multi-purpose' venues were the usual locations, and therefore the equipment and infrastructure associated with film exhibition had to be easily portable. This militated against the use of acoustically recorded sound, live orchestras or performances which required the use of sets or lighting.

This period pre-dated the production/distribution/exhibition model which characterises today's global film industry and which started to become established with the emergence of Hollywood towards the end of World War One. Copies of films were sold outright by producers to exhibitors, and one of the key ways in which they advertised their products was through the use of extensive printed catalogues. These often contained elaborate descriptive prose relating to each title offered for sale, which in turn would form the basis for 'ad-libbed' lectures and commentaries performed by the exhibitor. As Richard Crangle points out, these were not usually as crucial to understanding the action in a film as lantern lectures were to the narratives in slide shows, and were therefore not usually scripted and performed verbatim.[12] But as the desire to convey more complex narratives through film (both fictional and non-fictional) developed during the 1900s, lecturers did sometimes find themselves in the position of having to deliver information to audiences which the embryonic grammar of film editing was then unable to. For example, the British crime drama *A Daring Daylight Burglary* (1903, dir. Frank Mottershaw), is described by one writer as 'remarkable for its precocious editing experiments':[13] so remarkable, in fact, that a crucial plot point (that one policeman has telephoned another, several miles away, to arrange for a criminal's apprehension upon his arrival), fails to be conveyed by the film itself at all. The audience was reliant on a commentator to provide this information (who in turn got it from the catalogue description), without which the action on the screen makes little sense.

Lectures and commentaries became less common and had effectively disappeared by the outbreak of World War One. Purpose-built cinemas became the norm, in which an orchestra, organ, sound effects equipment and eventually audio playback equipment could be permanently installed. The main precedent for film commentaries, the magic lantern, died out rapidly during the early 1910s as film replaced it as the dominant visual entertainment medium. And the films themselves changed. Editing became more sophisticated. Filmmakers learnt to convey specific information or meaning through the juxtaposition of different sorts of shot, and audiences learnt how to 'read' it. The intertitle (a short section of prose or dialogue, rarely longer than twenty words, shown momentarily against an opaque background in between shots) gradually replaced the commentator in situations such as the example given above. The length of films increased substantially, and the early 1910s saw the emergence of the 'feature' film, usually lasting significantly over an hour, as the main component in a cinema presentation. Sound, therefore, no longer had to fulfil a direct narrative function in the way that it did during the lecture and commentary period. The result was that a transition took place, at the end of which the live musical performance had emerged as the primary form of film sound in cinema exhibition.

As with the provision of lecture notes, the nature and extent to which the form and content of musical performances was determined at the performance stage, relative to the evolution of practices within individual cinemas, varied considerably from film to film and from exhibitor to exhibitor. By the 1920s the major Hollywood and European studios often provided full-scale orchestral scores, complete with detailed instructions as to the projection speed needed in various different sequences of the film (as has been noted in chapter two, camera and projection speeds were not standardised until the conversion to sound) in order to synchronise them to a performance at a given tempo.[14] Performance in the cinema ranged from large orchestras in prestigious, city centre locations to an individual pianist or organist in suburban or rural second-run venues. Depending on the musical ability and preferences of individual performers, the accompaniments were frequently improvised. In addition to musicians, a variety of machines were marketed which produced specific sound effects, either mechanically or pneumatically. The 'Allfex', sold in Britain from 1910, claimed to produce over fifty effects, including a steam engine, rain, hail, smashing china, a machine gun and a fire alarm.[15] However, their reception by audiences was problematic, with many complaining of inappropriate use or over-use of these effects.[16] As with first generation synchronised sound (see below), it seems that their use had all but died out by the end of the 1910s.

More elaborate modes of live performance designed to accompany a film projection were much rarer in the West. Where they did occur they were usually the work of experimental or avant-garde filmmakers, and did not represent the mainstream form of the medium. For example, *Entr'acte* (1924, dir. René Clair) was made to be shown during the performance of a ballet by the avant-garde composer Erik Satie. In other film cultures they were far more widely used, however, most notably in India and Japan, where cinema initially served to augment other, established forms of narrative culture. For example, while the lecture/commentary mode was essentially a primitive device in Western cinema – a throwback from an earlier form of entertainment, reluctantly used only until film could stand on its own two feet – in Japan, it was a deeply-rooted form of cultural expression to which film was effectively regarded as an added extra, as the director Akira Kurosawa recalls:

> The narrators not only recounted the plot of the films, they enhanced the emotional content by performing the voices and sound effects, and providing evocative descriptions of events and images on the screen – much like the narrators of the *bunraku* puppet theatre. The most popular narrators were stars in their own right, solely responsible for the patronage of a particular theatre.[17]

It is perhaps because North American and European filmmakers and technologists had a cultural view of cinema as a unique, self-contained medium which should not need to interact with other, established forms of entertainment in the way that it did in many Asian societies, that the push to develop audio recording and reproduction into a technology which was specifically suited for use in synchronisation with moving images originated primarily in the USA, and was based to a certain extent on European expertise. In other words, an underlying assumption existed to the effect that if the picture could be canned, so could the sound.

Film sound before the conversion: recorded sound

In the section on Edison above, we have seen how the nascent audio industry discovered by a process of trial and error that the most economically successful way of exploiting its technology was in selling mass-duplicated copies of music recordings for reproduction in the home. Equipment which enabled consumers to create their own recordings, or which facilitated their use in a business context, found virtually no market.[18] The film industry, too, evolved an economic model which was predicated on exploiting multiple copies of a single production, though a combination of technical (for example, the fire risk from nitrate film, the cost of equipment and the expertise needed to operate it) and cultural (initial film exhibition practices emerged from other forms of theatrical performance) factors dictated that consumption took place in a communal, rather than a domestic setting. Beginning at around the turn of the century and continuing well into World War One, there were a number of sustained attempts to combine the two technologies. While they enjoyed limited but notable popular success, they are of interest mainly because they illustrated clearly what had to be done in order to truly converge the technologies. These embryonic attempts to link moving images and audio recording, therefore, set the cultural agenda for a process of research and development which culminated in the 1920s, which made its first significant impact with the 'key moment' conversion process and which has essentially continued to the present day.

Edison's failure to establish a market for synchronised sound films in the 1890s did not mean that they disappeared from circulation for the following three decades. On the contrary, they found a niche market which exploited their capabilities and minimised the impact of their drawbacks, and which continued until World War One. The systems developed all used acoustic recording and amplification, mostly with the Berliner-type discs which had largely eclipsed Edison cylinders in the domestic audio market. The overwhelming majority of 'sound films' made were of musical or theatrical performances, ranging from popular songs to opera. In some ways they could be compared to modern music videos, and in the decade-and-a-half before feature films became the norm, their length, style and format was probably not radically different from those of other films which were produced and exhibited at the time.

In France, then the world's most economically advanced film industry, the company founded by Léon Gaumont began research in 1896, which culminated in the launch of its cylinder-based 'Phono-Cinéma-Théâtre' system at the Paris Exposition in 1900. This synchronised a musical accompaniment to footage of a number of well-known theatrical performers. It did not catch on with audiences, and the venture lost 150,000 Francs during the two months of the exhibition.[19] Four years later, in Germany, the movie pioneer Oskar Messter launched his disc-based 'Biophon' in 1903, and is said to have produced over 500 three-minute musical shorts which were shown successfully in Berlin during the following decade.[20] Britain was represented principally by Cecil Hepworth, whose 'Vivaphone' system of 1911 introduced a novelty: not only were musical numbers filmed and recorded, but he also produced footage of campaign speeches made by Lord Birkenhead and Andrew Bonar-Law, which must be among the earliest examples of political propaganda in synchronised film.[21] The following year, Gaumont was back; this time with the 'Chronophone' device, which mechanically interlocked a phonograph disc player and a film projector to ensure synchronisation. Like its other European predecessors Chronophone was exported to the US on a limited scale, where sound films were also regularly shown during this period, either for novelty value or as a component of variety hall or theatrical performances.

All of these first generation sound systems shared five key characteristics which defined the extent of their use in conjunction with film. Firstly, acoustic recordings (both cylinders and discs) were what would now be termed a 'write once' medium: after recording, the signal characteristics could not be altered in any way. Secondly, no editing of the recording's content was possible, analogous to the way that films could be cut and spliced. Thirdly, they were time-limited to (approximately) two minutes in the case of cylinders, and three for discs. Fourthly, there was the issue of synchronisation in playback. As all these systems relied on two separate carriers for the picture and sound, a means was needed of regulating their playback speed. They ranged from sophisticated to effectively non-existent. Chronophone, for example, used a crude form of mechanical interlocking which enabled the projectionist to adjust the relative speeds of the projector and turntable. Vivaphone only enabled the projector speed to be adjusted in order to match that of the record, but did provide a visual indicator in the projection box which enabled these adjustments to be made with some degree of accuracy.[22] At the other end of the scale were systems which, to all intents and purposes, ran the records 'wild', with consequent catastrophic results for the accuracy of synchronisation.

And finally, the acoustic nature of the medium imposed severe limitations. Recording had to take place close to a large horn – so close, in fact, that it would be visible in the frame if any attempt were made to actually record the picture and sound simultaneously. Therefore, almost all of these 'synchronised' films were produced by a performer lip-synching in front of a camera while a copy of the record (which had been recorded earlier) was played. In the case of the American 'Cameraphone' and the British Walturdaw 'Singing Pictures' series, these were not even specially produced for the film, but rather were off-the-shelf commercial recordings that were already on sale to consumers.[23] The other restriction imposed by acoustic technology was that of amplification. As has been mentioned above, playback of acoustic recordings was accomplished by passing a needle through the groove, which vibrated in response to the indentations made by the recording needle. The changes in air pressure generated by these vibrations were amplified by resonance within a metal horn, thereby replaying the recording. However, the extent of this amplification was inherently limited by the width of the needle, which in turn was limited by the width of the groove (or, in an Edison cylinder, its depth). Acoustic reproduction offered no means of increasing the playback volume without practical limit, something which was essential if recorded sound in large auditoria was to become a regular fixture. A number of experiments were attempted in order to overcome the problem, most notably that of pneumatic amplification, in which the vibrations of the playback needle operated a valve which released large quantities of compressed air under high pressure into the playback horn, in proportion to its vibration in the groove. While they succeeded in increasing the capacity of acoustic amplification, this was, according to many contemporary accounts, at the expense of sound quality (a subtle indication of this can be found in the acoustic method being described as the 'siren type' of amplifier by one technical historian).[24] The Chronophone playback system was succinctly – if a little unkindly – described by Eyman as 'jerry built'.[25]

In a last-gasp attempt to launch acoustic film sound as a viable commercial proposition, Edison briefly re-emerged on the scene in 1913 with the cylinder-based 'Kinetophone' system, which attempted – unsuccessfully – to address some of these issues. The mechanical synchronisation was more sophisticated than with any of its predecessors, the maximum running time was increased to approximately six minutes and the pneumatic amplification tweaked to improve the playback quality.[26] But after a six-month run at a New York theatre, acoustic recording – and Edison – disappeared from the film industry; this time, for good.

Classical Hollywood and electric recording

In order to explain why the wholesale adoption of synchronised sound happened when it did, it is necessary to identify and analyse two factors: the technologies which were appropriated for the purpose, and the cultural and economic factors which precipitated their use in synchronisation with moving images. As has been mentioned above and in previous chapters, the (approximate) period of 1910 to 1917 witnessed a wide-ranging process of change in the North American and Western European film industries. Itinerant exhibition and temporary venues gave way to purpose-built cinema theatres; the practice of producers selling copies of films directly to exhibitors was replaced by the rental-based distribution system, very similar to the one in place today; the beginnings of the studio system were established; and Donald J. Bell's 'fondest hope' of technological standardisation started to take shape. The form and content of the films which were produced and shown changed, too. The industry's first decade-and-a-half was characterised by short films, usually based around real events, special effects, or simple and embryonic fictional narratives. By the early 1910s, the films were getting longer and the stories they told were becoming more complex. A self-contained 'language' of cinema began to emerge, based on a combination of acting, directing, visual and editing devices. This has been termed the 'continuity system', or 'classical Hollywood', and was developed largely in America during this period.[27] It was a language that proved straightforward for filmmakers to implement and easy for audiences to understand: as one cinemagoer noted in 1917, 'ninety-nine picture fans in every hundred can instantly tell whether the continuity in a picture is good or bad'.[28] With the benefit of hindsight, it is not at all difficult to understand why acoustic recording declined and disappeared during exactly the same period as the continuity system took hold: the technical limitations of the former were simply incompatible with the requirements of the latter.

In an essay on sound reproduction in cinemas, Rick Altman argues that the decline of acoustic recording in the 1910s took place because it 'fell prey to a systematic producer campaign to feature continuous musical accompaniment' in preference to recorded sound.[29] Whether this transition was due to a systematic campaign or simply organic market forces, the reason is clear. Recording technology which offered very poor reproduction quality in large auditoria, did not allow editing or mixing and had a maximum playing time of three minutes that clearly could not be used in conjunction with films which by now were frequently over an hour in length and told complex stories through editing and a whole raft of other devices; ones which would require any synchronised soundtrack to be manipulated with an equivalent level of versatility.

The conversion eventually took place when a form of sound technology emerged, and was adapted to the film industry's needs, which did more or less fit that bill. Unlike the acoustic audio used in conjunction with early films, this technology did not come primarily from the record industry; rather, it drew on research carried out by telephone, radio and consumer record industry engineers. What the various components of technology all had in common was the use of electricity to capture and amplify the audio signal, both for storage using an offline carrier (be that a disc, photographically, or later, on magnetic media and as digital data) and for amplification in post-production and playback in the cinema auditorium. It would prove to be ideally suited for overcoming the limitations of acoustic audio.

Electrical audio works by converting the changes in air pressure which the human ear detects as sound into variations of the characteristics of electrical energy over time, which are then recorded. The device which detects those changes in the first place is known as a microphone (or 'mike' for short). The earliest known working example was patented by Alexander Graham Bell – inventor of the telephone – in 1876, and consisted of 'a wire that conducted electrical direct current, with audio signals generated and received via a moving armature transmitter and its associated receiver'.[30] Microphone technology was developed and improved substantially during the following four decades, with increasing sensitivity to changes in air pressure and increases in the signal to noise ratio (the proportion of 'good' recorded sound to distortion in a given recording or transmission). Telephones remained the only application for which microphones were used on any significant scale until the early 1920s, the decade in which radio became established as a mass medium. In the US, for example, regular broadcasts began in 1920 and radio sales 'took off' in the spring of 1922, with over $60 million in sales of receivers recorded during that year.[31] By the end of the decade, most of the populations of North America and Europe either owned a radio or were able to listen regularly to broadcasts.

By contrast, the recorded music industry stuck entirely with acoustic technology until 1925, and even after electrical recording was introduced, playback in the home remained primarily acoustic until the introduction of microgroove records in 1948. The reason for this was partly cultural and partly technological. Unlike the close ties which exist between the recorded music industry and broadcasting today (for example, music broadcasting and the role of disc jockeys), the two were seen as oppositional in the 1920s. Radio sold itself on being a live medium, distinct from both the 'canned' media of film and recorded sound. Commercial entertainment broadcasters, therefore, rejected the use of recorded programming as inferior.[32] When the recorded music industry did begin to switch from acoustic recording using horns to electric recording with microphones in the spring of 1925, the impetus came from a sector about as far removed as possible from the popular cultural remit of the now established Hollywood film industry, that of classical music. The improved reproduction quality of electrical recordings was initially exploited in this genre: upon hearing an experimental test record the legendary classical record producer Fred Gaisberg immediately announced his intention to record Wagner operas.[33] As David Morton describes the phenomenon, 'high culture = high fidelity'.[34]

The main technical reason why the manipulation of sound as an electrical signal was restricted to telephones for so long was the problem of amplifying that signal to a level at which it would power a recording device or loudspeaker (i.e. be audibly played back within a large space). Once again, it was the telephone industry which set about trying to solve this problem, in an attempt to extend the limits of its geographical coverage. Most of the main research was carried out during the 1910s: beyond the telephone network, it was commercially exploited by radio broadcasters initially (1920), followed by the recorded music industry (from 1925), and in the cinema last of all (systematically from 1926).

The individual who made the first key discovery and who was instrumental in its subsequent application to film sound was Lee de Forest, an electronics engineer born in Iowa in 1873. On 20 October 1906 he demonstrated a device which he called the 'Audion tube', a vacuum-sealed glass receptacle containing three electrodes, through which a source signal flowed and could be varied (what would now be called a 'triode valve'). Although at the time, he was thinking in terms of using it purely as part of a radio receiver, subsequent research revealed that triode valves could amplify the voltage of an input signal almost five times, while retaining the original pattern of modulation. The resulting electrical current could then be fed to a device which converted the current into audible changes in air pressure – a loudspeaker, in other words – thereby playing back the signal generated by a microphone. The impact of this discovery on all forms of audio technology, not just film sound, was enormous: as Kellogg puts it, de Forest's Audion tube 'unlocked the door to progress and improvement in almost every phase of sound transmission, recording and reproduction'.[35]

De Forest himself was keenly interested in synchronised film sound, and eventually produced the short-lived but technically very successful Phonofilm system (see below). But he could not have adapted his valve amplification technology into a workable system on his own. Instead, the patents to his Audion tube were bought by the American telecommunications giant Western Electric in 1913. Over the following years Western Electric developed amplification technology for use in long-distance telephone transmission, and following the end of World War One further work was done on producing microphones and loudspeakers of sufficient sensitivity and power for use in the then embryonic radio industry.

The chain of events which culminated in the research and development that would lead directly to the mass-rollout of synchronised film sound, together with extensive analyses of the individuals and businesses involved, their roles and their motivations, has been covered exhaustively, painstakingly and authoritatively elsewhere.[36] There would be very little point in my trying to provide a detailed summary of this research here, since this book is principally concerned with the evolution of moving image technology as a whole and with drawing parallels between different branches of it, not with analysing the minutiae of one chapter in that history in order to draw broader conclusions. Rather, we should take the following salient points from that chapter as having a specific bearing on the development of film sound technology thereafter:

• by 1926, four distinct methods – which would subsequently be marketed under the trade names Vitaphone, Movietone, RCA Photophone and Western Electric – had been successfully demonstrated for electrically recording and reproducing audio in synchronisation with a real-time film transport mechanism (i.e. a camera or projector);

• the first significant commercial launch was of the Warner Bros./AT&T/Western Electric 'Vitaphone' system, used in the feature films *Don Juan* and *The Jazz Singer* (1926 and 1927 respectively dir. Alan Crosland). These combined a recorded orchestral score with synchronised effects and, in *The Jazz Singer*, one scene containing a brief exchange of recorded dialogue (which included yet another apocryphal line: 'You ain't seen nothing yet!'). Although the release of these films has been characterised as 'the birth of the talkies' by a number of writers, it is important to note that by this stage, the competing systems were in the final stages of development and almost ready for commercial launch.

• these early experiments were deemed to have made both a cultural and a business case for integrating film with electrical audio recording. In the years 1927–30, the output of film studios in the US and Western Europe gradually moved to including soundtracks. Actuality and other short subjects came first (for example, Fox Movietone News), but by February 1930 95 per cent of Hollywood's output had synchronised sound. The only significant remaining 'silent' areas of production were in documentary, industrial, educational and amateur films.

• the installation of reproduction equipment in cinemas took significantly longer, because of the volume of equipment which needed to be manufactured and shipped, and the financial investment in it was so great (tens of thousands of cinemas, as distinct from ten to twenty studios). In this sector, the conversion process was not totally completed until the end of 1933, and will be covered at greater length in the following chapter.

Within the details of this process, however, can be found some important clues as to the way in which film sound technology would develop during the latter half of the century. Acoustic sound had failed because (i) the continuous playing time was severely limited, (ii) editing was impossible, (iii) synchronisation was unreliable and (iv) it could not be amplified sufficiently for playback in a large auditorium. The form of electrical audio which eventually established itself in the film industry of the 1930s – optical sound-on-film – successfully overcame all these issues.

The transition process – from experimentation to standardisation

Of the four systems which had been developed in the years leading up to 1926 it was, interestingly, the one which was least compatible with these criteria that was used for Warner Bros.' initial launch. Vitaphone represented a huge step forward from the acoustic technology of the 1900s and 1910s in that it used electrical amplification both to record and reproduce the signal. The signal carrier, however, was still a disc. There were some refinements. The discs were 16 inches in diameter, and rotated at a significantly slower speed – 33rpm compared to the 78rpm which had been universally adopted by the recorded music industry as the standard speed for records for domestic sale in 1915. To compensate for the consequent loss of dynamic range, a thicker and softer shellac compound was used together with a larger groove pitch, which enabled wider and deeper modulations to be inscribed. This produced a comparable signal quality but a longer continuous playing time than the discs' domestic counterparts. However, as a result, Vitaphone discs tended to wear out far more quickly than domestic shellac records. They could be played 'between 18 to 22 times with fairly good results'; according to a projectionists' manual from 1929, cinemas needed to have three sets of discs to last for a full week's run.[37] From Warner Bros.' standpoint, the decision to use disc technology represented a significant time and cost saving, as Western Electric had already invested in the research and development which had led to the introduction of electromagnetic disc cutters – in which the cutting needle is driven by an amplified electrical signal rather than acoustic vibration – in 1925, and these could be adapted for recording synchronous film sound at very little extra cost.

Vitaphone did, however, suffer from some of the same drawbacks as acoustic recording. Each disc surface carried about 10 minutes of playing time (which corresponded to a reel of release print), which all had to be recorded in a single take. Any mixing had to be done 'live' during the recording itself and the disc could not be edited thereafter.[38] As a former radio engineer who moved to Hollywood in the wake of the conversion recalled, 'it was one thing to put a recorded song on a disc, with an ad-lib by Al Jolson at the end of it; it was quite another to edit a fast-moving melodrama, in which there might be a dozen short scenes in one minute of film'.[39] And there were also synchronisation problems. Again, the use of electricity achieved a vast improvement over any of Vitaphone's acoustic predecessors. In the studio, cameras and disc cutters

Fig. 4.1 The projection booth of a cinema in Tooting, South London, circa early 1930s. The projectors are equipped both with optical sound-on-film reproducers and turntables (underneath lamphouse) for Vitaphone records. Courtesy of BFI Stills, Posters and Designs.

were driven by self-synchronous (selsyn) motors, in which the frequency of the AC source power supply ensured that their speed remained identical. In the cinema, 'the turntable carrying the disc is on the same motor shaft that drives the mechanism of the projector',[40] theoretically ensuring perfect synchronisation. But though it was much improved, this arrangement was not foolproof (especially at the exhibition end, which will be covered in the next chapter).

In order for synchronised sound to be fully compatible with the production techniques which had been evolved by Hollywood during the 1920s, a system was needed which, like Vitaphone, exploited the possibilities of electrical recording and amplification; but which, unlike Vitaphone, did not suffer from the drawbacks of using discs as the carrier (unreliable synchronisation, limited running time and inflexibility in editing). This technology was already in an advanced stage of development by the time Vitaphone was launched to the public, and would push it out of the marketplace altogether by the early 1930s. This was 'optical' sound, a technique which worked by creating and reproducing a photographic record of the electrical impulses produced by a microphone.

The technique of representing soundwaves photographically (yet again) was not new. Possibly the first person to have discussed the idea in print was Professor E. W. Blake of Brown University. In 1878 he produced 'photographic records of speech sound on a moving photographic plate, using a vibrating mirror',[41] though there is no evidence that he managed (or even attempted) to reproduce them as audible sound. As with electrical disc recording, it was the discovery of Lee de Forest's Audion tube and the subsequent development of microphone technology which really opened up the possibility of using photographic film to record sound as well as pictures. In crude terms, this technique works as follows: as with telephone or radio transmission, and disc recording, sound is captured by a microphone and the resulting electrical impulses are amplified by electrically, by means of valves. However, instead of feeding this signal to a disc cutter to produce a groove in a record, it is used to control the flow of light to a strip of unexposed film, which passes in front of the light source at a constant speed. The stronger the signal, the brighter the light, and vice-versa. When processed, the exposed film therefore carries a permanent record of the modulated signal over time, one which is theoretically capable of being reproduced as audible sound.

In practice, working systems for both recording and reproduction did not become available until the early 1920s. This was because, even though it would be theoretically possible to regulate a light aperture controlling exposure acoustically, optical sound was wholly dependant on electrical amplification for playback. As there were no indentations in the processed sound film, the only way it could be played back was to find a means of converting the processed photographic sound record back into an electrical impulse for amplification. A key discovery toward making this possible was demonstrated by the British engineer Willoughby Smith in 1873, which was that the metal selenium generated small amounts of electrical energy in proportion to the intensity of light it was exposed to. This eventually led to the production of the 'photoelectric cell', which forms the basis of the optical sound reproduction head in all cinema projectors manufactured from the early 1930s onwards. However, it was impossible to exploit this discovery in the absence of any means of electrical amplification for playback. Though de Forest's work cleared the way for optical sound to become a reality, systematic research did not get underway until the late 1910s, primarily because the telecommunications industry initially had other research and development priorities for the Audion tube, and thereafter World War One put a stop to most technical research and development in the US and European film industries.

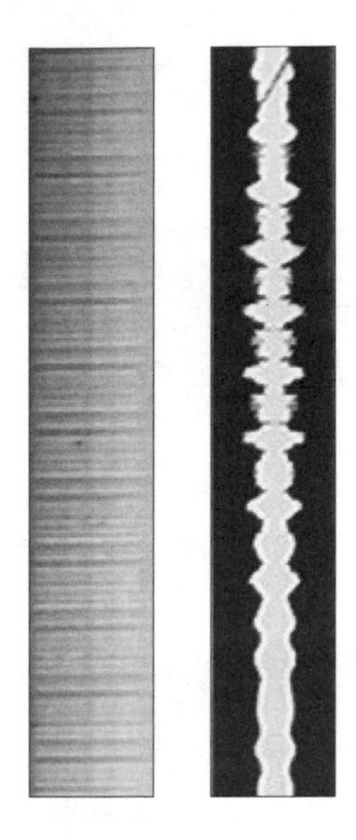

Fig. 4.2 Variable density (left) and bilateral variable area (right) optical sound records.

By the start of the 1920s, a number of inventors were working on different systems, backed by a combination of industrial capital and private investment. As we have seen above, the Warner Bros./AT&T Vitaphone system was the first to bring electrical recording and reproduction to the market, primarily because it represented a cleverly integrated combination of two 'off the shelf' technologies initially developed by separate industries – the amplification technology found in telecommunications and radio and the audio carrier used by the retail music industry – rather than a fundamentally new technology which was specifically tailored to the needs of film studios and cinemas. This gave it a slight head start. There were also no less than three optical systems being developed simultaneously, the most successful of which (albeit in a much modified form), remains in widespread use well into the twenty-first century. These were, briefly:

• a system which resulted from work carried out by Lee de Forest and another American, Theodore W. Case. In 1917 Case had discovered a significantly more sensitive photoelectric compound than selenium, and by 1922 had developed the 'AEO-light', a hydrogen-filled bulb which responded to variations in electrical current. Combined with de Forest's amplification technology, the package was christened Phonofilm and was first publicly demonstrated in April 1923 in New York. In the Phonofilm sound camera, the intensity of exposure through a fixed aperture varied according to the current applied to the light. As the developed sound record was thus of fixed width but with varying modulation according to the density of silver salts remaining in the emulsion, it was known as a *variable density* soundtrack. For the next few years Phonofilm was used on a small scale to make a number of musical and comedy shorts ('about 34 theatres' in the US-installed playback equipment, according to Coe,[42] and the system was also used on a limited scale in Britain), until the Fox studio eventually bought the rights in July 1926. By the following year Fox was starting to release full-length Movietone features on a regular basis, and from autumn 1927 a bi-weekly newsreel with synchronised sound (something which the logistics of Vitaphone recording would have made impossible).

• a system which recorded optical sound by varying the size of the aperture rather than the intensity of the light source. This produced an exposure of equal intensity, but with variations in the signal modulation being registered as movement of the boundary between the exposed and opaque areas of the recording. This, therefore, was known as *variable area* sound. The invention which facilitated it was a modified galvanometer which caused a small mirror to vibrate in response to the input signal. When a uniform light source was projected onto this mirror, it was reflected through an aperture and onto the emulsion of the passing unexposed film stock in proportion to the source modulation. This technique was first demonstrated by the scientist Charles Hoxie in 1921 and subsequently refined by researchers working for the Radio Corporation of America (RCA), which bought the rights. At the time RCA was effectively a front organisation for two large American industrial combines, Westinghouse and the General Electric Corporation, both in exploiting new audio technologies and also as one of the emerging broadcasting majors. The problem RCA had with Photophone was that between them, Warner Bros. and Fox had effectively saturated the (then nascent) film sound market. RCA's solution was to start its own studio and cinema infrastructure, formed by buying out a large theatre chain (Keith-Albee-Orpheum, hence 'KO') and with RCA providing the sound technology. RKO, therefore, is unique among the Hollywood majors of the 1930s and 1940s in that it was established with the sole aim of commercially exploiting sound technology. Initially only RKO used Photophone among the Hollywood majors, although the system quickly gained a growing market share among producers of B-pictures, newsreels, documentaries and animation. Ironically, the second major studio to adopt it for feature film production was Warner Bros. itself, having finally abandoned Vitaphone in the early 1930s and variable density recording in 1936.

• a refined variable density system developed by Western Electric as it became clear that the technical advantages of sound-on-film would push disc recording out of the market. In this method the modulation was not recorded by varying the brightness of the lamp making the exposure, but by means of a 'light valve', which varied the width of the slit aperture which regulated it. As it was lighter than the RCA mirror and worked at a much lower voltage than the AEO light, it was far more sensitive to changes in modulation – especially at lower volume – than either of the other two optical systems. In 1928 the five major Hollywood studios agreed to adopt this new standard for subsequent feature production.

Of the four systems which had been successfully demonstrated by 1926, the years between 1928 and 1932 saw two (Vitaphone and Movietone) fall by the wayside and the other two (Western Electric and RCA) establish market shares which they would maintain and develop for two decades subsequently. Vitaphone discs remained in use as a playback medium in cinemas for 2–3 years after the system had been abandoned as a production medium, largely because of the investment which cinemas had already made in the equipment (this issue will be covered in greater detail in the next chapter). Compared to the Western Electric method, the sound quality recorded by the AEO light was 'poor, and the Fox-Case system soon went the way of Vitaphone and the carrier pigeon'.[43] Of the two technologies which remained, Western Electric dominated Hollywood throughout the 1930s. RCA, however, established far more of a foothold in Western Europe (especially in the UK and France, where variable area recording, either using RCA equipment itself or sub-licensed forms of it, soon dominated those studios) and South America. It soon became apparent that variable area recording offered significant technical advantages over variable density, of which more below.

The 'key moment'

The historian Thomas M. Cripps notes that:

> The chronicle of soundfilm was no more risk free than the stories of other technologies. But the manner of its success differed from almost all other cinematic achievements. Neither the adventurers in the banks nor the visionaries in the lab could have predicted soundfilm's tremendous success after a vaudeville-like snippet of it premiered in the summer of 1926.[44]

He is certainly correct to identify the speed and universality of the conversion process as being unique among the complex and interrelated instances of technological, economic, industrial and political forces determining the when, where and how of technological change in moving images. Here are some examples by way of comparison: camera and projection speeds took nearly three decades to standardise. Photographically recorded and reproduced colour took almost four decades to make the jump from a prestige, niche-market technology to being almost universal in film and television production. The 'Academy' aspect ratio was abandoned as a universal standard for film in the early 1950s, and over half a century later, no single widescreen standard has emerged to replace it. In almost eighty years of regular broadcasting there has never been a single, worldwide standard for the signal format of an analogue television transmission. Compared to these timescales, the practice of producing a synchronised and mixed soundtrack supplied alongside the moving image, and recorded in a way that was universally replayable on a wide range of cinema equipment, took place almost overnight. The only other conversion processes to have happened anything like this quickly were the introduction of panchromatic film and the nitrate to acetate conversion (see chapter one). There are some notable similarities and differences between the two, of which more below.

As stated in the introduction, the 'key moment' has been heavily researched, discussed and mythologised; probably more so than any other process of technological change in the history of the film industry. In fact, it is probably the *only* process of technological change which the lay cinemagoer is likely to be aware of, in any meaningful historical sense. Support for this contention can be found in the fact that of ten people I recently spoke to at random in the bar of my local arthouse cinema, eight had not even heard of the term 'CinemaScope' (and the other two were not able to define it to any degree of accuracy), while all but one knew that sound had been 'invented' in the late 1920s. Seven respondents gave their source of information as the film *Singin' in the Rain*. Even at the time that film was set, one of the proliferation of technical manuals for projectionists published in the wake of the conversion noted the extent of the public reaction it had generated, commenting with frustration that 'every now and again the newspapers and the trade publications hail some individual as having been the first to have invented "sound pictures"'.[45] So why is this the case, and do the unique aspects of the 'key moment' set it totally apart from other instances of technological change in the history of moving image technology?

The first unique factor to note is that the 'key moment' was highly visible and heavily publicised. Hollywood had spent the previous decade developing the 'classical' continuity system of shooting and editing, the main aim of which was to make the role of technology invisible. As a member of the cinema audience you were not supposed to know or care whether a scene had been shot using orthochromatic or panchromatic stock, how it had been lit or even if it had been shot on location or a studio set. Sound, on the other hand, was a rare example of technology – and specifically the idea of technological change – being heavily promoted and marketed to consumers of the film industry's output. True, this also happened to a limited extent with Technicolor and early widescreen as well. But the publicity associated with these technologies was aimed at identifying the films and cinemas which offered them as prestige, niche market products, as exceptions which proved a rule. This was very similar way to the phenomenon David Morton identifies, of marketing of hi-fi classical music recordings by the retail music industry when electrical recording was introduced in the mid-1920s and then long-playing records in the early 1950s. In contrast, the promotion of movie sound was explicitly pitched as a new technology which was here to stay and which would engender permanent change.

To a certain extent, accounts of the conversion such as the one found in *Singin' in the Rain* represent the end result of a legend becoming fact and then being printed as such. It will already have become apparent by now that most published research on the 'key moment' – including this summary of it – deals almost exclusively with events that took place in the United States. A demonstration of the extent to which the Hollywood 'key moment' myth has come to dominate can be seen by looking further afield, where the conversion process starts to extend and become more complicated. In Western Europe, for example, patent and licensing disputes inhibited both the conversion process itself and the evolution of film sound technology thereafter. Eventually, compatibility issues and licensing restrictions associated with the German 'Tri Ergon' variable density system were eventually used as a political weapon by the Nazis, as a way of restricting the import of Hollywood films.[46] In the Asian subcontinent, where the American giants could not enforce their patent rights as effectively as they could in the developed world, the conversion process happened a lot more slowly, as indigenous companies developed their own equivalents of the Hollywood technologies: in Japan, for example, 14 per cent of feature films were still were still being shot silent by 1942.[47]

A second reason for the 'key moment' being understood as a unique phenomenon was its sheer cost in relation to the speed at which it took place. As argued in chapter two, a combination of the depression and the time taken to amortise the investment in sound almost certainly inhibited the commercial development of widescreen for two decades subsequently, and may well have done likewise both with colour (see chapter three) and stereo (see below). This investment was not so much in the research and development of the technologies themselves: in this modifications, the Paris agreement continued to operate until the outbreak of War Two.[53]

the Hollywood studios, gradual refinement of the Western Electric and RCA ns continued. It was noted earlier that one of the main reasons for Vitaphone's e was inflexibility in editing. The first generation of Movietone and Photophone ing equipment was not much better: it was 'single system', meaning that both age and the optical sound record were exposed simultaneously, in the same a and onto the same strip of film. The advantages were foolproof synchronisa- nd a reduction in the amount of equipment needed for location work (hence tone's initial launch in newsreels), but the possibilities for creative editing were nited. As editing optical sound film was accomplished in exactly the same way ting the picture – i.e. by cutting and joining it – any cut made to the original ve would affect both the picture and the sound. Add to that the 20-frame and it quickly became clear that for everything except footage that would eed minimal sound editing (for example, actuality footage), single system was too restrictive. Possibly the most widely known single system feature film ne of the most heavily analysed early sound films in existence) is *Blackmail* dir. Alfred Hitchcock). The fact that the single-system Photophone camera post-synching impossible was exacerbated by the fact that the lead actress the part of a London shopkeeper's daughter, but was a Czech who spoke no h! The production had started life as a silent in which this obviously was not a m. When the studio took delivery of some of the first RCA equipment to reach midway through shooting, a last-minute decision was made to release both m as planned and a talkie version, the latter containing a number of scenes lly reshot for sound.[54] As the use of single system made it virtually impossible t-dub an English-speaking actress (to achieve this, all the other components of undtrack would have had to be recreated in real time along with her voice), she the lines into a microphone off-screen while they were mimed by the Czech s on camera. As a result of logistical problems such as these, one of the many diting manuals to have dissected *Blackmail* in minute detail observes that 'the nce between the two versions is palpable. The freedom available to shoot dit without constraint is manifest in the variety of structuring possibilities that ock exploits in the silent version.'[55]

Vithin a few years single system had been superseded by the use of sepa- sound cameras' (devices which expose an optical sound record onto negative independently of the camera used to shoot the picture), with clapperboards ding a reference point for synchronisation (see chapter two). The separate negatives could be cut and joined far more flexibly and in perfect synchroni- with the picture, as there was no fixed offset in the recording. Furthermore pportunities for mixing were to all intents and purposes unlimited, as two or reproducers (each running film which had been edited as necessary) could be ed to differing volume levels, with the combined output from all of them fed sound camera which exposed a new negative carrying the mixed recording. this did introduce the generational fading which is common to all duplication

regard the film industry was following one of a long line of precedents of cashing in on technical innovation which had been generated either by small scale 'cottage industry' research by private individuals, or by adapting technology produced by big business in other related industrial sectors, or – as in this case – by a combination of the two. The big bucks were spent on installing equipment in cinemas. I shall discuss this process at greater length in the next chapter; it will suffice to note at present that by 1929, there were four systems on the market being used by Hollywood to record sound, and over 300 for reproducing it in the cinema. The asset base of Warner Bros., for instance, grew from just over $5 million in 1925 to $230 million by 1930.[48] It was a similar story with the other vertically integrated majors, not to mention the thousands of independent exhibitors which found themselves having to buy equipment in order to stay in business. Most of this infrastructure was paid for by speculative investment, and it all happened at the height of one of the biggest economic booms in America's history, when such investment was easy to come by. As Douglas Gomery notes, 'Warner Bros. 'had the support of America's most important banks throughout this period of expansion. And sound was only one part, albeit a very risky one, of the investment surge [of the mid-1920s]'.[49] Despite the overlap in technology with other allied industries, this electronic hardware and the manpower installing it could not be removed and put to other use. Once the conversion bandwagon had gathered a critical mass of momentum, there was simply no going back (apart from writing off a colossal investment) even if the industry and/or its bankers had wanted to do so, especially after the Wall Street Crash. By contrast, the technical perfection of widescreen and stereo did not happen until shortly after the bust which followed. In economic terms, the commercialisation of these two technologies was set back to where sound had been two decades earlier, even if, by the early 1930s, the technologies themselves were considerably more advanced.

However, there are some other aspects of the 'key moment' which do have similarities with other examples of technological change, most notably the nitrate to safety conversion. The first – and this flies in the face of a lot of 'key moment' mythology – is that it was achieved largely on the back of off-the-shelf technology. With the sole exception of the hardware for recording and reproducing optical sound (and that was only developed to the point of marketability as a direct response to the obvious shortcomings of the alternative), virtually all the technology used to make the 'key moment' happen had been developed for use in telecommunications, broadcasting and/or the retail music industry. It was adapted for use with films when it became clear that it would enable the restrictions of acoustic disc-based recording to be overcome and thereby allow recorded sound to meet the requirements of the 'classical' production conventions being evolved by the mainstream studios. Safety film had followed a similar pattern. Attempts to inhibit the flammability of nitrate went back at least to 1904, but for over four decades subsequently it made more economic and technological sense to manage and minimise the risk of fire than to adopt an alternative which was more expensive, less reliable and to all intents and purposes just did not do the job. With both sound and safety film, there were important points along the way: de Forest's invention of the triode valve, improvements in microphone technology, the mass-manufacture of diacetate and butyrate for use in small-gauge stocks, or the introduction of propionate in 1938. It is less clear with safety film than with sound as to what precisely tipped the scales; but as with sound the conversion happened astonishingly quickly once that point was finally reached. There are other examples of similar chains of events, but none which took place in anything close to the time scale as that of the 'key moment' or the introduction of safety film.

The second aspect of the 'key moment' which bears similarities to the mass-rollout of other moving image technologies is in the role of standardisation, which enabled sound to integrate with dominant business models. Barry Salt expresses the rationale for Bell's 'fondest hope' in somewhat more pragmatic terms:

> The almost inevitable sequence of events is that one company chooses a standard for its own use, and then, either because the company concerned was first in the field, or because it establishes economic superiority, or because the standard is obviously a sensible and practical one, the other companies adopt it as well. Only after this has happened is the standard ratified by an industry body.[50]

In other words, market forces do the job, the end result of which is then enshrined by formal technical standards, which in turn derive their authority from an already-established mass acceptance. In the case of sound the pioneer systems – Vitaphone and Fox/Case – ultimately failed, for exactly the same reasons that (for example) Technicolor and the CDS sound system system (see chapter 5) eventually followed suit: they proved the business case for the deliverable, but then someone else came along with a more efficient way of delivering it. Initially, the four sound systems were totally incompatible with each other at the exhibition end. A film recorded using Vitaphone equipment could not be shown in a cinema which used Western Electric reproducers, and vice-versa. The studios very quickly realised that if this situation had been allowed to continue, it would push costs up and reduce revenues unnecessarily. Cinemas which could afford them would install multiple sound systems, which would benefit the electronics manufacturing industry at the expense of the Hollywood studios, but other cinemas would not be able to show films which they otherwise might have done, thereby reducing the earning power of the studios' product. As with 35mm film itself, it was for these reasons that the five major studios agreed in 1928 to standardise certain technical characteristics of optical sound implementation between the rival products available to studios, principally the film transport speed (24fps), the position and width of the soundtrack and the 'offset', i.e. the distance between the synchronisation points for picture and sound (for 35mm film the sound was offset 20 frames ahead of the picture). This ensured that, while there may be differences between the cost and quality of the various systems available, any soundtrack which adhered to the standards would be playable in any cinema throughout the world. That is not to say that compatibility issues never surfaced thereafter. In particular, there is evidence to suggest that the uptake of specific sound systems was used by the Hollywood studios to gain economic advantage. In

the early 1930s, for example, allegations surfaced
were being refused access to American-owned dist
recording equipment imported from the US.[51]

By the early 1930s, therefore, moving images a
become two inextricably linked technologies. An ir
quickly the 'key moment' had taken hold can be four
es (though admittedly, not many) who still believed
inextricable, that silent cinema had not gone forev
forward to a long period of co-existence. One exam
Betts, who, writing in the highbrow journal *Close U*
come the established norm for newsreels, actualitie
'because of their novelty and magnetism of the hum
have a place in our film programmes',[52] but that syn
for features, the majority of which would continue
his were increasingly becoming exceptions which p

Consolidation in the 1930s and 1940s

The following two decades present a similar patter
they consist primarily of the consolidation of existin
industrial practice, and the beginnings of the rese
which would eventually bring new ones to the mark

As has been mentioned above, the early part of
and Movietone disappear and Western Electric va
area sound (and derivatives thereof) become the pri
use worldwide. Although Vitaphone had effectively b
duction medium during 1929 in favour of Western E
soundtracks continued to be released to cinemas u
phased out (more on this in the next chapter). With
studios also adopted Western Electric. RCA Photoph
nating market position in Britain, while the rest of E
a German variable density system known as Tri-Erg
consortium of European investors, the Tonbild-Synd
witnessed what was almost a full-scale trade war v
systems. It was eventually resolved at a conferenc
between key US and European representatives, who
cartel for the export markets of sound technology. M
the exclusive territory of Tobis; though once again, a
was that of universal compatibility between playbac
was unrestricted, with the result that a number of h
Notable among them were British Acoustic Film, la
variable area technology developed by the Danish e
nold Poulsen, and Visatone-Marconi, which was in us
used extensively in newsreel, documentary and low-b

of analogue signals, various methods were evolved for keeping the loss of signal to noise ratio (i.e. sound quality) to a minimum. Most notable among these were that film manufacturers developed stocks which were specifically designed for optical sound mastering and duplication. While the introduction of panchromatic film in the late 1920s (see chapter one) greatly expanded the creative opportunities available to cinematographers, it was not the ideal type of film emulsion for optical sound recording. As the exposure in a sound camera came from an artificial light source, a more accurate sound record could be obtained by using film stock which was photosensitive only to those areas of the colour spectrum in which the sound camera generated light. This obviously was not an option with single system, because the same strip of film had to carry both the picture and the sound. By 1932, both Eastman Kodak and Du Pont (the two major American stock manufacturers) were marketing fine-grain stocks optimised specifically for variable area and variable density sound mastering.[56] The sound negatives these produced could be copied optically (i.e. by contact printing) with negligible loss of signal, or electronically if mixing was needed.

Of the two methods of optical sound recording which became established during the 1930s, variable area gradually emerged as the technically superior system. Initially the RCA sound cameras had suffered from what was termed 'volume expansion' by the sound engineers who worked with it. The mirror which directed the light beam in response to the input signal was significantly heavier than the equivalent light regulation device (the 'Carolus cell') in the Western Electric variable density sound camera. It failed to respond effectively to low signal levels (meaning that quieter sounds were less audible), but over-reacted to higher ones, resulting in overmodulation. This was gradually overcome by progressive refinements to the RCA galvanometer system throughout the 1930s and early 1940s, eventually leading to the use of ultra-violet light to make the exposure in the RCA sound cameras marketed in the late 1930s. Its much lower wavelength than that of light within the visible spectrum vastly increased the strength of the recorded signal and its durability through successive generations of copying.[57]

The main flaw with variable density was not so easily overcome. It depended on variations in the chemical composition of the emulsion across the topography of the film to accurately represent the sound record. Variable area, however, only needed the film to register the difference between opaque and transparent – it was the accuracy in where that boundary lay which determined the quality of reproduction. Therefore, a density track was far more likely to be significantly degraded through successive generations of copying than an area one, because quality control in the lab was far more critical. As one archivist who has spent a significant proportion of his career working with variable density soundtracks (and who describes himself as 'possibly the last person on this planet to re-record using a variable density sound camera') stresses: 'control, or, if you wish, grading of variable density sound is both vital to both quality and volume of duplicated sound materials. Signal to noise and distortion can become enormous problems if the closest of attention is not paid to sound control.'[58] Furthermore, a number of noise reduction techniques for variable

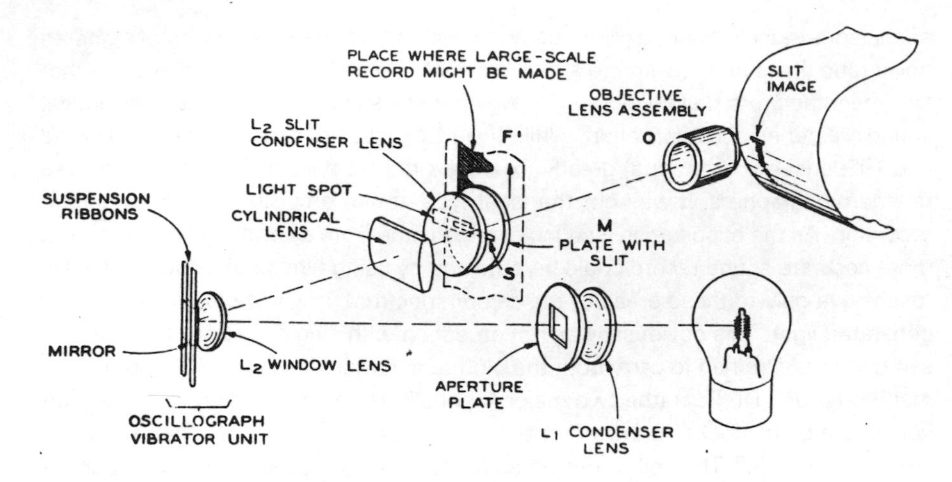

Fig. 4.3 Schematic of an early RCA variable area recording mechanism.[59]

area systems were introduced throughout the 1930s which further enabled almost lossless copying (thereby increasing the creative possibilities for editing and mixing), notably the 'push-pull' soundtrack, while their variable density equivalents proved to be far less effective. However, due to patent restrictions and contractual arrangements between studios and the owners of sound technology, Western Electric maintained its dominating market share as Hollywood's principal film sound system until the advent of magnetic recording in the late 1940s. Optical sound would, however, remain the primary method of reproducing film sound in the cinema up until the time of writing. The main developments in this technology in the post-war period were concentrated almost exclusively in the area of release formats for cinema exhibition, and it is for this reason that they will be discussed in the next chapter. For the purposes of this one, it is enough to note that as the technical superiority of variable area became apparent, the RCA system and derivatives thereof gradually increased their market share. The use of variable density declined significantly in the late 1940s, and had all but disappeared from mainstream use by the end of the 1950s.[60]

Magnetic sound in the 1950s

The use of magnetic tape to record and edit film sound was the next major technological development to significantly affect the production and exhibition of film and television. Optical systems captured a record of the modulating input signal using photographic emulsion. Magnetic sound does so on a metallic compound which is sensitive to magnetic fields (initially iron oxide, but subsequent refinements introduced magnetic media of greater sensitivity, such as chromium dioxide) coated, like film emulsion, on a flexible base. The input signal is fed to an electromagnetic 'head' while the coated base passes in contact with it at a constant speed. The head converts the input signal into magnetic energy, which in turn causes the oxide particles to change their relative positions (i.e. some particles are positively charged while others are not) in a way that creates a record of the input signal. Playback is simply the reverse of this process: the coated tape or film is passed across the head a second time, only without any signal being fed to it. This time the head responds to the changing patterns of magnetised oxide on the tape, and the resulting signal is electronically amplified in order to play back the recording.

The development and commercial rollout of magnetic sound followed a similar pattern to that of the other technologies discussed in this chapter. Early attempts by engineers to put theory into practice stretch back well into the nineteenth century. The first one which technological historians generally cite as having been successful was a magnetic recorder demonstrated by the Danish engineer Valdemar Poulsen at around the turn of the twentieth century. His 'Telegraphone' used steel wire as the recording medium. It was of limited use because the absence of any means of amplification meant that the quality of the recorded signal was low and that it could not be played back at any significant volume. There is no evidence to suggest that magnetic recording technology underwent any major developments between then and the 1930s, though a number of scientists and engineers worked on it experimentally.

Once the conversion to sound was complete, the film industry realised early on that this method offered several potential advantages over optical sound. Although the experimental magnetic recorders built before the 1940s generally produced much poorer sound quality than their state-of-the-art optical equivalents, research in the post-war period quickly enabled magnetic sound to record a much wider frequency range and with much lower noise than any optical system. And just as optical sound had provided a number of significant advantages over acoustic and electromechanical disc recording, so did magnetic over optical. Although optical sound could be cut, edited and mixed without practical limit, the process of doing so was costly and time consuming. This was because, like discs, optical sound was also a 'write once' medium: the film stock could only be exposed once, and the sound record could only be played back after it had been sent to a lab for processing. Although a defect in an optical recording could be corrected by replacing only the affected element or section (unlike with discs or single system, in which all elements of the affected reel or shots – including the picture – had to be retaken), playback or mixing was not instantaneous. With magnetic sound it was: as with a modern tape recorder, all the sound engineer had to do was to rewind the reel and press the play button. Furthermore the tapes could be erased and re-recorded ad infinitum. Noting these points, one American writer suggested in 1931 that:

> What seems to be the most practicable method thus far suggested is to record the electrical variations as changes in magnetism. If it could be perfected, some apparatus such as the electromagnetic recorder, known as the Telegraphone, invented years ago by the scientist Poulsen, could be used.[61]

In fact, the next major developments in magnetic sound appear to originate (as with safety film and tripack colour) with the Nazis, who seem to have been more interested in its potential for use in radio broadcasting than as a means of adding sound to moving images: as Winston argues, 'it was not until after World War Two that American interest in magnetic recording was revived with the importation of Nazi "Magnetophon" recorders'.[62] In contrast to the reluctance of American broadcasters to use recorded sound because they promoted radio as being a primarily 'live' medium, the Nazis found it ideally suited to many of their propaganda requirements. One interesting footnote to this story is that the Allied intelligence services were persistently unable, throughout the war, to identify the geographical origin of many 'black' propaganda broadcasts to the UK: they were unaware of the advances which had been made in magnetic sound and therefore assumed the broadcasts to have been live because their quality was too high for the source to have been a disc recording. While the early models of recorder (such as the 'Blattnerphone' of 1929, which was also used on a limited scale by the BBC in the 1930s) used by Nazi broadcasters also recorded onto steel wire, the development of a flexible base was quickly identified as a priority, mainly because of the relatively low sensitivity of 'raw' steel, but also because of the health and safety risks of using sharp steel wire in a high-speed transport mechanism. In 1934 the German chemical giant Badische Anilin und Soda Fabrik (BASF) started to manufacture tape on a cellulose diacetate base (with the oxide being supplied by another infamous name in Nazi heavy industry, I. G. Farben), and in 1938 German state radio began using modified versions of a later Poulsen-designed recorder, labelled the Magnetophon.

As with the Agfa tripack colour system, this technology was appropriated by the Americans in the immediate aftermath of the war. The Ampex corporation (of which more in chapter six) quickly established itself as the US's key manufacturer of equipment, while the Minnesota Mining and Manufacturing Company (later 3M) was the first industrial-scale producer of tape in the West, initially on a triacetate base but subsequently on polyester. Even as late as the early 1950s, the embryonic Sony company in Tokyo was still experimenting with paper as the base,[63] while in Britain magnetic recording was not established among broadcasters and the music industry until the mid-1950s.[64]

The film industry moved quickly. By 1948 both Western Electric and RCA were marketing magnetic recorders which were designed to integrate with the existing production practices of studios that already used their optical equipment. The initial recording, editing and mixing took place on magnetic stock. At the end of the whole process, an optical sound negative was exposed of the final mix, which was used to add a combined optical soundtrack to the release prints. The efficiency of this process was further enhanced when the studio engineer Norman Leevers came up with the idea of coating conventional 35mm raw stock with magnetic oxide and making the recording at the same speed as photographic film passed through a camera (i.e. 1½ feet per second, the equivalent of 24 frames assuming a four-perf pulldown), meaning that the sound record could be run, cut and joined in synchronisation with the picture film during editing, just as optical soundtracks had.[65] The disruption to studio techniques and the retraining needed by recordists and engineers was therefore minimal.

By the late 1950s magnetic recording had become standard practice in all forms of production, from Hollywood blockbusters to home movies, and would remain so until the film and television production industries turned to digital sound recording in the mid- to late 1980s. Soon, ¼-inch or ½-inch tape recorded on portable machines quickly became the norm for location recording, while an updated form of single system was adopted for television news and documentary shooting: the 'combined magnetic' track, in which a 'stripe' of oxide was applied to unexposed 16mm film stock in the space normally occupied by the optical track. Since the content of this track could be copied onto another magnetic tape or film for editing after processing and then the final mix rerecorded to the original element after it had been cut if necessary, the magnetic single system was a lot more flexible than its optical predecessor.

The rollout of magnetic sound as a playback medium for cinemas followed a significantly different pattern. In fact, the introduction of this technology broke a long-standing commercial link between the manufacture and supply of sound-recording equipment for studio production and for cinema reproduction. The two were no longer packaged in the same way as had been the case during the period when optical recording was in mainstream use, and the cinema exhibition industry proved increasingly reluctant to invest in buildings and equipment upgrade. Where new technologies were rolled out, backwards compatibility became an absolute prerequisite, as will be discussed further in the next chapter.

Stereo sound in the film industry: 1952 to the present day

Multi-channel or 'stereo' sound for films was also nothing new by the time of its first commercial rollout in the early 1950s.[66] Examples of early research into the idea date back to October 1911,[67] though it is generally agreed that the first successful demonstrations took place during the 1930s. In Britain, the telecommunications and radar engineer Alan Blumlein produced a sound camera which contained a 'totally new' galvanometer. It enabled two variable area optical tracks to be recorded in the space previously occupied by one, and Blumlein carried out extensive tests with various techniques for positioning microphones.[68] The subject matter of his 'binaural' films was clearly designed to reproduce a sense of spatial positioning when played back through loudspeakers placed to the left and right of the auditorium. Of the surviving films, the two most effective (both shot in July 1935) show steam trains passing through a suburban station in West London, and a horse-drawn fire engine travelling across a field, the bell on which can clearly be heard panning from left to right. Similar experiments were carried out by Bell Telephone Laboratories in the US, supervised by the engineer Harvey Fletcher, eventually leading to what is generally believed to have been the first mainstream feature film to be recorded and presented in stereo. *Fantasia* (1940, dir. Ben Sharpsteen *et al.*) was a feature-length cartoon, the soundtrack of which consisted mainly of orchestral music. The music was recorded using eight channels and then mixed down to four for exhibition, with the stereo mix being played in a small number of 'road show' venues which had installed the necessary equipment.

This appears to have set a precedent for stereo in more ways than one. The 'high culture = high fidelity' phenomenon identified by David Morton in relation to the introduction of electrical recording in the 1920s repeated itself with stereo, both in the consumer audio and film sound markets. In the former, the 'long-playing' microgroove record (LP), introduced in 1948, was 'envisioned with classical music in mind'[69] while stereo recordings, first sold as ¼-inch tapes from 1955 and, following the launch of the Western Electric stereo disc cutter in 1958, as LPs, were marketed aggressively at middle-class consumers with high disposable incomes. Once again, classical music was mainly responsible for the sector's growth, in the form of a 25 per cent surge in LP sales between 1959 and 1961. Unlike the 1920s 'key moment', when electrical recording had been redefined from an elitist consumer technology to a mass medium in the form of film sound, stereo sound in films would remain a high value niche market in films as well, until the introduction of Dolby stereo variable area optical tracks in 1976 (which is covered in the next chapter) shifted the economies of scale to enable a mass-rollout of stereo in the cinema.

The systematic introduction of stereo recording and mixing within the film industry took place in the early 1950s. While the techniques and technologies for studio production practices were gradually evolved and standardised, two key factors prevented any form of it from becoming a universal standard. Firstly, all the delivery systems for playback in the cinema were magnetic, and therefore required significant additional investment at the exhibition end. Secondly, all the early stereo systems were packaged with widescreen. Cinerama, CinemaScope and Todd-AO all included their own reproduction formats, which required dedicated equipment and which were incompatible with any of their rivals. While the 'Perspecta' pseudo-stereo optical system, which extracted directional sound information from a single channel, was used to a limited extent in conjunction with the early widescreen systems (most extensively with VistaVision), stereo remained an exception which proved the rule until the latter quarter of the decade. Following the lead established by *Fantasia* and the popular perception which was established in the 1950s of stereo being first and foremost a medium for high fidelity music reproduction, most of the 'road show' prestige feature films released during this decade and featuring stereo sound (many of which were musicals anyway) used the stereo spread predominately for music, and in some cases only for music. As John Belton notes, 'critics, industry personnel and audiences accepted stereo scoring, while rejecting directionality for dialogue'.[70] Over the past five decades the use of a centre channel, specifically for dialogue and nothing else, has gradually become an almost universal convention, both in studio mixing and cinema reproduction. Its use became widespread following the introduction of Dolby SVA in 1976, which standardised four channels (left, centre, right and surround) as the industry norm for the subsequent two decades. Rick Altman argues that:

> Listening to the centre channel is like listening to a telephone during a music concert, simultaneously satisfying our expectations for music reproduction (large room with high levels of long, slow reverberation and a wide frequency range) along with the standards that we

> have learned to apply to dialogue transmission (spacelessness and no reverb, with a relatively narrow frequency range).[71]

With stereo sound, therefore, we have witnessed the same process of evolution which characterised the 'key moment' and then the move to magnetic: from a form and use of technology which was either experimental or borrowed from other industries (in this case, the retail music industry) to one which addressed the production and reception context of film specifically.

Developments in film sound technology from the mid-1970s to the turn of the century are concentrated mainly in the area of cinema exhibition, and so will be covered in the next chapter. In the context of production, the 1970s and 1980s saw stereo become the norm as cheap, reliable and backwards compatible equipment for cinema playback became available to support its use in shooting and post-production. The number of channels and the dynamic range possible from magnetic recording gradually increased, and a number of new developments further cemented the use of analogue magnetic recording throughout the 1970s and 1980s. A number of very high quality and easily portable ¼-inch magnetic tape recorders were developed, mainly in Europe, including the Swiss Nagra and the German Uher. Although these machines were initially used mainly in television production and among documentary and experimental filmmakers, these and other advances in tape technology enabled ¼-inch to supersede 35mm magnetic film for many applications in the mainstream film industry. Notable among developments was the introduction of a technique invented by the physicist Ray Dolby which significantly reduced noise levels on magnetic recordings. In crude terms, it worked by passing the input signal through an electronic device which increased the modulation level in weaker areas of the signal's dynamic range (i.e. boosted the recorded volume of quieter sounds), in order to raise the signal to noise ratio. In playback the process was reversed, thereby restoring the balance of the original input signal. The first commercial application for Dolby noise reduction was in the consumer audio market, first in LP mastering and then as an add-on device to the 'compact cassette' format, launched by the Dutch consumer electronics giant Philips in the early 1960s. Cassettes had originally been envisaged as a low cost, easy to use format for consumer recording and playback, designed specifically for office dictation and educational use.[72] Philips and its licensed manufacturers soon discovered that customers were actually using it to play music recordings (either made from radio broadcasts, purchased in the form of pre-recorded cassettes or copied from LP records), and Dolby therefore produced a version of his noise reduction technology that was specifically designed to reduce the 'tape hiss' which characterised the cassette as a lower-quality medium than either LPs (which could not be recorded at home) or ¼-inch tape (which was significantly more expensive). As with virtually all the other forms of sound technology covered in this chapter, Dolby noise reduction was appropriated by the film industry after its benefits in other sectors had been established. Following a series of experiments, the first major feature film to use Dolby in post-production audio mixing was *A Clockwork Orange* (1971, dir. Stanley Kubrick).[73]

This pattern repeated itself yet again with digital recording. Given the extent of the issues and techniques around it which are common to both moving images and audio, the 'd-word' will be covered in one fell swoop in chapter eight. As far as its application to audio is concerned, the technology derived primarily from the computing and IT industries. The use of computers to represent, store and manipulate sound as digital data on a commercial basis have their origins (as with Lee de Forest and electronic amplification) in the telephone industry, and experiments in the early 1960s to establish if digitising transmissions could significantly increase the 'bandwidth', or capacity, of long-distance telephone cables. The record industry began to take an interest in the mid-1970s, when computer processing power and magnetic data storage increased to the point at which the sound quality of digital audio became comparable to that of analogue tape. The first generation of studio digital recorders, however, were not suitable for film industry use. Their mechanisms were those of helical scan Umatic video cassette recorders (see chapter six), adapted to read and write digital audio data rather than an analogue video signal. At that time, a helical scan tape mechanism was the only method available of reading and writing the volume of data necessary for digital audio and in real time. They were many times more expensive than ¼-inch tape technology, could only be used in fixed installations (i.e. they were not portable) and were not perceived by film studios to offer a significant improvement in the final mixed soundtrack as it was delivered to cinemas. Although the compact disc made digital audio playback in the home available from 1983, sales of the new format did not start to rise significantly until the end of the decade, and it was around this time that the use of digital sound first appeared on the film industry's agenda. One landmark development was the launch of the Digital Audio Tape (DAT) format by Sony in 1987. It recorded two channels of sound in an almost identical data format (and therefore quality) to that of the compact disc, but was much smaller and more rugged (despite also being based on a helical scan mechanism) than any previous digital recorder. Its use followed exactly the opposite trajectory to that of the analogue cassette: although it was originally intended as a consumer format, it is believed that opposition from the music industry, based on fears that the format would mainly be used by consumers to produce digital 'clones' of compact discs, kept hardware and media prices high.[74] In the event, DAT began to be used extensively in the early to mid-1990s by broadcasters and filmmakers as an origination medium (the portability of the equipment made it ideal for location use), as an effective replacement for ¼-inch, even though subsequent mixing and editing was still generally done using analogue magnetic tape. During the latter half of the 1990s, personal computer and audio technology began to integrate with a vengeance, as processor speeds and hard disc capacity increased to the point at which audio could be captured and manipulated as easily, as quickly and with as much versatility as other forms of data, such as text in a wordprocessor. A proliferation of stand-alone, hard-disc based audio recorders entered the market at around the turn of the century, and by the time of writing hard discs were being used almost exclusively for recording and post-production mixing in the film industry. Due to concerns over the long-term preservation of digital data (see chapters seven and eight), however, high-quality magnetic analogue master elements are still frequently made, especially for higher end production materials, as archival storage masters.

Conclusion

With the one exception of analogue optical sound, perhaps the most startling conclusion which can be drawn from this narrative is that none of the technologies which have been successfully used to add synchronised sound to film-based moving images were initially developed specifically for that purpose. Furthermore, and in each and every case, there was a significant time lag (ranging from a few years in the case of the 'key moment' constituent technologies to over four decades in the case of stereo) between that initial development and/or commercial exploitation elsewhere, and its mainstream appropriation by filmmakers. Acoustic recording could perhaps be characterised as an exception to this rule – the embryonic Victorian audio-visual industry in general and Edison in particular had been trying to synchronise it with moving images almost since the word go – but, compared with the sound technologies which later arrived and stayed, it could not be described as successful. Recordings were only ever produced to accompany a tiny fraction of the total films released during this period (1895 to 1926 approximately), and the emergence of the 'classical' model of film production in the aftermath of World War One killed off acoustic sound completely. Of the technologies which facilitated the 'key moment', microphones and electrical sound amplification were initially developed by the telephone and radio industries, and electrical disc cutting by the radio and retail music industries. The one significant exception to this rule – optical sound – was only developed to the point of commercial viability at all because of the obvious shortcomings of the alternative (discs).

Stereo sound had both its developmental and its commercial origins in the retail music industry: Alan Blumlein worked for EMI (Britain's largest record company) at the time of his experiments, and stereo LPs went on sale to consumers almost 30 years before stereo sound was widely available to the majority of cinemagoers. Magnetic recording was an experimental cottage industry without any obvious application until the Nazis got hold of it, and was then primarily a broadcasting medium before the film industry took it up soon after World War Two. And digital audio spent the best part of two decades (about the same time as electrical analogue amplification) being developed by the telephone and music industries before the film studios started to take an interest.

Why, therefore, has this area of moving image technology depended so heavily on the products of other industries and in most cases waited a considerable length of time before starting to use them? Some of the answers can be found in the unique relationship between culture, technology and economics which cinema has always embodied. As with the MPPC experiment of making sales of equipment conditional on sales of film stock, Edison believed that by synchronising acoustic recordings to films he could boost the revenue-earning potential of both. When that proved not to be the case (i.e. feature-length 'classical' silent films became the economically dominant format, with which his sound technology was incompatible), he fell by the wayside. A widespread myth exists to this day that Sam Warner decided to launch Vitaphone simply because his company was losing market share to the other Hollywood majors, and therefore 'gambled on a new technology'[75] in order to try and win it back. Though the extent of Warner Bros.' financial dire straits has certainly been exaggerated in many accounts of the 'key moment', the fact remains that they would not have introduced sound when they did either (i) in the absence of a perceived business case, or (ii) if their system was not significantly more compatible than its predecessor with the cultural mode of film production which had by then become a mainstay of the Hollywood film industry.

Magnetic, stereo and digital sound were also introduced on the same basis: when it became clearly apparent that their use would enhance the production of the sort of films which the film industry knew would make money at the box office, and not before. If these technologies had been used in other industries for many years previously, then so much the better: there was less of a chance of mistakenly adopting a system which would ultimately prove unsuitable (as had been the case with Vitaphone) and less of the initial research and development costs to absorb.

Sound, therefore, has been always been treated very differently in connection with film than in connection with other moving image technologies. There was never any such thing as 'silent' television broadcasting on any significant scale (television grew directly out of radio in an industrial sense, in which audio was the mainstream, and the 'key moment' had already happened in film by the time regular broadcasting began); no mainstream videotape recorder was ever built which did not record and reproduce at least one soundtrack in synchronisation with the picture (apart from those designed for CCTV recording); and in the world of digital media, as with broadcasting, audio essentially came first. With film, moving images came first, and continued to be the core business. The technology was made to 'work' on a commercial basis over thirty years before audio technology became available which complimented and enhanced its cultural and economic role. For the first quarter of its life, therefore, the film industry did not integrate sound technology as a key component in its cultural norms and its business model, and has never wholly depended on it. Indeed, the first generation of sound technology was found wanting, and duly discarded.

Unlike the key developments in other areas of film-based moving image technology (e.g. film emulsions, formats, cameras and colour), sound remained an accessory to its key business and culture, right through to the introduction of digital audio in the 1990s. Whereas developments within image technology either came from within the industry itself or from its allied service sectors (for example, film stock manufacturers), sound technology was taken off the shelves of other cultural, communications and mass-media industries as and when it was perceived to deliver a benefit, and its integration then enshrined through the use of standardisation. It is unlikely to have escaped your notice that I began this chapter by suggesting that film historians tend to dwell too much on the 'key moment' of the 1926–32 conversion to sound, and then proceeded to spend over half its total word-length doing precisely that. The 'key moment' is simply the most dramatic and wide-ranging example of this phenomenon, which can also be identified in every other significant development in the history of film sound. It is hoped that its coverage in this chapter has demonstrated that fact, and thereby contextualised the role of sound in relation to the other constituent technologies of film-based moving images, the last of which will be discussed in the following chapter.

chapter five | cinema exhibition

'When he saw the living pictures, he jumped up like a hare; and it took nine ushers just to hold him in his chair.'[1]

The first half of this book has examined the evolution and implementation of the various technologies used to record and manipulate moving image content with photographic film as the primary medium. The issues covered have included materials chemistry (film bases), photographic chemistry (emulsions, dyes and colour processes), physics (light and its manipulation in order to record a photographic image), mechanical engineering (cameras, printers, lenses and acoustic sound recording), electronics (electrical sound recording), economics (the business models through which technologies have succeeded or failed in the marketplace) and politics (the legal regulation of technology and the role of technical standards).

In the following two chapters we turn our attention to the other side of the fence and consider the technological factors which shaped and influenced the way this content has been delivered to its consumers. These are cinema exhibition – the projection of films in front of a mass-audience – and the allied technologies of television and videotape. While the latter are also a significant primary production medium, it is estimated that some 80 per cent of the total footage broadcast on public television stations over the history of the medium was originated on film, and over 90 per cent of the world's videocassette recorders are to be found in private homes. For the purposes of this discussion, therefore, we will consider television and video primarily as delivery systems.

At first sight, the rationale for treating the technology of cinema exhibition as a separate issue from those of film production, post-production and distribution might not seem obvious. There are, however, two powerful reasons for doing so. Firstly, although it is now a relatively small industry, for over half the total lifetime that moving image technology has existed as a mass-medium, it was a major industrial and cultural force and the only significant means through which content was delivered to its customers. Simply put, until the late 1950s (in the developed world; even later in other regions), going to the cinema was the one and only way in which most people saw any form of moving image. Secondly, the sheer infrastructural size of this sector of the film industry had a fundamental effect on the economic model through which it operated, one which had a very specific effect on both its own technologies and those of the studios which supplied it. At the sector's economic peak of the mid-1940s, there were probably several thousand cinema projectors in use for every one studio camera. The technological costs at the exhibition end, therefore, were a far bigger factor affecting development and change in this area than they were in studio production. The role of standardisation was thus especially significant (because economies of scale made the stakes higher).

This chapter, therefore, will examine the ways in which cinema projection technology has been established, evolved and changed in the 110 years since what is believed to have been the first public film projection took place. In doing so, I hope to show that the capabilities and limitations of this technology has been a fundamentally important influence in all areas of film's use of a mass-medium, and one which is a key factor in the use of moving images within society and our relationship to them.

The beginnings: 1889–1914

For the first six years during which film was being used to record moving images, there was no known method in widespread use of substantially enlarging the picture and displaying it before an audience. When the combination of the Edison/Dickson camera and Eastman's film was combined to produce a viable means of recording moving images in 1889, the initial expectation (as discussed in chapter one) was that it would be marketed primarily as a technology for individual viewing. It is worth bearing in mind that Eastman Kodak had hitherto been mainly producing materials and hardware for amateur still photography, and that Edison's only other audiovisual interest before he became involved in film – the Phonograph – was also designed and marketed for home use. Cultural technologies which involved communal viewing (e.g. lanterns and slides) were not a market in which either company had taken any systematic interest.

The only form of moving image 'playback' which was used on any significant scale during the early 1890s was the Edison Kinetoscope (designed by Dickson and patented in 1891). This consisted of a wooden cabinet in which a 50-foot endless loop of film was transported by an electric motor over a series of rollers between an incandescent light bulb and a rotating disc shutter. Above the shutter was a lens, similar to that in a magnifying glass, through which the user saw the image. From chapter one it will be recalled that the key discovery which enabled moving images to be photographed was the ability to create the illusion of continuous movement and that the invention of a flexible, transparent solid (i.e. film) cleared the way for this to be exploited in a camera mechanism. All successful motion picture camera mechanisms, from Dickson's onwards, were designed to stop the film movement completely during the moment of exposure. The Kinetoscope did not: it was a continuous motion machine, perhaps as a result of continuing difficulties which Dickson had experienced with the intermittent mechanism on his 'projecting Kinetograph' in

1889–90.[2] In order to minimise visible flicker, a film speed of 40fps was found to be necessary, almost double what would eventually become the industry standard of 24fps and almost three times the effective minimum needed for flicker-free projection using an intermittent mechanism (16fps). A Kinetoscope had a film capacity of 50 feet, giving a total viewing time of just over 20 seconds.[3]

For a few years from April 1894 a number of 'Kinetoscope parlours' opened in American cities, in which these machines were used on a coin-operated basis. But apart from the illusion of movement, they did not offer customers any real advance on the optical toys of the late Victorian period. By the turn of the century they had been superseded by the method of film exhibition which, in the developed world, dominated the industry until the mass-rollout of television in the 1950s: projection. The event which many historians argue marked the start of the film industry as we now know it took place in a Paris café on 28 December 1895, when two brothers who had previously run a business manufacturing glass plates for still photography, Auguste and Louis Lumière, demonstrated their 'Cinématographe' to a paying audience. This screening set both a technological and an economic precedent.

The Cinématographe served three functions in one machine – it was a combined camera, printer and projector. Projection was achieved by removing the back cover from the mechanism and placing a light source behind it, which passed through the exposed film and the lens. About 30 feet in front of the Cinématographe was a matt white screen, approximately 10 feet by 7 in size, from which the light source was reflected, thereby enabling an audience sitting in the room to view the photographic image on the film. The reproduction of movement was very straightforward. The mechanism which enabled the exposure of still images in rapid succession when the Cinématographe was used as a camera served exactly the same function in projection, only with light travelling in the opposite direction. A claw-and-cam system, similar to that used to insert and retract the needle in a modern sewing machine, was mechanically linked to a rotating disc shutter similar to the one used in the Kinetoscope. While the shutter blade obscured the flow of light, a claw would be inserted through perforations in the film and pull it down through a gate for the length of one frame. The claw would then retract as the shutter blade rotated away from the light path, thereby enabling the stationary frame to be projected through the lens. The process was then repeated when the shutter blade completed a revolution and returned to block the flow of light.

The economic precedent was that of exhibiting the same film before a gathered audience. Only one person could view each Kinetoscope screening, meaning that substantial numbers of machines had to be installed in each parlour in order for the operation to generate any significant revenue. The Cinématographe, however, was able to service a much larger customer base (it was estimated that the audience for the December 1895 show numbered about 100) for each screening, and thus recoup the investment in manufacture and research and development far more quickly. This principle turned out to be one of the main drivers behind the growth of film as a mass medium during the following three decades: by the time the film industry's 'classical' period was established in the mid-1920s, ever increasing audiences for each screening ensured that the cost of the technology which recorded, duplicated and reproduced the moving images themselves had become a relatively small component in the industry's overall running costs. In other words, it would be impossible to pay Hollywood stars vast amounts of money for each appearance if there were no cheap and efficient means available of screening it to thousands of audiences of hundreds, spread over a large geographical area.

The following spring the Cinématographe made its London debut, showing, as it had in Paris, a 20-minute programme of actuality shorts. With the business case for projected film screenings thus established, the design and manufacture of purpose-built projectors began very quickly, by several companies and individuals, and on an industrial scale. The next significant development came in the United States, where the engineer Thomas Armat began public screenings using his 'Vitascope' projector in 1896. Later models of this projector incorporated a technique developed by another American, Woodville Latham, at around the same time. The 'Latham loop' was a length of film positioned between the continuously-moving sprockets which transported the film through the mechanism, and the gate in which it was intermittently held stationary.

The loops (above and below the gate) acted as a shock absorber, minimising contact between the film and any hard surfaces as it was advanced intermittently through the gate, and has been incorporated into virtually every cinema projector built to this day. The final core element of technology which formed the projector mechanism as it would enter mainstream use for the following century was an intermittent mechanism that was capable of withstanding the heat generated by the light source needed to illuminate a large screen, while at the same time inflicting minimal wear and tear on the film's perforations. This was the 'Maltese cross' movement, a device which in essence converts the continuous drive from a cranking handle or motor into the intermittent movement of a shaft with a sprocket at one end, which engages the film's perforations.

Fig. 5.1 The 'Latham loop' below the intermittent sprocket on a modern cinema projector.

This device, which was developed during the 1890s and early 1900s, gradually replaced earlier forms of mechanical intermittent and by the mid-1900s had almost totally superseded them in mass-manufactured 35mm projectors. Unlike any of the claw-based alternatives, the sprocket teeth inserted and retracted through the projectors at a far gentler angle, and the downward force was exerted by at least two sets of teeth at any one time. This minimised perforation damage and enhanced picture stability by providing a more accurate length of pull-down than the claw-and-cam based alternatives. A technical manual from the 1920s noted that 'no other intermittent movement has yet been evolved which compares with the Geneva [Maltese] movement for accuracy',[4] a statement which, for projection, holds true to this day. It appears to have emerged in Europe and America more or less simul-

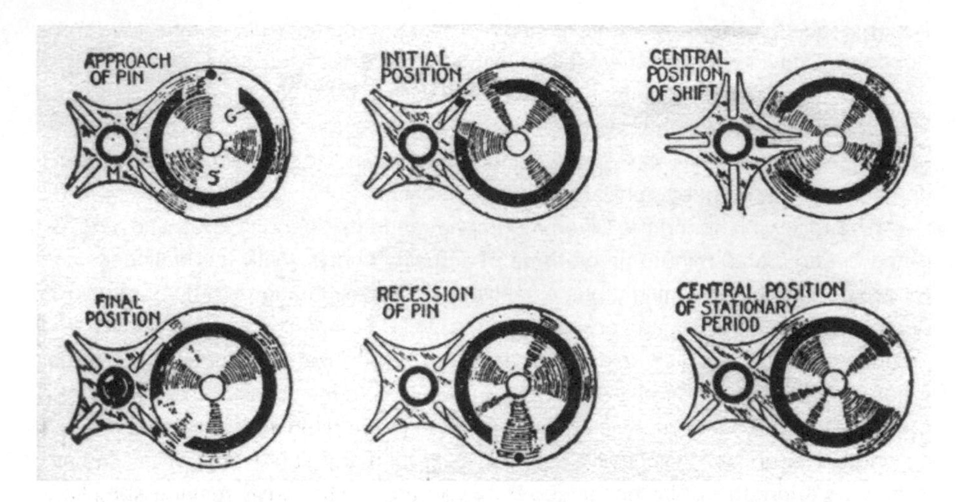

Fig. 5.2 A schematic of the Maltese cross intermittent mechanism.[5]

taneously: in New York, Thomas Armat's 'Vitascope' projector, based on a Maltese cross intermittent, was used in America's first significant commercial film projection, when a programme consisting mainly of Kinetoscope shorts was shown at Koster and Bial's Music Hall on 23 April 1896. In Europe, the earliest known patent for a Maltese Cross mechanism was granted to the German film pioneer Oskar Messter in 1902,[6] although projectors using this technology were being manufactured some years earlier. Until the turn of the century, however, the claw-and-cam system remained the most widely manufactured form of intermittent, not least because this had now become an 'off the shelf' technology which could be lifted from a range of popular camera designs.

In the Cinématographe and the first generation of purpose-built projectors, the light source was identical to that used to project lantern slides. Indeed it was often the same lantern, with the operator mounting a slide or film projector mechanism in front of it as required. This was a form of lighting which was used extensively in theatres and other auditoria throughout the second half of the nineteenth century, and was produced chemically. Known as limelight, it was generated as follows:

> An intensely hot flame was directed against a stick of unslaked lime. The flame was produced by burning oxygen and hydrogen gases in proper combination. One tank, known as the saturator, supplied the hydrogen gas by vapourising sulphuric ether. Oxygen was generated in a larger tank, where water dripped into an oxygen-generating compound.[7]

It is worth noting at this point that at around the turn of the century, mains electricity was not universally available even in many city centres of Europe and North America, let alone in suburban or rural areas. Furthermore, until the late 1900s, cinema exhibition was essentially an itinerant business – in Britain and America, cinema buildings designed exclusively for the purpose did not start to go up on any significant scale until 1908–10. The most usual venues for film projection were music halls, variety shows, fairgrounds and 'nickelodeons' (or 'penny gaffes' in Britain – both epithets refer to the price of admission) – hastily converted premises, usually shops, which entrepreneurs would turn into ersatz cinemas for the smallest possible up-front investment. The projection equipment, therefore, had to be easily portable and not dependent on external services such as an electricity supply.

It does not take a PhD in chemistry to work out that the combination of nitrate film, the intense, localised heat from a limelight flame and several gallons of highly volatile substances – all in an unprotected auditorium containing an audience of hundreds – is not exactly conducive to health and safety. The risk from fire became a major influence in the design of cinema projectors within a few months of their commercial use and the first fires resulting therefrom. The earliest is believed to have been at an auditorium operated by the British film pioneer Birt Acres on 10 June 1896: no one was hurt and the damage was minimal, probably because the quantities of film being handled were not sufficient to cause a serious conflagration. Later that year a temporary pavilion in Berlin being used to show Edison films with a Cinématographe was burnt to the ground, though it is believed that an electrical fault rather than a film fire was to blame. The following year came an incident which placed health and safety at the forefront of cinema exhibition practice, legislative regulation, building and equipment design for the next fifty years. On 4 May 1897 a fire broke out in a projector in Paris, it is believed as a result of its operator using a match to illuminate the refilling of the volatile ether in the darkened room; 125 people were killed and many more were seriously injured. The event had been a charity screening, and many of those killed were the wives and children of prominent French politicians, industrialists and society figures, with the result that the fire received extensive media coverage.[8] Over the next decade, fire prevention moved gradually up the agenda, though until mass-manufactured projectors and the buildings used for film exhibition started to incorporate safety features (such as segregated projection booths), cinema fires with three-figure death tolls happened regularly during the 1900s.

The highest casualties in a US cinema fire came in Boyerstown, Pennsylvania on 13 January 1908 when 167 were killed, and although it was subsequently established that the projector was not the cause, the incident is generally credited with providing the impetus for systematic regulation of the American film exhibition industry. In the following year came the incident which caused the highest known death toll in a cinema fire started by igniting nitrate, when over 250 lost their lives in Acapulco, Mexico. By the middle of the following decade, film fires that claimed lives in double figures had become very unusual (especially given the vast increase in cinema buildings and ticket sales compared to a decade earlier), but by no means unheard of: Britain's worst incident came as late as 31 December 1929, when a nitrate fire which started during a childrens' matinee at Paisley, Scotland, killed 69 and seriously injured a similar number (see also chapter 1).[9] As early as 1901 an advertisement for the British Prestwich model 5 projector stressed the safety benefits of its 'automatic light cut-off' feature (i.e. the most probable cause of ignition would be removed in the event of a film jam),[10] and within a decade this and other safety features would be a legal requirement for public film screenings in many countries.

Towards the end of the 1900s the design and manufacture of equipment and buildings used for film exhibition gradually evolved from a cottage industry onto an industrial footing. This mirrored the trend seen across all sectors of the industry as a whole: technical standardisation began to impose economies of scale, the techniques for editing individual shots in order to form extended narratives were perfected, the outright sale of film prints was replaced by the rental distribution system and the economic interests which controlled film production, distribution and exhibition began to consolidate. The development and production of the technology used in these processes gradually came to be supplied by an independent sector, functioning on a service provision basis for producers, laboratories and exhibitors.

Establishing the standards: 1914–26

The second main phase in the development of cinema exhibition technology consisted of two major processes. The production of projectors and related equipment shifted from a large number of designs manufactured in small quantities to the production line manufacture of a much smaller number of models (the functionality of which was to all intents and purposes identical). Secondly the venues in which films were shown began to be designed and built specifically for the purpose. The latter trend would culminate in the 'picture palaces' of the 1920s and 1930s, built with the unique requirements of the audience and the technology in mind.

The projector designs which emerged during the 1910s and which were in mainstream use throughout America and Western Europe by the early 1920s had a number of standardised features which had evolved to meet the needs of film exhibition in the emerging 'classical' period and which would remain largely unchanged until the arrival of sound. In essence, these were:

- the use of a Maltese Cross mechanism to advance the film intermittently;
- the use of 'Latham loops' to prevent film damage as its motion changed from continuous to intermittent and then back again;
- a concentric disc shutter consisting of two or more blades;
- a spool capacity of between 1,000 and 2,000 feet (considered to be the safest maximum quantity of nitrate film which could placed at risk of accidental ignition);
- a range of fire safety precautions, including enclosed spoolboxes for the feed and take-up reels, spring loaded blades which would automatically cut the film above and below the mechanism in the event of ignition, liquid-cooled gates (a system of pipes positioned around the gate area through which water was pumped, removing heat by radiation) and built-in fire extinguishers.

In the United States, two manufacturers emerged during the 1910s which, between them, would account for the bulk of projector sales throughout most of the twentieth century. In 1898 Alvah C. Roebuck, of the Sears-Roebuck mail order empire, was making a healthy profit supplying magic lantern equipment, but soon turned his attention to the film business, believing that this was where the future lay. He founded the Enterprise Optical Manufacturing Company, which, using the trade name Motiograph, progressively refined its line of products, culminating in the launch of the model 1A in 1908. The model E in 1916 and model F in 1921 introduced improvements to the intermittent mechanism, shutter, safety features and picture stability of the mechanism.[11] During the early 1900s the pioneer filmmaker Edwin S. Porter (of *The Great Train Robbery* fame) and two engineers designed the Simplex projector, first sold in 1909, and which sold thousands of units each year until its replacement with the Super Simplex in 1928. Although the company was taken over in 1983, projectors continue to be sold under the Simplex brand almost seventy years later. In Europe, the German textiles manufacturer Ernemann diversified into the film industry, launching its first model in 1904. The following year, Bauer of Stuttgart entered the business, initially servicing and repairing Pathé machines, then producing its own models from 1910.[12] Kalee in Britain, Philips of the Netherlands and Cinemecannica and Prevost of Italy were all marketing cinema equipment by the late 1920s.

During this period electricity replaced limelight as the main light source for cinema projection, and elbow grease as the means of driving the mechanism, as electric motors increasingly replaced hand-cranking during the 1920s. The basic principles of carbon arc illumination were first demonstrated by the British engineer Sir Humphrey Davy (better known as the inventor of the safety lamps used by coal miners) in the late nineteenth century. He discovered that if an electrical current is passed between two thin rods of carbon, combustion of the resulting gases produces light. For use in cinema projection the two rods of copper-coated carbon electrode are mounted on stands in front of a mirror which focuses the beam of light on the projector gate. The light is 'struck' by placing the electrodes in contact with each other and applying the current. The positive electrode is then 'trimmed' by retracting it a short distance from the negative. This creates an electrical spark which crosses the resulting gap (the arc), which in turn ignites the vapours produced by the heated carbon. It is these burning vapours which produce the light. The carbon electrodes are gradually consumed in this process and must then be replaced.

Fig. 5.3 The British Carey-Gavey 'flickerless' projector – a typical (apart from the four-blade shutter) design of the early 1920s.[14] Note the hand-cranked mechanism, enclosed spoolboxes and film cutter (for protection from burning nitrate), and shutter positioned in front of the lens. By the early 1930s, the majority of projector designs positioned the shutter between the mechanism and light source, in order to reduce heat exposure to the film.

Carbon arc lighting, which was also used in studio production until superseded by incandescent light in the late 1920s,[13] offered significant advantages for projection over limelight. The running costs were lower, and it was a lot safer than storing large quantities of combustible materials close to the projector. A separate machine, known as a rectifier, was needed to convert the high-voltage, low-current mains supply to the high-current, low-voltage power used to feed the arc. Although these generated a significant amount of heat and also contained highly toxic chemicals, including mercury, they could be (and usually were) placed in a separate room and connected to the projector's lamphouse by heavy duty cables running under the projection booth floor.

This was one example of how, by the 1920s, fire safety was not only a consideration in the design of projection equipment itself, but also of the buildings in which it was operated. In the US, state statutes and federal regulations usually decreed that any film which was not actually being projected had to be stored and worked on in a separate 'rewind room' fitted with a fire-retardant door, as did rectifiers and transformers. Two separate exits, one of which led directly to the exterior of the building, were required, as were iron shutters mounted on a system of runners covering the projection portholes: in the event of any film fire, they would automatically fall into place, like a guillotine blade, in order to minimise the risk of fumes or flames entering the auditorium.[15] The standard length for each reel of a release print remained 1,000 feet (which was shipped to and from the cinema in a fire-retardant solid steel container) until after the conversion to sound, when 'double' reels of 2,000 feet started to be introduced. From the 1910s onwards, almost all purpose-built cinemas were equipped with at least two 35mm projectors (usually more) which were aligned to the same screen, which allowed a continuous programme of any length to be shown. At first changeovers were carried out manually, with an operator working on each projector. As the outgoing reel finished, a cue sheet provided by the distributor described the scene in the finishing reel at which the projectionist operating the incoming machine was to start his motor and open the 'douser' blocking the flow of light from the carbon arc. The conversion to sound required changeovers to be carried out with a greater degree of precision, due to the risk of cutting dialogue if the timing was not almost perfect. A system of electrical interlocking between the two projectors combined with visible cues on the screen was therefore introduced. Standardisation was fast becoming a necessity anyway: in 1927 a projectionist writing in the *Transactions of the SMPE* complained of the 'punch mark nuisance', noting that home-made cue marks consisting of 'stickers of all shapes, sizes and descriptions' were becoming a major distraction for cinema audiences.[16]

By the early 1930s the latter had been standardised as a mark which was printed on four consecutive frames of each print (usually by punching holes out of the internegative used to strike it), positioned in a corner of the picture. The reel in the incoming projector would be threaded to a pre-determined point. When the projectionist saw the first cue mark, he would start the motor and strike the arc. On seeing a second mark, which appeared approximately six seconds later, a changeover switch would be operated. This had the effect of opening an additional shutter, simultaneously closing another one on the machine projecting the reel which had just finished, and switching the sound source between the two projectors. If this procedure was carried out successfully, the audience would be totally unaware of it having taken place (the cue marks appear too briefly for anyone who is not aware of their existence to notice them).

The other major developments in projection technology which took place during the late 1910s and early 1920s was precipitated by the big increases in the size of auditoria which were typically built during the decade. These were a gradual increase in projection speeds and a move towards reducing the typical number of shutter blades from three to two. The increased 'throws' (distance between projector and screen) and screen sizes of the new cinemas resulted in a consequent increase in the light output needed from a projector, and of maximising its efficiency. This was because of the increased light absorption resulting from the thicker lens elements needed for longer throws and the large, smoky auditoria in the 1920s picture palaces. As a technical manual from the period notes, 'projecting conditions have changed. Whereas 20 to 30 amperes was considered sufficient for projection, owing to longer throws necessary in the larger theatres, many houses are now using considerably over 100 amperes.'[17]

The design of shutters was a crucial factor in the light efficiency of a projector mechanism, and one which in turn was affected by another parameter which would eventually be standardised – the film (transport) speed. As explained in chapter one, the recording and reproduction of 'moving' images exploits the ability to create the illusion of continuous movement. In relation to film-based moving images the describes the effect whereby if a sequence of still images is captured and displayed in quick enough succession, it will appear to the viewer as being a continuously moving image. The function of a shutter, in both a camera and a projector, is to separate the recording and reproduction of each image, or 'frame', by blocking the flow of light as the film is advanced. Through a process of trial and error, the first generation of filmmakers established that in order for the flow of movement to appear fluent, a speed of 16fps is, for most people, a bare minimum. Any less than this and the amount of time which elapses between the exposure of one image and the next, especially if rapid movement is present within the subject (e.g. a shot of someone running), will be so great that a 'jerk' or uneven movement, will be perceived when the sequence is reproduced. There is also a maximum period of time which may elapse while the light flow from a projector is obscured during the intermittent mechanism's 'pull-down' cycle. If the light source is restored within that time, the human brain will not have time to register the fact that the screen has gone dark. If it is not, a gradual fading of the image will be seen, followed by a reciprocating increase in light intensity when the following frame has been pulled into place. This will be perceived as a flicker, and from the 1920s onwards one of the major problems facing projector designers was to maximise the use of available light while minimising the appearance of any flicker.

The issue of flicker reduction in itself went back even further, as this description of a projector sold from 1897 makes clear:

> The movement is a six to one, i.e. the film is stationary for five-sixths and is being changed in the remaining one sixth, so that the shutter used is but a very small one, and only one-sixth of the light obstruction (sic), which reduces flicker to a minimum.[18]

Yet before the post-World War One cinema building boom the problem was vastly simplified in that light intensity was not a major problem, given the short throws to small screens which characterised the first generation of temporary auditoria. The only crucial requirements were to minimise the time taken by the pulldown and to ensure that the flow of light was even. The appearance of a perceived flicker was not affected by the total light output over the complete duration of the shutter cycle, only by the time of each 'obscure' phase within it. It did not matter if, during an hour-long film presentation, there was no light being projected on the screen for over thirty minutes of the total, just as long as each individual period of darkness did not exceed a certain limit. In order to produce an acceptably consistent picture with a film speed of 16fps, it was eventually established that a concentric shutter with at least three blades was needed. Two of those were 'cycle' blades, and the pulldown took place while the third covered the light source. The greater the number of blades, the more even the screen illumination appeared – hence the projector shown in fig. 5.3, which was advertised as being 'flickerless', featured a shutter with as many as four blades.

The *quid pro quo* for increasing the number of shutter blades was a reduction in light output, as the greater the total length of the shutter cycle is spent with the light obscured, the less bright the image will appear to the human eye. When the 1920s picture palaces started to be built this became a major issue. In order to maximise the economies of scale for their operations, exhibitors put up cinemas with higher and higher capacities: by the end of the decade, auditoria containing 2,000 seats were not uncommon. This necessitated a far greater throw than had ever been needed in pre-classical times, with the result that even the output capacity of the newly developed carbon arc lamps was being stretched, not to mention the consequent increase in film fire risk from the additional heat they generated. Reducing the number of shutter blades from three to two could potentially increase the amount of light on the screen at no extra cost and without making the projector any hotter. In 1921 another projectionist training manual advised that, 'the three equal sector shutter is scientifically most perfect', although other designs were being developed to boost light output.[19] The problem was that without a compensating increase in the film speed, a two-bladed shutter would introduce flicker.

By the end of the 1900s, 16fps had effectively established itself as a default speed for studio production, one which was consolidated after the introduction of the Bell and Howell 2709 camera (see chapter two), which exposed 16 frames for every two revolutions of the cranking handle. A projection manual from 1922 describes 60 feet per minute (16fps) as the 'standard speed'.[20] But until the introduction of sound forced the issue, there was no formally standardised shooting and projection speed; and nor could there be, while the majority of cameras and projector mechanisms were hand-cranked. As the perceived accuracy of reproduction did not depend upon standardisation until sound came along – as an engineer quoted in the discussion of format standardisation in chapter two noted, shooting and projection speeds 'can vary over wide limits without apparent falsity' – introducing it was not believed to offer any economic benefit. Indeed, not only was there no standardised speed but both producers and exhibitors exploited the leeway this offered within their own industrial practices. In the studio, for example, a camera would deliberately be 'undercranked' (operated at a lower speed than was intended for projection) in order to heighten special effects or stunt scenes. In other cases, the shooting speed was dictated by technical considerations:

> ...the cameraman will, under adverse light conditions, use as large a lens opening [aperture] as is practicable and slow down as much as he can, in order to obtain sufficient exposure. Conversely, when the light is strong, the tendency is to speed up.[21]

Although the major studios usually supplied information to cinemas making it clear what speed a feature was supposed to be projected at, this was often ignored, especially in the cheaper 'fleapit' houses, where films would be shown as fast as the manager thought he could get away with, in order to maximise the number of screenings that could be fitted into a day. 'Picture racing', as it was termed, was said to be a significant problem by the mid-1910s: 'an evil existing mainly in cheap, poorly-run theatres, but which once in a while pokes its sinisterly rapid head among the seats that retail at a quarter or half a dollar'.[22] One projectionist gave this example:

Fig. 5.4 A British Kalee 11 model projector, typical of those in use during the 1920s. Note the manual cranking handle linked to the lower sprocket, and the speed indicator above the gate. Picture courtesy of BFI Stills, Posters and Designs.

> I remember running 1,000 feet in 12 minutes in the old days of hand-cranking at the 8 o'clock show, and in the afternoon I used to project the same reel so slow that it took Maurice Costello ages to cross the set. Those were my manager's orders.[23]

Around the late 1910s to early 1920s, therefore, it became apparent that a significant increase in the routine cinema projection speeds in use was becoming necessary. It did not help that studio practices appeared to be determined by an *ad hoc* combination of achieving certain artistic effects within specific technical limitations, nor did the fact that exhibitors appeared to be deciding projection speeds according to local economic factors. Nevertheless, by 1918, a prominent Hollywood projection engineer was calling for an increase:

> Our present rate is too slow with present powerful illumination and semi-reflective screens. It is not sufficiently high to eliminate flicker under those conditions, especially if the local shutter conditions be bad. Seventy [feet per minute, i.e. 18.6fps] will, on the other hand, place no unduly heavy burden on the film itself, or upon projection machinery, and will eliminate flicker in all but the very worst cases.[24]

Implementing this suggestion was not as simple as instructing projectionists to crank a little faster, however. It was soon discovered that the mechanical tolerances of most projector designs then in widespread use could not withstand a speed increase significantly in excess of 16fps without affecting picture stability and causing excessive wear.[25] Fears were also expressed that the increased mechanical stress caused by higher speed could exacerbate the risk of film breaking, and consequently catching fire.[26] Nevertheless, projection speeds did gradually increase throughout the decade, prompted by the refinement of mechanism designs featuring electric motors and two-bladed shutters, and which were specifically designed for use with high-current carbon arc lamps.

Sound and consolidation: 1926–48

The standardisation of identical shooting and projection speeds was the first significant effect of the conversion to sound to be felt. Whereas the reproduction of movement can, in some cases, vary over quite wide limits without the effect being perceived by the untrained eye, the reproduction of analogue audio cannot, for the simple reason that varying the speed of playback of a recording also varies its pitch. In September 1927 the SMPTE's Standards and Nomenclature Committee undertook a fact-finding exercise in order to establish what speeds the emerging sound systems were using. The two which were entering commercial use (Vitaphone and Movietone) both used 24fps. The RCA variable area system, still in development at that point, used 22fps, while De Forest Phonofilms (which had virtually ceased production by that point) ran at 20fps. Accepting that trends in exhibition practice over the previous decade and decisions made by the designers of the two most successful sound systems had effectively standardised 24fps by default ('we may as well accept facts and acknowledge that the speeds were adopted in the days of our ignorance'), the committee proposed that it be enshrined as a formal standard, which took place shortly afterwards.[27]

In the first phase of the conversion to sound, the equipment for playback in cinemas was designed and manufactured by the same companies which produced the studio recording equipment. However, when it became apparent that a full-scale conversion was taking place towards the end of the decade, the supply and installation of sound equipment evolved into a service sector along similar lines to that of projector manufacture. Many of the key components of the sound reproduction technology were also used in a variety of other electronic products (such as radios), thereby further encouraging the trend. Although both RCA and Western Electric supplied and serviced optical sound reproducers, amplifiers and loudspeakers across North America and Europe throughout the following four decades, a wide range of other manufacturers designed equipment specifically for cinema use. As early as 1929, a training manual for projectionists and engineers, written specifically as a guide to operating and maintaining the new equipment, provided a detailed technical description of over twenty product lines of cinema sound equipment, and by the middle of the 1930s it was estimated that over 300 companies were selling equipment

Fox Movietone and Vitaphone Installations

1. Place the film in the upper magazine, emulsion side toward light.
2. Be certain that the main blade of the shutter is *up* in position and that the intermittent has just completed one full movement.
3. Thread the mechanism in the usual manner.
4. Be sure that the STARTING MARK on the film is *in frame* at the aperture.
5. Make the loop between the intermittent and the lower sprocket, so that the film rests against the index finger held across the opening of the lower film shield. It is important that this loop be of proper length so that the film between aperture and the sound gate will measure 14½ inches.
6. Thread the film to the left lower idler under the lower sprocket. Draw tight to the sound gate sprocket. Then raise the film two holes before closing the idlers.
7. For film-recorded sound pictures, it is important that film be perfectly centred on the slit in the sound gate and that the gate be tightly closed.
8. Select the disk corresponding to the number of the film.
9. Place the disk on the turntable and clean it with the especially provided record cleaner.
10. Select a perfect needle. Insert it in the reproducer securely and place it exactly on the starting mark indicated on the disk.
11. Check all loops, idlers, and sprockets. Make sure that the number on the disk and the film correspond. Be certain that the starting mark on the film is in perfect frame and that the needle is fastened securely, tracks correctly, and is placed exactly on the starting mark indicated on the disk.
12. Turn down the record the required number of full turns indicated on the cue sheet, making sure that the film and the needle are tracking properly during this operation.
13. Strike the arc and start the motor on the cue. When all is up to speed, raise the dowser and on the proper cue bring the fader up to the required mark as per cue sheet.
14. Give the second projection the cue to strike the arc on the second machine, and stand by for change-over.
15. Stand by for the cue indicating the end of the record. Make the change-over on the fader to the second projector at the cue.

Fig. 5.4: The operating procedure adopted in the projection booths of a chain of cinemas on the West Coast of America.[29]

in the United States alone.[28] The conversion also necessitated an additional range of technical skills on the part of projectionists. During the silent period, the job had consisted primarily of film handling and operating mechanical equipment. With the arrival of sound, a substantial knowledge of electronics was also needed. Initially, the projection of sound films also called for a not inconsiderable amount of manual dexterity.

Sound on disc took longer to disappear for cinema playback than it did as a primary production format, for the simple reason that a significant number of venues worldwide had invested in disc-based reproduction systems, but were not equipped for optical sound-on-film reproduction. For several years after the system's inflexibility and lower sound quality had forced it out of production use, the main Hollywood and European studios had to operate a 'dual inventory' system for supplying release prints. A Vitaphone wax master was cut from the final mix optical sound negative for each reel, and discs pressed for supply to those cinemas which had not installed optical reproducers. There were other reasons why Vitaphone was also doomed as a reproduction format. While the use of electronically interlocked turntable and projector motors should theoretically have ensured perfect synchronisation between the two sources, the synchronisation was lost if the record stylus jumped during playback or if film prints were not repaired properly following any damage which resulted in lost footage. Technical journals and trade papers from the period give the impression that these were both significant problems, and in any case the records themselves had a much shorter lifetime than the film prints, thereby adding to the costs of continuing to support the format. Vitaphone had to all intents and purposes disappeared from the US by the end of 1931, although the distribution of discs continued in Europe for a little longer: in Britain, for example, Warner Bros. ceased supplying them in February 1932.[30]

One other significant area of standardisation was established as a direct result of the conversion to sound, which was the 'Academy' aspect ratio of approximately 1:1.38. The reasons for this are explained in chapter two, and it will suffice to note here that this was largely precipitated by the methods used by projectionists to get round the problem of Vitaphone and sound-on-film prints having different frame dimensions. This was the first occasion in which projectionists had ever needed to deal with more than one ratio interchangeably. As a result, most projectors built from the early 1930s onwards featured lenses and aperture plates (a steel plate with a hole punched out of it corresponding to the frame area on a release print, and which was positioned within a projector's gate assembly) which could easily be removed and replaced.

By about the end of 1932, therefore, the technology which could be found in a typical cinema projection booth in North America or any European country was in roughly the form it would remain for the following two decades. As was noted in the previous chapter, the conversion to sound had necessitated an unprecedented level of investment within the exhibition industry, and it is interesting to note that almost all the significant developments and advances in film-related technology which took place during the late 1930s and 1940s were either not rolled out or could be delivered without any significant cost to exhibitors. Refinements to the RCA and Western Electric optical sound systems during the 1930s did not affect the ability of release prints to be played back on existing equipment. Although the experimental use of stereo variable area tracks began during this decade, multi-channel sound was not introduced on any significant scale until the mid-1950s, and did not become standard until the late 1980s. The brief widescreen launch of the early 1930s did result in some notable developments in projector technology, most importantly the 70mm variant of the Super Simplex designed for the Fox Grandeur process; but again, this research and development ended up in mothballs for the following two decades. Three-strip Technicolor release prints, introduced in 1935, were projected in exactly the same way as their black-and-white counterparts – in fact, many other systems which had competed with Technicolor for market share during the 1910s and 1920s (e.g. Kinemacolor and lenticular Kodacolor) fell by the wayside precisely because they depended on non-standard equipment for projection. As the main cinema building boom in North America and Europe had taken place during the 1920s, the acoustic properties of auditoria were not a significant architectural consideration in the majority of cases. Unlike the introduction of widescreen a generation later, the conversion to sound had very little impact on the auditoria themselves, as cable runs were usually concealed in pre-existing false ceilings and loudspeakers were usually positioned beside or above the screen. Perforated screens, which allowed speakers to be placed directly behind them without any loss to sound quality were not commercially produced until the 1970s. Acoustic panelling and fabric was used to some extent in auditoria where excessive reverberation was found to be present.[31]

The only other major development in film exhibition during this period was the introduction of the 16mm film soundtracks in 1933, which marked the start of this format's use as the primary 'non-theatrical' exhibition medium for the following half-century. Portable projectors which used incandescent light (i.e. generated by a filament inside a sealed glass tube, similar to a domestic light bulb) and with a self-contained optical sound reader and amplifier began to appear, and were used extensively in schools, businesses, community groups and even some wealthy homes: in short, anywhere films needed to be shown which was outside a purpose-built cinema. As 16mm was a 'safety only' gauge from the outset (see chapter two), it represented the ideal medium for such screenings. Although 35mm projectors designed for non-cinema use had previously been marketed on a limited scale (a 1931 research report on the use of films in schools described a 'portable' 35mm sound film projection kit, occupying five trunks!),[32] their use had proven logistically impossible in the majority of situations, not least because the volume of 35mm safety stock manufactured before the 1948–50 conversion to triacetate was negligible.

Adjusting to industrial change: 1948–76

The three decades following the end of World War Two saw a number of wide-ranging changes in the technology of film exhibition, ones which addressed and reflected the changing social and economic demands of cinemagoing. During the 1930s and

1940s, cinema buildings consisted mainly of single auditoria seating a thousand customers or more. They were usually located in town and city centres, and reached on foot or by public transport. By the 1950s, major changes were underway. In America and Europe, the working-class population in the teens and twenties before the war, who had tended to live close to urban centres and who constituted the bulk of the film industry's customer base, started to move out to newly-built suburbs and start families. The pattern of leisure activities began to change, prompted not least by the growth of television (though many social historians believe that this effect has tended to be overstated) which is discussed in the following chapter. The 'classical' model of film industry vertical integration began to fall apart due to a combination of political change (in particular the 'Paramount case' of 1948, in which the US government successfully argued that the practice of studios, distribution infrastructure and cinemas being owned by the same company was unfair and monopolistic) and reduced consumer demand for its output.

The technological model of film exhibition which had been stabilised by the mid-1930s, therefore, was no longer sufficient to safeguard the industry's long-term economic future. As a result, two trends of technological change began to emerge and, eventually, to converge. The first was an attempt to reinvent the prestigious city centre venue in a way which foregrounded hitherto unmarketed image and sound technologies, in a deliberate form of product differentiation both from television and from the cinema of the previous decade. This was represented principally by the commercial rollout of widescreen and stereo sound in the mid-1950s. The second was the introduction of technologies which reduced the running costs of cinemas and increased their operating efficiency. Chief among them were the introduction of safety film, xenon illumination and 'long play' film transport systems, which enabled existing cinema buildings to be divided into multiple smaller auditoria without significant extra running costs, and eventually the multiplex boom of the 1980s.

As will be recalled from chapter one, safety film had been around in one form or another since the mid-1900s, and had been mass-produced for use in amateur film formats since at least 1923, when the 16mm gauge was launched. The discovery of 'high acetyl' cellulose triacetate in 1948 eventually offered a product which offered similar mechanical and tensile strength to that of nitrate, and which could be manufactured without significant extra cost. By film industry standards the ensuing conversion took place very quickly: Eastman Kodak ceased to produce nitrate in February 1950, and the base had effectively disappeared from release print circulation within a couple of years. This had major implications for projection practice, in that the extensive and expensive fire safety precautions needed within projection booths and equipment used in handling nitrate could potentially be abolished, including the use of separate 'rewind rooms', fire extinguishers built into projectors, costly fire-retardant fabric within auditoria, and – most significantly – the level of projection booth staffing needed to comply with nitrate handling regulations in most countries. As with the obsolescence of Vitaphone, the advantages of safety film were felt last of all in the cinema. The information pamphlet on safety film reproduced as fig. 1.6 on page 21 advises that the safety precautions associated with nitrate must not be relaxed until 'every last foot' was removed from routine circulation, a process which was not complete until the mid-1950s.

The next major developments were widescreen and stereo sound. Despite two distinctly negative precedents – Edison's Motion Picture Patents Corporation in the 1910s, and attempts to package widescreen with the conversion to sound in the early 1930s – the two were initially sold as a package. The technical principles behind three of the four widescreen systems launched during the 1950s (Cinerama, CinemaScope, VistaVision and Todd-AO) had been established at around the time of the conversion to sound, but attempts at commercial introduction proved unsuccessful. The sound systems that were designed to go with their widescreen counterparts were, however, based on a fundamentally new technology: magnetic recording. It would be another two decades before Alan Blumlein's work on stereo optical sound in the 1930s would be developed to the point of commercial rollout; in the meantime magnetic sound, which had entered studio use as a production medium soon after the first tape recorders had been liberated from Nazi Germany, was also being seen as the future of sound playback in the cinema. Leslie Knopp, technical advisor to the British Cinema Exhibitors' Association, spoke for many:

> It is now recognised that magnetically recorded sound is superior in quality to the optical soundtrack and I think it is the view held by the majority of technicians that the future system of sound recording and reproduction will be by means of the magnetic track or tracks.[33]

In particular, magnetic tracks were seen as offering two distinct advantages. Firstly, there was the perceived quality of the sound itself, as even the first generation of magnetic sound technology used in the West was capable of recording and reproducing a greater frequency range than any optical technology then in use. Secondly, this method could easily be adapted for mixing and playing back multiple 'channels' synchronised to the same picture (i.e. stereo sound), without the costly modification of existing reproducers. Therefore, all the three widescreen systems which incorporated a stereo sound process used magnetic reproduction – a separate strip of magnetically coated 35mm film carrying six channels and synchronised electronically in the case of Cinerama, and a number of oxide 'stripes' on the release prints in both the CinemaScope and Todd-AO formats. All three combinations achieved notable success in the prestigious first-run city centre market where admission prices were high and the technical quality of the projection and sound was an explicit selling point. Indeed these systems were consciously marketed as distinct theatrical events which were intended to represent a break from the standard, mass-produced technical medium which cinema operated as during the 1930s and 1940s. To start with, only a relatively small number of auditoria were fully equipped with the new technologies, and in the case of the inaugural Todd-AO release of *Oklahoma!*, seats had to be booked weeks in advance and customers were required to attend the screenings wearing full evening dress.

Both widescreen and stereo eventually made the transition from a novelty to a technological standard. The ways in which they did so upheld existing precedents for the rollout of new technologies, and re-established them when it came to subsequent change. Whereas the introduction of sound had necessitated investment in new equipment, widescreen also required significant architectural modifications to existing cinema buildings. It is therefore hardly surprising to note that the systems which needed the highest equipment investment and structural alteration of auditoria achieved the lowest market saturation.

Cinerama, with its very large, deeply curved screen and three separate projection booths was arguably the least successful. The number of auditoria was believed never to have exceeded thirty, and by 1959 (seven years after the format's launch), only 22 Cinerama venues were operating worldwide.[34] The Todd-AO principle of using 70mm film to enhance the picture definition on a large screen and to carry multiple, high-quality magnetic soundtracks survived, and to the present day remains a niche format used in a small number of venues. Unlike the modified Simplex projectors used for the Fox Grandeur system, the new generation of 70mm machines, starting with the Philips DP-70 (sold as the Norelco AA in North America) which was used to launch Todd-AO (of which more below), were backwards compatible: that is, they could also be used to project conventional 35mm film.

CinemaScope is perhaps the most interesting case in point, as the way it produced a wide picture (the anamorphic lens) and the precedent it established of using an aspect ratio significantly over twice the width to height did become a widespread technology within two decades of its launch, accounting for approximately a third of Hollywood features by the mid-1970s. But this was not in the form originally envisaged. Its promoters, Twentieth Century Fox (TCF), soon discovered that many exhibitors were not prepared to invest in stereo sound at the same time as having the stages of their auditoria rebuilt and buying the new projection lenses, just as they were similarly disinclined to invest in widescreen in the immediate aftermath of the conversion to sound. Indeed, there were many instances recorded of cinemas being wired up to mix all four channels of the CinemaScope magnetic prints into a single mono one for playback, in order to avoid having to buy additional amplifiers and speakers. However, unlike the companies behind Cinerama and Todd-AO, TCF and Paramount adapted their systems to meet customer demand rather than dogmatically continuing to promote them as roadshow-only formats. TCF produced a modified version of CinemaScope which was compatible with existing mono sound systems,[35] while Paramount developed a means of producing release prints from VistaVision originals which were fully compatible with existing cinema projection equipment but which offered a significantly higher image quality. In both cases, the only significant items of capital equipment which the cinema needed to purchase were an extra set of projection lenses.

Fig. 5.6 A small cinema in Cardiff, Wales, following its conversion for widescreen in 1954. The original stage area has been left largely unchanged, with wider ratios obtained by lowering a vertical masking system (author's collection).

Furthermore, it soon proved possible to adapt smaller auditoria for widescreen projection with only minimal rebuilding, although the resulting image was frequently a lot smaller than its Academy ratio predecessor. Therefore, through an emphasis on backwards compatibility and reducing the level of investment needed at the exhibition end, widescreen successfully made the transition from a high-end roadshow format to an industrial norm, while preserving one of the key reasons for introducing it in the first place – using technology to market cinema as a distinct product from television. Stereo sound would eventually follow it, but not before a system became available which also fulfilled these criteria.

In notable contrast to the American-led development of film-based technologies generally, the next two important products that would have a major influence on cinema exhibition practice both had their origins in Europe. When it was introduced in the 1920s, the carbon arc lamp had offered significant advantages over its chemically-lit predecessor: it was safer, cheaper, easier to operate and produced light at a more even colour temperature. There were, however, two key drawbacks. As the flame which produces the light in a carbon arc is caused by using electricity to burn the substance of the positive electrode, eventually the carbon rod which forms that electrode will burn away completely, and need replacing. During the burning process itself, the distance between the two electrodes needs to be constantly adjusted, or 'trimmed', to prevent the gap between them from becoming so great that the flame goes out. Although some mechanisms were devised for automating this process towards the end of the period when carbon arcs were in everyday use, the bottom line was that these lamps required frequent attention from the projectionist. Furthermore, most designs of lamphouse enabled the use of carbon rods which had a maximum burning time of approximately 30 minutes before they needed to be replaced, thereby making it impossible to show anything approaching an entire feature film using only one projector. During the nitrate period this did not matter, as fire safety considerations dictated that 2,000 feet (about 20 minutes) was the maximum length of film of which it would be safe to risk igniting. But when the conversion to acetate removed this restriction, cinema owners began to think about the reduction in staff costs and the possibility of constructing multiple auditoria within a single building which could potentially be enabled by technology that projected an entire programme unattended.

With the risk of fire greatly dimished by the conversion from nitrate to acetate (cellulose triacetate is said to have similar burning characteristics to those of paper), there was no theoretical reason why a single reel of up to several hours (or 10,000 feet plus) could not be assembled, if equipment could be devised to handle it. In reality, two problems needed to be overcome: the limited burning time of a carbon arc lamp, and the fact that the vast majority of projectors on the market still had a maximum spool capacity of only 2,000 feet.

The limited continuous burning time of a carbon arc was overcome with the introduction of xenon arc illumination in the 1950s, a technology which succeeded in combining the high light intensity and colour temperature of a carbon arc with the flexibility and ease of use of incandescent light. Incandescent illumination – in which a thin wire, or filament, is enclosed within a glass bulb and produces light when electricity is passed through it – had been in widespread use for domestic lighting since the late nineteenth century. This technology was unsuitable for use in film projection due to the limited amount of light which each bulb could reliably generate, and its relatively low colour temperature. In crude terms, incandescent bulbs produce a 'yellowish' light which would distort the appearance of the silver image or colour dyes in projection. From the 1920s onwards it was found that the colour temperature could be improved by filling the sealed bulbs with an inert gas (tungsten at first, now more usually halogen), and this enabled the use of incandescent light for studio use, and in projectors lighting smaller screens (such as slide projectors and small gauge projectors for home and small hall use). But so far, it has proven impossible to design an incandescent bulb which can generate enough light to illuminate a full-sized cinema screen.

The xenon arc lamp provided a compromise which overcame this problem. Like a carbon arc lamp, it generates light by producing a spark across two electrodes which in turn causes a gas to glow. But unlike in a carbon arc, that gas is not created by burning away a carbon rod. Instead, the two alloy electrodes are inside a quartz bulb which is filled with xenon gas under high pressure. The arc is struck by momentarily applying a very high voltage signal (usually of several thousand volts) across the electrodes, which overcomes the resistance produced by the gap between them, thereby establishing a flame. The electricity supply then changes to a low-voltage, high-current signal, which maintains it. The heat generated inside the bulb causes the xenon gas to glow, thereby producing light of a similar intensity and colour temperature to that of a carbon arc. The lamp will then continue to operate for as long as is needed until the power supply is switched off. The earliest models of xenon lamp had service lives of 100–200 hours, while the ones on sale at the time of writing are typically guaranteed to between 1,500 and 2,000 hours. Apart from the use of protective clothing when fitting and removing them (the pressure of the xenon gas inside the bulb creates a small risk of explosion), no special maintenance or safety precautions are needed. Originally developed in Germany, the first xenon arc lamps for projector illumination were demonstrated by the Zeiss Ikon co. at the Photokina trade show in Berlin, 1954. By the end of the decade, they were reportedly in use in 50 per cent of German screens.[36] Interviewed in 1957, the chief engineer of a British projector manufacturing firm noted that 'the new xenon light sources now under development will make the light on the screen independent of the operator',[37] acknowledging that one of the key advantages of this new technology would be that the cinema industry could potentially use it to cut staff overheads.

But although there was now a light source available which removed the time restriction of a carbon arc, the spool capacity of projectors still had to be overcome. In the wake of the conversion to sound, 2,000 feet had been established as a standard reel-length for release prints, meaning that two projectors interlocked for changeover operation were needed in order to present a complete feature film without interruption, and that a projectionist needed to be present throughout the screening. During the nitrate period this was not seen as being in any way restrictive, as the safety regulations in most countries required a higher level of staffing than was strictly needed to present the film, anyway. With the introduction of safety film and xenon illumination this was no longer the case, and therefore exhibitors started looking for ways of projecting entire programmes using a single machine.

The first systems to enter the market, from the mid-1950s onwards, consisted mainly of high capacity spools of up to 12,000 feet, which were mounted on heavy-duty spindles. These were either built into an enlarged pedestal underneath the projector mechanism and lamphouse, or positioned vertically on a separate structure known as a tower, situated behind or to one side of the projector. These spool carriers usually contained separate motors to power the take-up and rewind. Although release prints were still supplied to cinemas in 2,000-foot reels, the projectionist would 'make up' or assemble a complete programme onto a single large spool, consisting of the feature film and any other material which was to be shown (such as advertisements or trailers). Using a xenon arc lamp, this programme could then be shown using a single projector, completely unattended.

Fig. 5.6 A film transport platter system, as installed in a modern cinema. The film is paid out from the feeder on the top plate, and taken up around the 'collar' at the bottom. The centre plate can be used to assemble and take down programmes while the film is running.

The film-carrying system which would eventually make it economically viable to construct multiple auditoria within a single cinema building is known as the 'platter' (or sometimes the 'non-rewind' or 'cake-stand'). The first prototype was developed in 1964 as the result of a collaboration between Willi Burth, a cinema owner in southern Germany, and H. P. Zoller, an engineer, and was manufactured commercially from 1968.[38] The device consisted of two or more horizontal turntables which held an assembled programme of up to three hours in length. Film is drawn from the centre of the roll through a regulating device as the feeding plate is rotated by an electric motor, and after passing through the projector is wound around a central support, or 'collar' on the take-up plate. At the end of a show, the collar is pulled out of the taken-up film and replaced with a feeder, with the result that the film can be shown again without the need to rewind. The time taken to rethread a projector and platter between screenings can be as little as two to three minutes, and, as with any other device which enables an entire feature to be screened using a single projector, such a system can be left unattended during each performance.

The final piece in the jigsaw of technology which would reshape cinema exhibition in the 1970s and 1980s was the introduction of devices which automated a number of projection functions that were previously under the control of human beings. The first electromechanical automation systems were developed in the mid-1960s, and consisted of a control unit and a reader which was either attached to the projector itself or placed elsewhere in the film path. The reader detected 'cues' which were placed on the film surfaces, usually in the form of reflective adhesive labels. In response to these cues the reader would generate an electrical signal which caused the control unit to carry out functions such as lowering the auditorium lighting, opening screen curtains or adjusting the sound volume, and some systems even allowed performances to be started on a timer. In less than two decades, therefore, cinema exhibition had been transformed from a dangerous, highly-skilled, time-specific and labour-intensive process to one which could be managed according to economies of scale and carried out with far lower staffing levels.

The multiplex and beyond: 1976–2005

The evolution in cinema projection technology that took place between the end of World War Two and the mid-1970s equipped it to meet the demands of a very different commercial reality to that which had existed during the 'classical' period following the conversion to sound. The combination of safety film, xenon arc illumination, the platter and automated projection had an impact on the cinema exhibition industry and the culture of cinemagoing in general which is difficult to underestimate. In the 1930s, cinema provided not just the only routine access to moving images on offer but, in the developed world, represented the largest individual sector of the leisure industry. Its technical standards were shaped by a combination of commercial forces within the industry itself and the technologies on offer through other media. Cinemas offered electrically recorded sound when the technology was available in a form which integrated with the film industry's existing norms, and because it had also become available to consumers in other ways, principally radio and records. Widescreen was rolled out in the 1950s when moving images in the Academy ratio became available through television and leisure time was increasingly taking place in a domestic setting. During the following two decades it gradually made the transition from a roadshow technology to a routine fixture in every high-street cinema, as that product differentiation was found to be economically necessary.

Three new cinema exhibition technologies emerged during the final quarter of the twentieth century, and, like all their predecessors, they became established by the same process of market forces. The multiplex cinema – that is, a building which, instead of containing one large auditorium seating 1,000 or more, was subdivided into several smaller ones, some housing as few as 50 – was essentially a 1980s phenomenon, but one which had its origins in the conversion of cinema buildings in the 1960s and 1970s. This period was not an encouraging time for the cinema business, as is graphically illustrated by the statistics.[39] British cinema admissions peaked in 1946 with 1,635 million ticket sales, declined gradually to 1,100 million admissions in 1956 and then began a relentless year-on-year fall, reaching a low of 54 million in 1984. Over two-thirds of the UK's licensed cinema premises were closed during this 28-year period. As has been noted elsewhere in this book, a widespread perception developed within the exhibition sector and beyond to the effect that television and changing leisure patterns were largely responsible for the industry's virtual collapse. One of the strategies used by the industry to fight back was the combination of new technologies and an attempt to reinvent the social context of cinemagoing, as embodied by phenomena such as Cinerama and Todd-AO (and subsequently and to a lesser extent, Imax). The other was to replace spectacle with consumer choice. This process began in the late 1960s, when the operators of town- and city-centre sites built during the cinema boom of the 1920s and 1930s began to 'twin' or 'triple' them. This usually involved enclosing an auditorium's circle or balcony to form a (relatively) large cinema with a capacity of 500 seats or so, and building a partition wall down the middle of the main floor space below to form two smaller ones. The use of xenon-lit projectors being fed by platters, and usually operated under the control of a cinema automation unit, enabled the reconfigured building to be operated at no significant extra cost to its predecessor. Although the screens were a lot smaller, programming was more flexible. This had become a necessity, because customers were increasingly attending cinemas to see a specific film rather than as the leisure activity of default.[40]

As the results of this restructuring started to filter though, the decline in cinema attendance slowed down markedly through the 1970s, before plummeting almost 50 per cent between 1980 and 1984. Interestingly, this period also saw an equivalent rise in the number of prerecorded videocassette rentals of feature films. During the period 1984 to 1994 this trend was reversed, largely thanks to the arrival of the multiplex cinema, which extended the principle of consumer choice to all aspects of the exhibition process. This new form of cinema, which began in the United States in the early 1980s before being rolled out to Europe and beyond in the middle of the decade, consisted of a large, aircraft-hangar type building, usually situated on a motorway or city ring road and with sufficient car parking space for all of its customers. The building itself was divided into a large number of auditoria (as many as 32 in the case of Britain's largest multiplex, at the time of writing), with a gallery along its centre forming the projection booth. Computer-controlled automation means that, whereas in the 1930s it was not uncommon for up to sixteen projectionists to be needed to maintain a picture on one screen, one projectionist could now maintain sixteen. The bottom line was that customers wanted the flexibility to see a specific film at a time which was convenient to them, and the multiplex concept proved very efficient at being able to deliver that. The extent to which the growth of the multiplex enabled a revival of the film exhibition industry is striking. During that 10-year period, ticket sales almost doubled, from 54 to 123.5 million. In 1984, the UK contained 660 sites and 1,271 screens (i.e. just under two auditoria in each building), but a decade later this had increased to 734 buildings and 1,969 screens (just under three).

The rising number of screens caused by the multiplex phenomenon also had a profound effect on laboratory practice, owing to the vastly increased number of prints necessary for each run. In 1945 it was estimated that an average of 40 prints for a newly-released Hollywood 'A' feature were needed for UK distribution;[41] by the late 1990s runs of over a thousand prints for a 'blockbuster' title were not uncommon. High-speed contact printers and developing machines were designed for the purpose, with some labs offering the capacity to print and process up to a million feet per day. The changed film handling environment in the multiplex projection booth also resulted in the almost complete conversion from triacetate to polyester stock for release printing in the early to mid-1990s, largely in the wake of campaigning by the National Association of Theatre Owners (see chapter one), as it was more resilient and repelled the dust which would tend to accumulate on a print when sitting for long periods on an exposed platter deck.

The other two developments during this period were both related to sound. The stereo film systems launched during the 1950s reached only a limited audience, due mainly to the industry's attempt to package it with widescreen being rejected by exhibitors. But as with electrical recording in the 1920s, consumers gradually started to experience it through other media; firstly in the form of long-playing stereo records (first sold in America in 1958) and later through VHF stereo radio broadcasting. Given that as late as the early 1970s, over 90 per cent of cinema auditoria were still only equipped to play the single-channel, limited range 'Academy curve' optical mono prints which had been the industry standard since the early 1930s, the market was ready for an upgrade.

Once again, such an upgrade materialised and was rolled out when it became available in a form which was compatible with pre-existing industrial practice. It was developed by Ray Dolby as a spin-off to his noise reduction technologies, and enabled four channels of soundtrack to be recorded onto two Blumlein-style variable area optical tracks, which occupied the same area of the film as a conventional mono one. A technique termed 'phase shifting' by Dolby enabled each variable area track to carry two discrete channels, which were decoded in a sound processor linked to the projector's optical pickup cell. The crucial advantage of this system over the magnetic stereo methods which had preceded it was that it was totally backwards compatible: a Dolby-encoded print could still be played as mono in cinemas which did not have the new processors. This avoided the problem of 'dual inventory' print distribution (i.e. a situation in which some sorts of print are incompatible with the equipment in some cinemas) which had ultimately sunk lenticular colour, Vitaphone, VistaVision as a release medium and CinemaScope in its original form, among other systems. Even if cinemas were slow to purchase the new sound equipment, this would not prevent the rollout of Dolby prints, thereby increasing the range of titles available and making the system more attractive to potential purchasers. The original Dolby system, known as 'A type', was launched in 1976 with the release of *Lizstomania* (1976, dir. Ken Russell). The 1980s multiplexes nearly all installed stereo playback equipment from the start (which, along with the computer designed sightlines and acoustic properties of the auditoria, was a major selling point), and by the end of the decade stereo sound in one form or another had become standard equipment in most cinema auditoria in Europe and North America. An improved version of the Dolby system, known as 'Spectral Recording', was introduced in 1988 which, while keeping the same four channels in each mix (left, centre, right and surround), increased the dynamic range available from each channel. Again, the prints were totally backwards compatible, both with mono systems and those which had A-type decoders.

Exactly the same system of precedents applied to the mass-marketing of digital sound for cinema playback in 1992. This had already been available to consumers in the form of the compact disc for a decade previously (and some LPs were advertised as having been 'digitally recorded' or 'digitally mastered' as far back as the late 1970s), and the three competing playback systems which were all introduced to the cinema market in that year offered ease of installation and backwards compatibility in the print format as key selling points. Eastman Kodak had made an abortive attempt to bring digital sound to the market two years earlier with its 'Cinema Digital Sound' (CDS) system, in which the conventional optical track on a 35mm print was replaced with photographically recorded digital data.[42] But with the analogue track removed the prints could only be shown in CDS-equipped venues, and CDS was ultimately undermined by two factors: the need for dual print inventory, and the absence of any analogue track meaning that there was no backup in the event of digital sound reproduction being lost.

The three systems introduced in 1992 were Dolby Digital (originally called SR-D), in which optical data 'blocks' were printed between the perforations on one side of the print; the Digital Theater Systems (DTS) method, in which an optical timecode on the film is used to synchronise the playback from data held on one or more compact discs read from drives inside the sound processor; and Sony's 'Dynamic Digital Sound' system, in which data is printed between the perforations and the outer edge of the film on both sides. While Dolby and DTS encoded six channels (directional surround and a separate sub-bass channel were added), Sony's method had the capacity to carry up to eight. All three systems initially established a market foothold. DTS is generally perceived to offer the highest quality sound, on account of the lower compression ratio (compression in digital audio and video is discussed in chapter eight) enabled by the decision to store the audio data separately on an offboard carrier, though the extra cost of producing and shipping data discs along with each film print means that it has never established the same market share as Dolby. The number of SDDS film releases never matched the level of either Dolby or DTS, and this was not helped by the fact that data tracks on the prints could be easily made unreadable by even slight edge damage to the film.

Almost two decades after the commercial rollout of the first mass-marketed cinema stereo sound system, therefore, this has become a standard technology in nearly all cinema auditoria in the Western world. Echoing the comments of those who believed that silent films had a future in the early 1930s, its introduction has caused some critics and spectators to believe that a perceived increase in the sound levels in cinema auditoria are actually detrimental to their experience of the films being shown:

> Once there was a time when the enjoyment of cinema could be compromised by a neighbour rustling his sweet packaging. During *The Lord of the Rings* in the seat next door someone could have been road-testing a jack hammer and you would have remained in blissful ignorance.[43]

a series of changing electrical modulations which can either be transmitted by radio or recorded onto magnetic tape. Later, the changing electrical modulations were replaced by representing the image as digital data which is encoded and decoded by a computer.

This had a number of implications. The chapters on film-based technologies have emphasised the role played by independently articulated technical standards which are observed by competing suppliers of products and services. This is because without these standards, the basic economic principle on which the film industry depends for profitability – that of reproducing a single recording in as many different locations and to as many paying customers as possible – becomes less effective or even impossible. Let us suppose, for example, that 35mm film was not a universal, worldwide standard, and that 28mm had survived to become a widespread release format alongside it. Assuming that each gauge achieved an approximate 50 per cent market share, any studio wishing to make a feature film available to all the world's cinemas would be forced to produce both types of print and to absorb the extra distribution overheads.

With television the stakes are even higher, because instead of the estimated 130,000 35mm cinema installations in use worldwide in 2003, there are literally billions of television sets. The technical properties of the broadcast signal therefore have to be standardised in order to ensure that all the receivers in a given area are able to display the images and play back the audio correctly. For this reason and because the broadcasting 'bandwidth' (the number of modulated signals or digital data streams which can be broadcast at any one time) is limited, broadcast standards and bandwidth use have been regulated by national governments from within a few years of sound-only radio starting to become established as an industry. Perhaps the best known of these government agencies in technical circles is the United States' National Television Standards Committee, of which more below.

One of the key roles played by governments in the technology of broadcasting is to allocate and regulate the use of bandwidth by privately-owned broadcasters,[5] though some countries have gone even further, by establishing state-run broadcasters financed by direct taxation. The United Kingdom (which was the first country in which regularly scheduled public television broadcasting took place) evolved its unique hybrid in the form of the British Broadcasting Corporation (BBC), a (supposedly) non-commercial organisation established by Royal Charter, but which is institutionally separate from the government executive. It is financed by a law which requires anyone in the UK who uses a television set in their home or workplace to buy an annual licence for the premises on which it is installed. This scale of political intervention has never been equalled in any other area of moving image technology in the developed world. Apart from the former USSR and Eastern Bloc, there have never been a significant number of state-owned cinemas, and nowhere near the same scale of state involvement in the production of content. The scale on which television operates as a mass medium therefore results in a very different model of regulation and standardisation to that of film, a model which is in evidence throughout virtually every stage of technological development and change. Hence, therefore,

Albert Abramson's remark that television was the first invention to be achieved by committee. In fact film had its fair share of committees too, and almost from the very beginning. But because these committees (or at least, the ones that mattered) were all industry self-regulation bodies (such as the MPPC or the SMPE), their role in developing and imposing standards was less obvious. In other words, the film industry was still essentially 'business, pure and simple', as D. W. Griffith put it, committees and all. The technical logistics of television initially ruled out standardisation through market forces. While a vertically integrated film industry could absorb the cost of Vitaphone proving not to be up to the job and of having to throw a few thousand turntables into skips – even in the depths of the century's worst economic depression – individual consumers would not (and, to a certain extent, still will not) risk a substantial capital investment on a piece of domestic equipment which is likely to go the same way, as Robson's research on British television in the 1930s reveals:

> The pace of sales was always sluggish, partly due to a lack of confidence in the technology that the BBC was using. The public had its major concern that the broadcast standard was still experimental and liable to change in the near future, and this apprehension was hardly allayed by government pronouncements at the time.[6]

Governments, therefore, had to start making pronouncements, and in regulating the television industry they have been doing so ever since. This makes it all the more ironic that when George Orwell wrote *1984* in 1948, just as television was on the verge of starting to grow from an experimental technology into a mass-medium, he settled on the 'telescreen' as his symbol of state intrusion into the private sphere.

Television – initial research, development and rollout

The timescale in which television evolved from a theoretical possibility into scheduled broadcasting and cabinets in living rooms was similar to that of film, but took place approximately half a century later. Both the illusion of continuous movement and the basics of photochemistry were understood and starting to be applied by the end of the 1830s; from there the research and development needed to put two and two together and produce film took a further half century. Likewise, the basic radio and electronic technologies needed to make broadcast television a reality were in place by the 1890s, and were evolved by a number of individuals and organisations over the first half of the twentieth century, to the point of becoming a mass-medium in the years immediately following the end of World War Two. This work took place predominately in the United States, and in Britain to a lesser extent. The only other country in which public television broadcasting took place before World War Two was Nazi Germany, and even then only on what was effectively an experimental basis.

The media technologies that were developed by the Edison/Eastman cartel in the late nineteenth century – i.e. still photography, audio recording and reproduction and film-based moving image technology – all had one crucial attribute in common: they were initially intended as a means for consumers to create and play back their own recordings. It was only in the case of still photography that this application proved to be commercially viable, even in part. Where audio and film were concerned, the law of diminishing returns came quickly into play and resulted in an industry based on the commercial exploitation of multiple copies of a single recording.

As we have seen in chapter four, the arrival of radio in the 1920s heralded a new variant of this business model. Broadcasters consciously differentiated themselves from the record industry by promoting the fact that almost everything they transmitted was live. As David Morton's research has revealed, the broadcasting of pre-recorded radio programmes was initially perceived as being culturally inferior, a perception that was deliberately encouraged by the radio industry of the 1920s and 1930s in order to distinguish their product from records and, to a lesser extent, films.[7] The radio industry was following the lead of record and film producers up to a point, in that it was making money by selling the same content to an almost infinite number of customers. But broadcasting had the added attraction of making the process time-specific.

The idea of television offered the same potential selling point, only with moving images into the bargain. The first generation of television inventors and pioneers were looking to exploit the same underlying technical principle as their counterparts in radio had discovered. This principle had been discovered as early as 1832, when the engineer Samuel Morse had discovered that electrical currents could be transmitted along metal conductors (wires) of almost infinite length. While the most widespread use for this discovery was the delivery of 'mains' electricity from power stations to homes and businesses, Morse also realised that by periodically interrupting the current to create a sequence of pulses, that sequence could be detected and reproduced at the other end of the wire, thereby forming a method of data transmission. This was the 'Morse code', in which the receiving end of the wire was connected to a buzzer or a light bulb, thereby enabling the sequence to be reproduced and interpreted as the 'dots' and 'dashes' which formed letters of the alphabet.

In the late nineteenth century the German scientist Heinrich Hertz established that it was possible to induce changes in the wavelength of an electromagnetic energy source, which could then be detected from a remote location without the need for any wires. While this did not enable the transmission of significant volumes of power, it did enable 'wireless' data transmission using a modulated signal, e.g. of Morse code. The invention of the first crude microphone by Alexander Graham Bell (see chapter four) enabled a representation of audible sounds to be modulated and transmitted, and it was the the Irish-Italian engineer Gugliemo Marconi who, in 1896, combined these technologies to carry out the earliest known voice transmission. Lee de Forest's triode valve marked another milestone (although, in chapter four, this is discussed primarily in the context of film sound, it is worth keeping in mind that valves were originally developed to enable the construction of long-distance telephone networks) in that it enabled the modulating signal to be amplified to the point at which broadcasting over a significant distance (i.e. radio as mass-communication) became feasible.

Just as records and films did not catch on as 'home media' to anything like the same extent that they became vehicles for mass-communication, radio was not developed and marketed as a form of wireless telephone. There were two key technical reasons for this, namely bandwidth limitations and the fact that no scrambling or encryption technology existed. In order to receive a wired transmission, it was necessary to establish a physical connection to the wire which carried it; but anyone with a receiver tuned to the appropriate wavelength can hear a radio transmission, whether the broadcaster wishes them to or not. It was precisely this characteristic which initially made radio a useless medium for one-to-one communication, but ideal for mass-broadcasting. As well as these technical reasons, records and films had, by the early 1920s, established the cultural precedent for electronically mediated mass-communication, even if the media were pre-recorded. Therefore, the broadcasting of moving images along with sounds was a logical next step in all the development which had been going on in the area of media technology during the late nineteenth and early twentieth centuries; and furthermore, a cultural framework already existed for its commercial exploitation by the early 1920s.

The process that would eventually lead to the emergence of broadcast television as a mass medium in the form we know it today involved two technological approaches and a number of individuals and organisations associated with each. To some extent, the two technologies can be analogised with the battle between sound-on-disc and sound-on-film in the 1920s film industry. The first system, which used an electromechanical approach to encoding and displaying the transmitted signal, established a cultural and business case for television as a medium, but eventually encountered technical limitations which caused it to fall by the wayside. The electronic method which emerged shortly afterwards would eventually provide the technological basis for television as it exists today.

Electromechanical television

This technology was associated first and foremost with John Logie Baird, a British engineer born near Glasgow in 1888 and who began his career in the electrical supply industry. After settling in south-east England in 1923 he began a series of experiments designed to exploit the potential of the 'Nipkow disc' in encoding visual images for radio transmission. Television, like film, relies on the illusion of continuous movement to electronically encode and reproduce a pattern of light which is detected by the human eye as the impression of a continuously moving image. But whereas film does this by regulating the rate at which whole images are recorded and reproduced, analogue television has to divide that picture up into a series of horizontal segments for transmission. The first practicable method of achieving this was devised and demonstrated by the German engineer Paul Nipkow in 1883. This consisted of a rotating metal disc (not dissimilar to the shutter in a film camera or projector), in front of which was a lens and behind which was a photoelectric cell (the same sort as would be found in the optical sound reproducer on a cinema projector). The disc contained a series of small holes around the perimeter, gradually spiralling inwards. When this apparatus was placed in front of a stationary subject and the disc rotated, the first hole in the sequence would move horizontally across the lens' field of vision; the second would do likewise on a line just below it, and so on until at the end of the revolution, the entire picture had been 'scanned', and the modulated signal for each line of the picture had been generated by the photocell.

Working largely in isolation and financed mainly by private collaborators, Baird undertook a process of research and development which culminated in a viable television system based on the Nipkow disc. Initially this took the form of a low-resolution, 30-line signal, but was eventually uprated to a 240-line Nipkow disc by 1935.

Electronic television

Baird had a north American counterpart, who was also a largely self-taught engineer working in isolation. Philo Farnsworth, born in Utah in 1906, was, however, working on a fundamentally different approach to transmitting visual images to that of Baird. His 'image dissector tube', first successfully demonstrated in 1927 (when its inventor was aged only 21!), consisted of a large glass bulb coated with a photoelectric compound. This surface was scanned continuously by a beam of electrons focused from the narrow end of the bulb, in order for the signal modulation generated by the photoelectric surface in response to light to be detected and transmitted.[8] The image was displayed by reversing the process in the 'cathode ray tube' (CRT) which, instead of being coated with a photosensitive compound, was coated with one that produced light in response to electrical stimulation.

Put like this, the technique sounds simple and straightforward. In fact, it was anything but, and Farnsworth and his competitors expended a great deal of time and energy attempting to improve the reliability and image definition of their tubes. One key problem was the relatively low sensitivity of the photoelectric compounds which were available in the 1920s and 1930s (as will be recalled from chapter four, the inventors of optical sound-on-film systems had exactly the same problem), which became more of an issue as the number of lines of horizontal resolution was increased. In particular, the phosphors in the receiving tube could not be made to illuminate for long enough to maintain the screen's brightness between scans, as a result of which the technique of interlacing was developed in 1929. In many ways this can be seen as television's equivalent of the developments in cinema projector shutters in the 1920s. Instead of scanning and broadcasting each line of the picture in succession (known as progressive scanning), lines 1-3-5 would be scanned first, followed by 2-4-6 (interlaced). This would maintain overall picture brightness, because as the odd-numbered lines started to fade, the evens would be freshly scanned and bright. Although today's CRTs are more than capable of displaying a progressively scanned image with uniform brightness, the television broadcast signals in use today still interlacing.

Interlacing was just one innovation produced by researchers working for the Radio Corporation of America (RCA): and is indicative of the point having been reached at which the development of broadcast television technology diverges radically from the romantic image of the innovations of lone inventors such as Baird and Farnsworth, and becomes a process driven by a unique combination of big business and political regulation. From the early 1930s onwards, RCA in the United States and the BBC in Britain became the key players – the 'invention by committee' described by Albert Abramson.

The formation of RCA was essentially a political act. At the end of World War One, the only significant provider of radio communications in the United States was a subsidiary of the British Marconi Company, which led to concerns in the US government and military for the security implications of the radio communications infrastructure being foreign-owned. On 17 October 1919, the Radio Corporation of America was formed, primarily as an offshoot of General Electric, which took over Amercan Marconi shortly afterwards. With the Marconi infrastructure at its disposal RCA soon established a dominant position in radio communications, and by the mid-1920s had also become heavily involved in broadcasting to the public, both as a manufacturer of equipment and an owner of radio stations. Throughout this period RCA enjoyed close relations with the US government, which were to prove especially important when it began to enter the field of commercial television broadcasting in the late 1930s.

There were two driving forces behind RCA's involvement in television, both of them Russian émigrés. David Sarnoff was born near Minsk in 1891 and emigrated to the US as a child. A self-taught radio operator during the 1910s, Sarnoff joined RCA as commercial manager upon its formation and became its president in 1930, a position he continued to hold until his death in 1971. He was essentially the commercial and political force behind the establishment of television in the US. The technology was supplied by Vladimir Zworykin, a scientist born in 1889 and who emigrated to the US in 1919. During his early career in Russia he had researched CRT technology thoroughly, though his initial interest was in its application for x-ray imaging rather than television. In 1920 he joined Westinghouse as a researcher working on photoelectric cells, and wrote a PhD thesis on increasing their sensitivity. By this stage he was actively interested in television, and in 1924 took out patents for both camera (termed 'Iconoscope') and display tubes. There were some subtle differences between Zworykin's tubes and Farnsworth's which resulted in litigious wrangles throughout the 1930s, but when Zworykin joined RCA in 1929 and persuaded Sarnoff of the commercial potential for television as a mass-medium, the ball started seriously rolling, one which would culminate in the launch of public television broadcasting in the United States in New York on 20 April 1939.

In Britain, the corporate impetus behind television was also a nominally independent organisation, but which in reality had close links to the institution of government. The BBC began life as the British Broadcasting Company in 1922 before being established under royal charter as a corporation in 1926. As it had a legally-enshrined monopoly in UK radio broadcasting, it was obliged, under orders from the Postmaster General (the government official responsible for allocating transmission frequencies), to co-operate with Baird when he approached them asking for infrastructural support to begin public broadcasting using his mechanical scanning system. The first experimental transmissions took place in 1930, and continued for some years, with Baird gradually improving the resolution of his system from 30 lines to 240 during the first half of that decade.

Electronic television was also being developed in Britain. Independently of Farnsworth or Zworykin, work on cathode ray tubes was also being carried out by Electrical and Musical Industries (EMI), based in West London, which had achieved an all-electronic television transmission and display system by 1932. By April 1934, EMI was able to demonstrate a 180-line, all electronic version based on its 'Emitron' camera tube, which already offered a number of crucial advantages over the Baird system, most notably a camera which could generate live images (at that point Baird's could only produce images by scanning film, i.e. it was a telecine device). In the summer of that year the government established what was to all intents and purposes a committee to invent television, i.e. to make recommendations on the feasibility of beginning regular public broadcasts and which system should be used.

Largely in order to protect its reputation for political neutrality, the Selsdon Report, published the following year, resulted in a competition. For a three-month trial period, between November 1936 and February 1937, the 240-line Baird system and the 405-line EMI system would broadcast in a series of alternate, scheduled transmissions, and the outcome of that trial would dictate which of the two systems the BBC would adopt for its 'high-definition' television service. The last 30-line Baird broadcast took place on 11 September 1935.

The outcome of the trial was an unqualified success for the electronic method, so much so that in its aftermath Baird's company sub-licensed some of Farnsworth's patents and attempted to market cameras and receivers based on them in the UK. The Emitron cameras could be used for live outside broadcasts, whereas the Baird system could only broadcast footage shot outside a studio which had been originated on film. EMI offered almost double the resolution, and the receivers proved to be far more reliable. From the BBC's point of view, the test period was an exercise to carry out the process of standardisation via committees and officialdom, the equivalents in film of which had been left to market forces. As Lord Selsdon said in a televised address in the opening public broadcast of the test period, on 2 November 1936:

> From the technical point of view, I wish to say that my committee hopes to be able, after some experience of the working of the public service, definitely to recommend certain standards as to the number of lines, frame frequency and ratio of synchronising impulse to picture. Once these have been fixed, the construction of receivers will be considerably simplified.[9]

The test certainly achieved that objective, with the 405-line, 50 hertz EMI system being adopted by the BBC as the television broadcast standard. EMI was declared the winner in February 1937, while Baird's Nipkow disc went the same way as Warner Bros.' Vitaphone disc – rapidly into obsolescence. Regular, scheduled transmissions on the EMI system continued for an average of two hours per day until the outbreak of World War Two in September 1939, when broadcasting was suspended for the duration of the conflict. Approximately 20,000 sets had been sold by that point, though Robson argues that the first two years of television broadcasting in Britain were broadly unsuccessful:

> The arrival of TV as we would recognise it – the most potent communication medium of the twentieth century – should have been an outstanding success, the 'must-have' for all middle-class Londoners in the years leading up to World War Two. In fact, the numbers of viewers never grew large enough for it to have any social force, and the short-lived experiment, lasting less than three years up to 1939, completely failed to achieve its potential.[10]

This judgement is surely on the harsh side, given that this service was the first of its kind anywhere in the world and that the cost of the early receivers was around half that of an average family car: its significance is more in the technical precedents set than as a 'social force'.

In America, RCA went through the same process, only using Zworykin's standards and technologies, approximately two years later. This time the inaugural event was the World's Fair in New York on 20 April 1939, when the RCA-owned National Broadcasting Corporation (NBC) transmitted images of David Sarnoff declaring that 'now we add sight to sound'.[11] As with Baird and EMI in Britain, a standards row was also underway in the US, the main protagonists being RCA and Farnsworth. In the absence of a BBC-type organisation, the US Government's Federal Communications Commission (FCC) had been established to regulate the radio industry, allocating frequencies and licensing broadcasters. In 1940 it established the National Television Standards Committee (NTSC) to do likewise with television, and one of its first tasks was to standardise the broadcast system. This was eventually established as the 525-line, 60 hertz interlaced system with which the term 'NTSC' is synonymous today, and the FCC authorised commercial broadcasting using this standard as of 1 July 1941. As in Britain, public service broadcasting and the commercial manufacture of equipment ground to a halt during the war, though in the US this was as of 7 December 1941 (the bombing of Pearl Harbor) rather than 3 September 1939 (Germany's invasion of Poland).

There is one footnote to add to the story of pre-war television, which is that Nazi Germany also operated a small-scale broadcasting infrastructure, using a 180-line mechanical system. The first transmissions took place in March 1935, and it is estimated that between 200 and 1,000 receivers were manufactured. Very little is known about the technical details of this system (except that it was totally incompatible with any of those being developed in the UK or the US), and it is not believed that the Nazi broadcasts were seen by anything like the number of viewers for pre-war UK and US programming.[12]

The post-war period: colour, PAL and the emergence of a mass medium

The two decades following the end of World War Two saw television make the transition from an embryonic prototype into the mass medium that succeeded cinema exhibition as the principal means through which consumers viewed moving images. Broadcasting in the UK resumed on 7 June 1946, and by 1949 150,000 licences had been sold.[13] In America the growth in sales was even more dramatic, including, for example, a 500 per cent increase in the number of receivers sold between 1947 and 1948.[14] In terms of the consumer hardware, economies of scale began to kick in with a vengeance as consumer confidence in the broadcast standard and an upturn in the Western economies, helped by an unprecedented level of political stability following the end of World War Two, encouraged broadcasters and consumers alike to invest. One key milestone was the launch of the RCA 630-TS television receiver, dubbed the 'Model T' of television, in 1946.[15] Priced at $385, it had sold 10,000 units by the end of that year alone. As the mass-manufacturing techniques (especially in relation to the cathode ray tubes themselves) evolved and became more reliable, prices of television receivers fell substantially in real terms.

Fig. 6.1 A British advertisement published in 1950. 'His Master's Voice' was a trade name of EMI, hence the claim 'made by the people who gave the world its first electronic television system'. The light output from the first generation of mass-produced cathode ray tubes was usually insufficient for viewing in anything other than subdued light, and 15 inches across was about the largest cathode ray tube it was possible to mass-manufacture at the time these receivers were sold.[16]

The next major technological change to be researched, developed and then rolled out was colour, and this followed a very similar pattern to that of the establishment of mass broadcasting: the basic techniques were established, shown to work reliably, proven to be economically viable on a large scale and finally subjected to the political processes which enshrined them through technical standardisation.

Unlike pretty much every film-based colour system used from the end of World War Two onwards, colour in television is exclusively additive (see chapter three for an explanation of additive and subtractive colour). Both Baird in 1928 and Herbert Ives of Bell Labs in the following year demonstrated crude mechanical colour systems based on multiple Nipkow discs. Instead of using a single sequence of spiralled holes in the disc which scanned a monochrome image based on the intensity of light which passed through it, the colour version had three; one for each of the primary colours (red, green and blue). The receivers contained gas cells which produced light of corresponding colours in the reproduction process. Although these systems suffered from all the same drawbacks as monochrome mechanical television, they established the principle of splitting the scanning process into three stages and 'adding' colour information to the existing monochrome scan that way.

As mechanical television became obsolete, engineers began looking for ways of capturing, broadcasting and reproducing colour information using the electronic scanning methods which had been established by EMI in the UK and RCA in the US. The first system to have been systematically demonstrated was in fact a mechanical/electronic hybrid, proposed by a scientist working for RCA's main commercial rival in the radio industry, Columbia Broadcasting System (initially the Columbia Broadcasting Company, but known by the initials CBS following a series of mergers in 1926), Peter Goldmark. He positioned a rotating filter wheel containing red, green and blue segments between the lens and a camera tube. When synchronised with the tube's scanning beam and the rate increased threefold, each frame was scanned six times in sequences of red-green-blue, the latter sequence being interlaced with the former. In the receiver tube the process was reversed, with the electron gun receiving the signal firing through a reciprocating filter to illuminate the phosphors at the end of the tube accordingly. In essence, the principle was the same as with the Kinemacolor film system (see chapter three), but with an electron beam scanning device rather than panchromatic film as the receiving medium.

When it was demonstrated in July 1940, Goldmark's system was generally acknowledged to produce an acceptable quality of colour reproduction. As a result, he faced intense opposition from RCA, which at the time was on the verge of rolling out commercial monochrome broadcasting. This opposition took two forms: firstly emphasising the mechanical nature of the system (i.e. trying to associate it with the proven drawbacks of the Nipkow disc, and characterising its claims to being electronic as 'counterfeit'), and secondly by highlighting compatibility issues. As the broadcast signal consisted only of the red, green and blue scanned colour records, there was no means of reproducing a monochrome image from it, and the signal could not be received on a set designed for the existing NTSC system. As RCA had undertaken a long process of intense, political lobbying to get its system adopted as the national broadcast standard, Sarnoff was not likely to sit around while a rival, incompatible system which also offered colour into the bargain was allowed to compete in the marketplace. The result was that the CBS/Goldmark system never progressed beyond a few experimental broadcasts.

Further significant research and development was interrupted by World War Two, after which the broadcasters' and electronics industry's emphasis – both in the US and the UK – was to increase economies of scale to the point at which monochrome television became a viable mass-medium. There was also the growth of TV in other parts of the world to consider, something which the American and British hardware manufacturers gave a high priority in order to recoup the large investment they had made in research and development since the late 1920s. Japan became the first country to begin regular broadcasting in the post-war period, starting in 1951, while the rest of the developed world followed gradually during the decade.

Work on colour did resume, with the start of a two-decade process that would culminate in the technology becoming universal. The next major advance was a purely electronic system developed by RCA and publicly demonstrated in June 1951. It was, in effect, an electronic, additive version of three-strip Technicolor. The camera contained three separate tubes and an optical system behind the lens which split the incoming light into its constituent primary colours of red, green and blue. However, unlike in Technicolor, the resulting images were not recorded on separate strips of monochrome film in the form of subtractive negatives – instead, the resulting electronic signal was modulated and broadcast. In the receiver tube, three separate electron guns reversed the process, illuminating phosphors which would glow with the reciprocating colour and intensity to those in the camera.

The superiority of RCA's system was not only in the absence of any moving parts. The Goldmark/CBS method was completely incompatible with the existing, monochrome signal standard which had been established by the NTSC in 1941: a CBS colour set could not receive the existing black-and-white service. The RCA system broadcast two separate modulated signals. The first was identical to the pre-existing black-and-white standard (the 'luminance', or 'Y' signal): in a colour camera, it was produced by combining the output of all three tubes to generate a single measure of brightness. The second signal contained the scanning information needed for the three electron guns in the receiving tube to reproduce the colour information captured by the camera, and was known as the 'chrominance' or 'C' signal. A black-and-white television set simply did not receive the 'C' signal, while a colour one combined both to reproduce the image as it had been shot by the three-tube camera. Conversely, black-and-white cameras could continue to be used, because the colour receivers could reproduce the images from them by simply firing all three electron guns with equal intensity (remember: in additive colour, combining all three primaries gives you black) in order to display varying shades of monochrome. The RCA system, therefore, could be introduced without any of the hardware which had been sold during the 1940s being rendered obsolete.

However, the successful conclusion of RCA's research programme came too late to prevent the FCC from licensing the CBS method, following a protracted legal wrangle between Sarnoff and CBS. The latter accused the former of attempting to monopolise the television industry, and pointed out that because the two systems were completely incompatible, the public should have their chance to decide between them in the marketplace. In May 1951, the US Supreme Court authorised CBS to begin commercial broadcasting, though in truth this turned out to be a purely symbolic victory. In the following months sales of CBS sets were negligible, and, at the request of the military, production was then suspended throughout the Korean War. In the aftermath of that and following another round of politicking, the RCA method was eventually adopted by the FCC, to be known officially as 'NTSC color'. The first sets went on the market in 1954, and commercial broadcasting on a small scale began soon afterwards.[17]

But it was to be another decade before colour television became widespread. Because RCA/NTSC was fully compatible with existing black-and-white hardware, there was little incentive for studios or consumers to make a fresh round of major investment – for all the same economic reasons as the film industry had ignored widescreen in the aftermath of the conversion of sound. By 1960, NTSC colour sets accounted for only one in fifty of total receiver sales in the United States.[18] A big part of the problem lay in the technology itself, as the three-gun colour tubes were expensive to manufacture, had a high wastage rate and were relatively unreliable (according to contemporary accounts, early models had the frequent and unfortunate habit of exploding). A useful comparison would be with the laptop computers sold in the early twenty-first century: even though they have been available for over a decade, they remain about 30 per cent more expensive than their full-sized counterparts, because manufacturing techniques and the availability of raw materials have not developed in order to reduce the unit cost of the high-power batteries and TFT screens which account for most of what a consumer pays in the shops. By the same token and until the late 1960s, colour broadcasting was limited and the cost of the receivers was substantially more than most consumers were able or willing to pay.

In the case of colour television, the development which broke this vicious circle came not from the United States, but from Japan. The Trinitron tube marked a huge improvement in colour resolution and picture definition and eventually proved a lot cheaper to mass-manufacture than the RCA tubes. It was the invention that made Sony a household name and marked the start of a process which would culminate in Japanese manufacturers dominating the consumer electronics market in the 1970s and 1980s. The major drawback with RCA receivers was the use of a focusing system termed a 'shadow mask', through which the three electron guns were discharged. This was needed to align the three beams to the viewing end of the tube accurately (and in some ways, can be considered analogous to the beam-splitting device in a Technicolor camera), but it absorbed a lot of light and tended to wear out after a few years' use, resulting in a blurred and indistinct picture.

By the early 1960s Sony had not entered the colour market at all, and its technical director, Masaru Ibuka, was determined that it should not do so until the company had developed a way of overcoming the drawbacks of the RCA tube design. Sony's first attempt was an adaptation of a US military radar display screen known as 'Chromatron': instead of using separate electron guns for each of the primary colours, the tube consisted of a single gun which illuminated three separate formulations of phosphor. In this way the need for a shadow mask was overcome, but the Chromatron tubes proved impossible to mass-manufacture reliably (initially, only around 0.25 per cent of tubes which came off the assembly line were usable, and this figure could not be improved substantially). The eventual solution, introduced in 1968, was the 'Trinitron' tube – a single cathode ray tube but which contained three cathodes (negative electrodes). The focusing method was known as an 'aperture grille', which absorbed only a fraction of the light of an RCA shadow mask. Both systems were able to receive the same NTSC broadcast signal. While shadow mask technology had also been steadily improving throughout the 1960s, the Sony method marked the breakthrough which would enable colour television to become a mainstream broadcast and consumer medium. This was partly due to the clarity and definition of the Trinitron's colour image, though the advances in tube design during the 1960s and 1970s were as much to do with process control and the ability to apply economies of scale to the manufacturing process as with the image itself.[19]

Both shadow mask and Trinitron-type display tubes continue to be manufactured today, although the use of CRT technology in television imaging has declined as new microelectronic and computer-based methods have emerged. In terms of recording electronic moving images, the charge coupled device (CCD) largely replaced CRT cameras for this purpose from the mid-1980s onwards. It was initially developed by George Smith and Willard Boyle, two scientists working for Bell Laboratories in the US in areas of optical communication devices, lasers and semiconductors. It is, in effect, an electronic memory which is 'written' by exposure to light and sequentially discharged in the form of electrical energy shortly afterwards. Although the first designs were not intended for television and video imaging, ongoing research into this technology resulted in its eventual use for this purpose. CCDs are a fraction of the size and weight of a CRT (the ones in today's cameras are around half the surface area of an average postage stamp), consume an insignificant amount of power and have an almost unlimited lifetime (although, in some circumstances, they can be damaged or destroyed by overexposure). CCDs record colour information in one of two ways: professional studio cameras and telecine devices usually contain separate chips for each of the primary colours, while the cheaper units used in domestic video camcorders (see below) and still cameras use a single CCD fitted with a filter device known as a 'Bayer mask'. This divides the individual photosensitive components, or 'pixels',[20] into a mixture of red, green and blue sensitivities.

Alternative technologies to CRTs have also been developed to display television and other types of electronic moving image. The liquid crystal display (LCD) has its origins in the discovery of liquid crystals in the late nineteenth century, when the Austrian chemist Friedrich Reinitzer (1857–1927) discovered a liquid substance similar to cholesterol which could be made opaque or clear by applying intense levels of heat (similar to the phenomenon whereby fat in a frying pan will become clear as it melts, and then opaque as it resolidifies after the heat has been switched off – only liquid crystals will not solidify). Research on the properties of liquid crystals developed slowly throughout the early twentieth century, but stalled in the 1940s due to the absence of any perceived industrial application. A resurgence in research activity in the 1960s resulted in the development of liquid crystals, the reflectivity of which could be permanently adjusted by applying an electric current. Most modern LCD devices are based on a method invented by the American physicist James Fergason and first publicly demonstrated in 1971, known as the 'twisted nematic field' effect. A more recent development in LCD technology was the invention of the 'thin film transistor' (TFT) display, which uses an 'active matrix' of microscopic transistors to form the image. Once energised, these will retain their chrominance and opacity for as long as is needed, effectively doing away with interlace issues, vastly improving the definition possible from an LCD and significantly reducing power consumption. For this reason, the most widespread use of TFT displays at the time of writing has been in laptop computers, though their use as a moving image display medium has been steadily growing and TFT-based television receivers are now readily available in the developed world (albeit at a substantially higher price than their CRT counterparts).

By the mid-1960s, NTSC was still the only nationally adopted broadcast standard capable of encoding a colour signal, used principally in the USA and Japan. Most of the rest of the world used the 405-line system adopted by the BBC in 1937, which was black-and-white only and of visibly lower definition relative to NTSC. This led to the rollout of the Phase Alternate Line (PAL) standard, which increased the definition to 625 lines but kept the 405-line scanning rate of 50 hertz (equivalent to 25fps). Some changes to the way in which the luminance and chrominance signals are modulated for transmission further improved the image resolution over that of NTSC. PAL was essentially a German invention, having been developed by engineers working for Telefunken, though the French, in the form of state broadcaster Radiodiffusion-Télévision Française, had also came up with a higher resolution broadcast standard: Système Èlectronique couleur avec memoir (SECAM), first demonstrated in 1957, in which each line of colour information was sequentially interlaced with a line of luminance.[21] As this is the only difference between PAL and SECAM, the two systems are compatible in black-and-white only (i.e. a PAL television can receive a SECAM broadcast, but will only display it in black-and-white, and vice versa). Starting from summer 1967 most of Western Europe adopted the PAL system (starting in Britain, when regular colour broadcasting began on 2 December 1967 following experimental transmissions earlier in the year)[22] while France, the Eastern Bloc and the Soviet Union converted to SECAM.[23] RCA had been hoping to further entrench its dominance in the television market by trying to establish NTSC as a global standard, exploiting the fact that the British 405-line system was not compatible with colour, and in a publicity brochure issued in 1964 (three years before the launch of PAL in the UK) had warned that:

> In Europe, government authorities concerned with telecommunications will endeavour, within the next few months, to reach agreement on standards of television for European countries. The responsibility is a serious one, for these standards must anticipate future as well as present requirements, and they must provide the best possible colour television system for all countries concerned.[24]

There were no significant changes to the broadcast standards in use during the last three decades of the twentieth century. In various examples related to film we have seen how, once established through a significant hardware and software base, technical standards in moving image technology become very difficult to supersede, even after advances in the state of the art have rendered them obsolete. This applies equally to television. In the UK, 405-line transmissions were not finally switched off until 1985 (although by that time, all but a tiny fraction of consumers had made the change to PAL receivers), while the US still uses the 'NTSC color' standard pretty much as RCA defined it in the 1950s. Digital terrestrial broadcasting has already began in some developed countries (and will be discussed in further depth in chapter eight), but even then the encoded picture is of identical characteristics and similar definition to its analogue predecessors (and therefore can be displayed on existing receivers in conjunction with an external digital to analogue conversion device). In most cases, the increased 'bandwidth' offered by digital television has been used to increase the number of channels being broadcast, not to improve the picture definition. Attempts to introduce 'High Definition [analogue] Television' (HDTV), principally by Sony in the late 1980s, ran aground. This was mainly due to consumer resistance at being forced to purchase more expensive hardware, which led to a chicken-and-egg situation in which manufacturers were consequently unwilling to invest in production lines and broadcasters were unwilling to incur the increased production costs.

Television and film

As with radio in the 1920s, television initially traded on being a live medium – literally, radio with pictures. It did not take broadcasters long, however, to realise that they needed a method of recording content 'offline' for subsequent transmission, and of recording live broadcasts for future use. For the first two decades of regular, scheduled transmissions, there was no effective means in existence for achieving this electronically, because the volume of 'bandwidth' (analogue signal information) consumed by a broadcast television signal was very much greater than with audio. Film, therefore, became the *de facto* recording medium, as the technology needed to encode film images electronically and capture television images on film had been developed to the point of reliability almost as quickly as television itself.

The 'telecine' device initially used an electronic camera tube in conjunction with a modified film projector. The scanning phases of the TV camera were synchronised to the shutter movement of the projector and its light output regulated in order to enable an even exposure. This process was formally known as 'photoconductive' telecine. It was in routine use both by the BBC and American broadcasters since the first scheduled transmissions: as the authors of the standard textbook on telecine technology note, 'the film and video industries have been intertwined almost since the beginning of television broadcasting'.[25] The earliest company which was formed specifically to market telecine equipment was Cinema Television Ltd., known as Cintel, founded in 1938 to service the BBC. The next significant development was the advent of 'flying spot' telecines, in which a small beam of light from a cathode ray tube progressively 'scanned' (i.e. projected through the film) the film surface, line by line, with the resulting flow of light being detected by a photoelectric device on the other side. This had one crucial advantage over the photoconductive method: it enabled the film to be transported in continuous motion rather than intermittently, as in a projector. Illumination was more even, and film damage or interrupted transmission due to a breakdown in projection was significantly reduced. From the late 1980s onwards, CCDs have largely replaced CRTs as the principal imaging device in use in telecine technology.

Two issues which have proven a continuous problem in the use of film as a storage medium for television have been that of speed compatibility and aspect ratios. By the end of the 1920s, 24fps was established as the standard shooting and projection speed for 35mm film worldwide. Film projection cannot be captured by any television imaging device at this speed, as the scanning rates for both 405-line/PAL (50hz) and NTSC (60hz) are determined by the frequency of the mains power supply in the countries in which they are used. In the case of 50hz systems the solution was very simple – the speed of the projector was increased to 25fps and each frame scanned twice per cycle. The slight speed increase is imperceptible and inaudible for most viewers, and in any case footage which is shot specifically for PAL television broadcast is usually filmed at 25fps in the first place. The only practical restriction in this method was that until analogue frame store and subsequently digital video imaging techniques became available in the 1980s and 1990s, pre-1930 footage (and amateur film taken since) which was shot at a significantly lower speed than 24fps had to be televised at 25fps. This technical issue is largely responsible for the popular image of silent films today as being exaggeratedly fast, as, for almost four decades, this was the only way they could be shown on television. With NTSC the problem was more complicated, as neither 24 nor any number remotely close to it can be divided by 60. The solution was the 3:2 (or sometimes 2:3) pulldown, whereby frames of film shot at 24fps were scanned alternately for three fields (1½ 24th-seconds of interlaced video) followed by two (1 24th-second). As with the speed increase necessitated by PAL telecine scanning, what should logically be a slightly uneven movement with the NTSC 3:2 method is to all intents and purposes invisible.

Aspect ratios were not a problem at first, because the Academy ratio was universal to all film and television. The advent of widescreen cinema introduced the problem of how material originated in this format could be made to fit the ratio of a televison tube, which could not be changed in the same way that cinemas and projectors could be modified. Two methods were developed. The first, known as 'letterboxing', simply displays the widescreen image in a section of the television screen, with the unused area appearing black. As this results in a substantially smaller image than would be case with Academy, a technique was evolved known as 'panning and scanning'. This involves selecting an Academy shaped section within a widescreen frame, using it to fill the television screen and cropping the rest. Paradoxically, although this enables the broadcast of material originated on widescreen film to take place without losing any of the television screen, it does mean losing a substantial amount of the original film image (almost half in the case of CinemaScope). For this reason, technically-aware consumers and enthusiasts generally prefer letterboxing; but when films are broadcast this way, TV stations inevitably receive a barrage of complaints, including those from people who needlessly called out repair men in the belief that the black bars indicated a technical fault with their receivers. While panning and scanning is disliked by purists and film industry professionals because of the image loss, broadcasters preferred it because 'the supposition was that viewers neither cared nor noticed'.[26] Given the relatively poor resolution of early generations of CRT displays and, in the United States, an FCC requirement that broadcasters fill the screen, panning ad scanning was, and by many broadcasters still is, considered the preferable option.

'Telerecording', or 'Kinescope' as the technique is known in US English, is the reverse of telecine: namely the capturing of electronically-originated images on film. The equipment consisted of a high-definition CRT display and a 16mm or 35mm film camera. The camera's shutter was synchronised to the scanning beam of the CRT, and during the three decades (mid-1940s to mid-1970s approximately) when the technique was in mainstream use, a number of film emulsions were developed specifically for the purpose.[27] As a BBC engineer noted, telerecording quality 'greatly improved' between 1946 and 1971. In later models, the two sets of interlaced lines were exposed onto the film separately in order to improve the definition of each individual line.[28] Before the use of videotape in broadcast studios became widespread from the mid-1960s onwards, telerecording was the only method available for making a permanent recording of moving image content originated by a television camera. Given the high cost of film stock, its use was restricted: even major national broadcasters would only telerecord broadcasts where repeat transmissions had been scheduled, the content was deemed to have significant commercial value (e.g. for syndication to broadcasters in other territories) or legal reasons dictated that a recording had to be kept (e.g. copyright registration requirements).

Video – initial development of broadcast systems

Both telecine and telerecording technology were initially developed because no reliable means existed of recording the broadcast signal electronically, hence the need to use film as a production and storage medium for television content. Although, for a number of reasons, attempts had been ongoing to develop the means of recording television electronically for almost as long as the medium had existed, they proved to be technically unsuccessful until the mid-1950s, and economically unviable (for use on a mass-scale) until the early 1970s.

'Video' – from the latin verb *videre*, meaning to see, is a term which has been variously used and abused over the second half of the twentieth century. For the purpose of this book, it will refer specifically to the technologies involved in recording television images on magnetic tape, both in analogue and digital form. The earliest attempt to carry out such a recording did not, in fact, use tape at all, but instead encoded the image as grooves on a record. This was the 'Phonovision' system, invented by John Logie Baird as a means of recording his 30-line mechanical images. Given the comparatively low bandwidth of the signal, Baird found that it was possible to modulate it as audible tones which were then recorded using a slightly modified acetate disc cutter.[29] Phonovision could not be considered successful in anything other than an experimental sense, though, as Baird never managed to reproduce the modulated signal as a moving image. However, experiments to play back the disc using computer software to analyse the signal on surviving discs were successfully carried out in the late 1990s, thereby proving that Baird was able to make an accurate recording.[30]

The capture of magnetic tape technology from the Nazis in the aftermath of World War Two stimulated research on video recording as with audio. Possibly the first working prototype came from an unusual source – the American singer and entertainer Bing Crosby. In the late 1940s he became one of the first major radio performers to pre-record his broadcasts on tape, primarily in order to reduce his workload and prevent the need to repeat live performances for broadcast in different time zones. When in 1948 he started to perform regularly on commercial television, Crosby asked the engineer and former Army Signal Corps officer John J. 'Jack' Mullin to produce a magnetic tape recorder for video signals. The result was the Mullin-Crosby video tape recorder (VTR), which recorded a black-and-white picture on ½-inch tape and was successfully demonstrated on 11 November 1951. The main technical problem in using magnetic tape to record video was to accommodate the vastly increased bandwidth. It was found that this could be accomplished in one of two ways: by increasing the tape speed to the point at which the entire modulation could be recorded as a straightforward linear signal, or by recording a number of parallel, diagonally-positioned modulated records, by means of rotating heads. The Crosby-Mullin VTR used the former method, as did the BBC's Visual Electronic Recording Apparatus (VERA) machine, first demonstrated in 1952 and used on a limited scale in 1958.[31] This method was soon found to be problematic, because the very high tape speeds it required (240 inches per second for Crosby-Mullin, 200 for VERA) severely limited the running time available from each reel, increased the cost of tape and resulted in high levels of mechanical wear.

It was the latter form of recording that would eventually enable videotape to become a standard production technology in television studios worldwide and subsequently to be adapted for consumer use. The American Ampex corporation[32] had been founded in 1944 specifically to commercially exploit magnetic audiotape technology, but by the 1950s was working on ways to reduce the tape speed and increase the image quality from that which could be obtained from the first generation of linear VTRs. The result was the model VRX-1000, first demonstrated at an American trade show in March 1956. This device used four rotating heads (hence the term 'quadruplex' used to describe this type of head assembly) mounted in a cylindrical assembly, across which 2-inch tape passed at a speed of 30 inches per second. The quality of television images recorded and broadcast using this machine were effectively indistinguishable from those broadcast live when viewed in the home, and over the following years it began to be used extensively in studios, primarily as a 'time shifting' device. Several drawbacks inhibited its use as a primary production medium, a role which continued to be filled mainly by 16mm film. Key among them were the cost of tape, the fact that a recording could only be played back by the same machine which had recorded it and difficulty in editing, which could only be done by physically cutting and splicing tape at the same angle to the modulated signal (which in turn required the use of chemicals to make the modulated magnetic oxide visible). Furthermore an edited tape could not be reused, so to start with, VTRs were only used to record complete broadcasts for subsequent transmission or repeat transmission. Writing as late as 1972, a BBC engineer stressed the restrictive nature of early broadcast videotape technology:

> Video-tape operation is not cheap. The machines themselves are costly, since they are built to a high standard of precision, and include complex servo mechanisms for maintaining the movement of the tape and of the head-wheel to the accuracy required for replaying in synchronism with other sources. Additional units are needed to maintain the even higher accuracy required for colour. The recording heads are expensive and have to be replaced after a few hundred hours. Running costs include the cost of tapes (which can, however, be reused many times), tape storage, maintenance and editing. At first editing was done by cutting and joining the tape, but electronic editing was provided by the second generation

> of video-tape recorders. This permitted the timing of cuts to be determined during the recording session, so that the producer was closely concerned with the editing.[33]

The key advantages of videotape were that the media was cheaper than film, recordings could be played back instantly (no need to send exposed film elements to a lab for processing) and the tapes could be re-recorded many times over. Sadly, this feature of the new technology led to the inadvertent loss of what many archivists would now consider to be culturally important programming as 'missing, believed wiped'.

Rotating head technology took another step forward with the introduction of the 'helical scan' method in 1961, which was a significant refinement to the quadruplex drum and reduced the number of recording heads needed to one. From then on the technology was gradually refined over the following three decades to add the functionality that users of broadcast video systems were taking for granted by the close of the century. NTSC colour recording followed with the launch of the Ampex VR2000, available from 1965,[34] with PAL equivalents coming on the market shortly after this standard's mass rollout in European markets. By 1972, the BBC had 42 machines.[35] Electronic video editing, which is achieved by selectively copying content from one tape to another, was also introduced by various manufacturers during the 1960s. This eliminated the need to physically cut and splice the tapes, and increased the volume of tape stock which a studio could recycle. The introduction of the 'UMatic' format by Sony in 1970 marked a further milestone, in that it was the first large-scale format to use cassettes as distinct from separate spools in the tape transport mechanism. From the 1970s onwards videotape technology became progressively more 'user friendly', with even broadcast-standard equipment being operable by non-technical staff, e.g. broadcast news journalists and schedulers. As one technical writer noted in 1981, the era of 'electronic news gathering' (ENG) was here to stay:

> The replacement of 16mm film for news gathering by U-matic cassette recorders had obvious advantages. The machine was portable, it eliminated high film developing costs, it could easily be edited using the new electronic control system and it was able to bring more immediate news coverage to the viewer. Cost savings were estimated at that time to be around 70 to 80 per cent. So began the ENG era.[36]

By the late 1980s, videotape had superseded film as the primary 'offline' medium in the television industry for everything except the broadcast of feature films and some high-budget, technically complex drama and documentary productions in which the 16mm was still used, either due to its superior image quality or technical versatility in other respects (for example, high-speed nature filming).

Consumer videotape systems

From the mid-1970s video cassette recorders (VCRs) designed specifically for use by domestic consumers began to be mass-manufactured and sold. Their ante-cedents go back to UMatic and the late 1960s, when a number of Japanese and American manufacturers sought to grow the market for videotape technology by extending it to homes as well as businesses. But, as with the Trinitron tube turning colour television successfully into a mass-medium, it was Sony which cracked the mass-manufacturing problems and found the application that would make the VCR an attractive proposition. Sony's founder, Akio Morita, came up with a catchprase for it: 'time shift', meaning to record broadcast television programmes off-air for subsequent viewing at a time more convenient to the viewer.[37]

The machine which Morita came up with to sell this concept was markedly different from its studio-based predecessors. Launched in September 1975, it used a format named Betamax. This consisted of a feed and take-up spool mounted inside a plastic cassette which contained a number of safety devices for preventing accidental damage to the tape caused by its owner mishandling it. The recording method was helical scan on ½-inch width tape, using a slightly more compact tracking geometry than broadcast standard formats. The result was a cassette around the size of a small paperback book which, initially, could hold an hour of footage. The VCRs which Sony produced for the format contained two features which were not included in studio versions of the technology: a television tuner, which enabled it to record off-air signals, and a timing device which allowed recordings to be made unattended.

To start with, Betamax was aggressively marketed as a 'time shifting' device. Advertisements were broadcast in which Count Dracula is shown using the machine to record primetime evening programmes ('If you work nights like I do, you miss an awful lot of programmes'), though it was never suggested – to start with, at any rate – that videotape was initially intended to allow consumers to accumulate their own permanent collections of material. This, however, was the concern of many broadcasters and film studios; with the result that Universal (which owned two hit television shows that were broadcast simultaneously on different channels and therefore considered itself a key target for 'time shifters') sued for breach of copyright. The case dragged on until 1981, eventually culminating in the 'Betamax ruling'. In it, the US Supreme Court ruled that the sale of equipment which could potentially be used to facilitate copyright theft cannot be restricted unless it was designed specifically and exclusively for that purpose. It is this decision which forms the legal basis for domestic VCR use worldwide, in effect ensuring that it is not an offence to record 'off air' material simply for the purpose of a single viewing later, as long as the intention is not to keep the recording permanently. Of course this has proven impossible to enforce, and in 1995 a prominent television industry body published its belief that the average VCR owner possesses a permanent library of between 100 and 200 hours of programming.

But, by the time the Betamax case reached the courts, the format itself was dead in the water. Two contenders emerged during the late 1970s. In one camp, led principally by RCA, were a number of short-lived disc-based systems. Two underwent an attempt at mass-marketing. RCA's 'Capacitance Electronic Disc' (CED), a vastly more sophisticated version of Baird Phonovision, was marketed from March 1981 to June 1986.[38] The modulated video signal was encoded as 'hill and dale' grooves which were scanned by measuring the varying resistance between the edge of the stylus and the groove's surface as the disc rotated. The 'laserdisc', initially developed by Philips and marketed from 1978, encoded the picture signal optically as indentations in a highly reflective surface. These were 'read' by bouncing laser radiation off it and measuring the time taken by a photoelectric device to register the reflection. The format was relaunched, with digital audio, under the trade name 'CD Video', in the early 1990s. It had the advantage over CED that as the reading device never made physical contact with the disc's surface, the signal quality would not erode with repeated playings. Film studios and broadcasters were initially very keen on these disc formats, and for one reason: as with records (and, for the first 15 years or so of their existence, CDs), they were 'read only' formats as far as the consumer was concerned, i.e. they did not allow time-shifting.

But, unlike with the audio technology of Edison's day (when there was no source material to record in the form of broadcast media), the inability to record was seen as a substantial disadvantage by consumers, and no videodisc system achieved significant sales until the launch of DVD in 1997 (see chapter eight). The medium which replaced Betamax as the principal domestic videotape format was the 'Video Home System' (VHS),[39] developed by another Japanese manufacturer, JVC (Japanese Victor Corporation), and adopted by all the Japanese majors except Toshiba and Sanyo. Crucial to VHS's success were two factors. Firstly, the format sacrificed picture quality on the altar of capacity, with running times of up to three hours for each cassette (thereby making it possible to view an entire feature film uninterrupted). Secondly, JVC persuaded a number of Hollywood distributors in which copies of feature films would be produced on VHS cassettes, either for rental or outright sale to consumers. As this British newspaper article from 1969, headlined 'A Sinatra movie on your shopping list', makes clear, it was an idea which had caught consumers' imagination since the first experimental domestic video formats materialised:

> On Friday night my wife goes to the supermarket with her weekend shopping list. She checks off what she needs: cereal ... eggs ... butter ... sausages. Then she says to me: 'What film do you want to show at home over the weekend: a Frank Sinatra or an Olivier?' And she'll buy the film, just the size of a jumbo postcard, along with the cigarettes at the check-out desk of the supermarket. At home I'll slip the film cartridge into the back of my TV set and we will see the film at whatever time we want.

Ease of use was considered another consumer 'plus':

> The real point about the cassette revolution is that for the first time it brings 'home movies' into real life – with none of that fiddling around with reels or lamps or screens.[40]

The VHS system could also be used for time-shifting, although this capability was not explicitly advertised to anything like the same extent as with Betamax. There was another, more subtle attribute of the format which, although not explicitly endorsed by major content providers, probably went some way to allaying their fears over unauthorised recording and copying: it was far less resilient and the picture quality was a lot lower than that of Betamax. VHS encoded a 'colour under' signal – that is, it combined the separate luminance (Y) and chrominance (C) information as broadcast into a single, phase-shifted pattern of modulation on the tape. Although this is capable of producing an acceptable picture on an average quality television set when a first generation recording is decoded and played back, any attempt to copy VHS tapes will result in a clearly visible loss of quality in each subsequent generation.

Thanks to Sony's high-profile advertising, time shifting was here to stay, though JVC had successfully managed to defuse the tensions which had caused the Sony lawsuit. In this way JVC had hedged its bets between the two potential applications for domestic video technology, and had secured the co-operation of the major content producers to use the medium as an extra revenue stream. By the early 1980s VHS had effectively pushed Betamax out of the domestic market, and by 1995, the rental and sale of prerecorded VHS tapes accounted for over half of Hollywood's revenue.[41] Ironically, Betamax turned out to have been of such high quality relative to the needs for consumer use that the cassette design and tape transport mechanism were recycled, in only slightly modified form, in three generations of broadcast VT format (Betacam, Beta SP and Digibeta), the latter two of which remain in mainstream use at the time of writing.

For almost two decades, VHS has dominated the market for consumer VCRs. For prerecorded content its market share is gradually being subsumed by the Digital Versatile Disc (DVD), which, being a digital medium, will be discussed in chapter eight. The speed at which domestic VCR sales accelerated is in many ways remarkable (from 1.8 million to 86 million units in use in the USA between 1980 and 1995, at which point 90 per cent of television owners also owned a VCR; in Britain ownership went from less than 15,000 units in 1976 to 3 million by 1983[42]), so much so that Brian Winston argues that they 'penetrated society more quickly than any other [mass communications] technology'.[44] Economies of scale have made the VHS format gradually cheaper over those decades, so much so that, in real terms, a VCR in 2000 cost 8 per cent of the retail cost of the first models to be sold in 1976. A startling illustration of the extent to which the accessibility of VHS has enabled its expansion can be found in the fact that it is apparently used as a primary production medium on quite a significant scale in some third world and developing countries. One example can be found in the so-called 'video films' produced in Nigeria. In the absence of a reliable broadcasting network and given that most Nigerians cannot afford to own television receivers in their homes, they consist mainly of drama series, not unlike the Western soap opera. They are produced, duplicated and distri-buted entirely on VHS:

> Aimed primarily at the urban youth market, these films are made for profit by small-scale video production companies and individuals and currently comprise an essential part of modern Nigerian culture which dwarfs any 35mm *film* culture as such. They are produced on VHS on very low budgets, and distributed through a thriving network of video rental shops and communal viewing centres in Nigerian cities.[45]

Despite what must be the very poor quality of an edited VHS master being duplicated hundreds of times over, this format has presumably found to be ideal for the purpose due to its low cost and reliability.

Piracy

The issue of piracy is one which has affected all commercially produced media content using every form of moving image technology yet invented. It is covered in this chapter because consumer videotape formats have been the principal focus of this activity above all others; but that is not to say that piracy has not affected either film distribution before it or the digital technologies which will supersede it (it does, with a vengeance – of which more in chapter eight).

Chambers' Dictionary defines the verb 'to pirate' in this context as 'to publish or reproduce without permission of the copyright owner'. It is not surprising that this should be a major issue where moving image content is concerned. The two crucial elements are reproduction and copyright. As I hope has become apparent thus far, the economics of moving image technology as a mass medium depend almost entirely on a single characteristic of that technology: the ability to produce multiple copies from a single original at little extra cost and with little significant loss of perceived image quality. This, of course, is a double-edged sword. One edge enables the individuals or companies who invested in an original product to distribute or broadcast it to a mass audience cheaply and easily. The other enables unauthorised third parties to do likewise.

This is really how the idea of copyright came to exist. Simply put, it accords legal protection to the 'creators' of media artefacts against third parties who seek to exploit them, by unauthorised copying and/or direct exploitation, for financial gain. By 'media artefact' I mean any means of mass-reproducing intellectual property, from Caxton's printing press onwards. The definition of 'creator' varies slightly between different legal systems. For example, in British and US copyright law, the copyright owner of moving image footage is generally understood to be the individual or organisation which put up the money that enabled a film or television programme to be made in the first place. The French, however, work mainly on the principle of the 'droit d'auteur' (right of the author), emphasising the right of a person who came up with the conceptual idea to control its commercial exploitation, even if he or she was only able to bring it to the cinemas or television screens with other people's money.[44]

However, just because laws exist which say that one must not do something, that does not mean to say that people will not try. And this is where the specific attributes of moving image technology – that double-edged sword – come into play. Piracy is nothing new. Its earliest incarnation took the form of duplicating release prints, either by using miniature 35mm cameras smuggled into cinemas, or by the owners of the original elements duplicating them by printing (something which was easier to do before the establishment of the production/distribution/exhibition system, when prints were sold outright to exhibitors). As a countermeasure, some production companies began to incorporate a logo or symbol into a costume or studio set (in much the same way that many television programmes are now transmitted with a 'spoiler', consisting of the broadcaster's logo visible in one corner of the screen) that would be visible on any unauthorised copies, of which surviving examples date back to the late 1890s.[46]

When the production/distribution/exhibition system based on the rental of film prints for projection in permanent buildings replaced the outright sale of prints to itinerant exhibitors, two new forms of copyright fraud emerged to exploit the new economic structure. The first was the outright theft of prints, most of which were exported and shown in foreign territories with less stringent copyright legislation (no cinema in the same territory as a film's production would be willing to book it from an unauthorised distributor, for obvious reasons). In America this problem grew to the point at which the Motion Picture Producers' and Distributors' Association of America (MPPDA) formed a film theft committee in 1922. Among its recommendations were that distributors should take greater care in ensuring that prints were returned at the end of their run and institute tracking procedures for ensuring that unwanted prints were destroyed. The other problem was of cinemas falsifying their box office returns. As the amount payable by cinemas to distributors was calculated as a percentage of the total ticket sales, rather than a fixed sum, an exhibitor could cheat a distributor out of revenue by claiming to have sold fewer tickets than they actually did.[47]

While these forms of fraud remained a significant problem throughout the period before consumer video technology became widespread, its extent was limited because 35mm film prints were large, heavy, expensive to produce, required technical expertise to handle and could only be projected in cinemas. No technology existed which enabled criminals to produce copies for direct sale to individual consumers. Any systematic box-office fraud could be detected relatively easily. In fact, a significant proportion of the widespread illicit copying and exhibition which took place in the developed world during this period was done in order to circumvent censorship rather than to make money.

The advent of domestic videotape systems let a technological genie out of its bottle as far as piracy was concerned, for the simple reason that video enabled film and television footage to be shown in a wide range of venues, using equipment that was relatively cheap and did not need any technical skills to operate. Furthermore, even the legitimate use of videotape had a negative impact on the film industry: as Kerry Segrave put it, 'Hollywood hit its roughest period as the VCR and the videocassette arrived and became ubiquitous'.[48] The VHS system had been specifically designed as a way for copies of feature films to be rented or sold directly to consumers for domestic viewing (the 'Sinatra on your shopping list'). Although these copies could not legally be used for any other purpose (i.e. in exchange for the rental or sale price, the copyright owner grants permission for private viewing in the home, only – hence the warning notice which appears at the start of commercially produced video recordings stating that they must not be shown in, for example, oil rigs, prisons and schools), there was no technical barrier preventing their use in other settings, as there was with film. This rapidly became a serious worry for the cinema industry, which feared that illegal forms of 'non-theatrical exhibition' using videotapes (i.e. screenings to an assembled group, but not in a cinema) posed a threat to their revenue. In 1981, the representative body for UK cinemas described the problem as being 'of very special importance', identifying two specific issues:

> The first relates to pirated material, that is, material recorded without the consent of the copyright holder and used without his permission. Into this category fall a large and increasing number of feature films which, in certain cases, have been shown in public houses at the same time or even prior to their local theatrical release … Secondly, but much more difficult to control, are the feature films made available for domestic use but which are improperly shown other than within the home.[49]

As with the 'taping' of commercially-sold music recordings and radio broadcasts which had been going on for the previous decade, the introduction of consumer videotape technology resulted in an ongoing level of piracy which the industry has been unable to eliminate. In the two decades during which the VHS VCR has existed as a mainstream consumer technology, the film and television industries have, in general terms, adopted a two-pronged approach. To combat professional piracy (i.e. criminals illegally duplicating recordings on a large scale and then selling them directly to consumers for profit) the response has been a combination of legal action (i.e. suing and prosecuting the people who produce the copies) and public relations measures. The latter have often taken the form of advertising campaigns trying to dissuade consumers from the idea that piracy is a victimless crime and/or suggesting that it is linked to more serious organised crime, including international terrorism. Against 'low level' piracy (i.e. individual consumers making a small number of copies which are usually given away to friends and relatives) the industry's defence has been technical rather than legal. As has been noted above, one such measure is actually incorporated into the design of the VHS system itself, in that any copy will inevitably be of much poorer quality than the original. Another widespread copy protection device used with commercial VHS recordings is called Macrovision, named after a California-based company which was formed in 1983. It works by manipulating the 'automatic gain control' (AGC), a circuit found in domestic VCRs which detects the optimum modulation strength for recording an incoming signal onto blank tape. The Macrovision signal on a source tape fools the AGC into thinking that the video and audio modulation it receives is weaker or stronger than it actually is, so that the resulting (copied) recording will be distorted and unwatchable. This does not affect playback of the original: it appears normally, because the television set to which the source VCR is connected does not have an AGC. Someone with a basic knowledge of electronics could easily defeat Macrovision by making minor modifications to a VCR being used to record the copy (instructions for which are readily available on the Internet). But the industry works on the assumption that the majority of consumers do not have such knowledge and that therefore, systems such as Macrovision will prevent most domestic piracy from taking place.

Conclusion

Television and video fundamentally changed the culture and economics of moving image technology, starting at the point at which film was the only viable medium. The initial commercial impetus was simple – radio with pictures. From the outset, the technologies used in producing, receiving and recording televisual images had to grapple with the conflicting demands of live and recorded content, both in terms of their capabilities and limitations and also of cultural expectations resulting from their use. The two decades during which broadcast television was a commercial reality but videotape recording was not established a complex relationship between film and broadcasting, which, although scaled down by the arrival of VT, remained in existence until the end of the century.

As with film, there was also the standards issue. Brian Winston argues that the videotape revolution was characterised by the 'proliferation of mutually incompatible boxes'.[50] In fact, commercial competition between incompatible technical standards shaped the evolution of practically all the technologies related to television right from the outset. Whereas the 35mm standard in film was determined in the absence of any serious alternatives and established itself as a bedrock of the industry during the first decade of its life, the birth of television was characterised by the BBC-mediated battle between Baird and EMI, promoting electromechanical and cathode ray tube image scanning respectively. As the subsequent fight between David Sarnoff and CBS over colour demonstrated, the establishment and operation of technical standards also acquired a political dimension where television was concerned. In this respect it followed the lead of radio, in which the political regulation of 'bandwidth' was a key influence in the development of the medium, as distinct from that of film, in which the only significant legislation affecting the technology was that which addressed the health and safety issues associated with nitrate.

Videotape technology was initially used only by programme makers and broadcasters, who had sought the versatility and economy of a recording medium which, unlike film, could be replayed instantly, rerecorded without practical limit and make permanent recordings of live transmissions without the expense of telerecording onto film. As VT became more reliable and easier to edit, a process began which saw it replace film for all except highly-budgeted productions which required its higher image quality as an origination medium. It was only a matter of time before the technology would be offered for sale directly to consumers.

The opening of the home video age was marked by yet another battle over standardisation, but on this occasion the impetus was economic rather than political. Sony's Betamax system was technically superior to JVC's VHS and marketed for a purpose which even required the invention of a new piece of technical vocabulary – 'time shifting'. VHS, on the other hand, was developed in conjunction with the media industry and intended primarily for consumers to rent or purchase prerecorded video copies of Hollywood films. Once that genie was out of the bottle, what had been the biggest economic strength of all moving image technologies to date – the ability to produce multiple copies quickly and cheaply from a single source – quickly became its worst enemy, as the illegal exploitation of 'official' recordings and the unauthorised production of new ones suddenly became very quick and easy.

We have seen in this chapter how the political and economic factors influencing technical standardisation shaped the emergence of the related technologies of television and video recording. The next chapter examines the issues these and other factors raise in an activity which no one who designed the mass-produced moving image technologies ever really considered.

chapter seven | archival preservation and restoration

> The paradox of Theseus's ship, which poses the problem of conservation and restoration, has always fascinated me. The ship in which Theseus, slayer of the Minotaur, returned home from Crete, was kept like a sacred relic by the Athenians for centuries. Over the years the old pieces of wood had to be changed to save the vessel from the ravages of time, to such an extent that after so many replacements and substitutions, it could quite legitimately be claimed that the original ship no longer existed.[1]

> Surely it is the images that are the important thing, not the base that they are on ... Give me acetate or polyester any day: I'll take my chances with vinegar syndrome – at least it is easy to identify – and I'll sleep easy knowing that my vault is not going to go 'wallop' in the middle of the night![2]

Throughout this book so far I have used the adjective 'permanent' in relation to a number of recording media. All these uses make the implicit assumption that as soon as a reel of film emerges from the fixing bath, or a Vitaphone master was removed from the electroplating machine, or a videotape is wound onto its take-up spool, a moving image or sound recording has been created which is indelible and will last forever, unless someone makes the conscious decision to dispose of it or (in the case of magnetic tape) overwrite the media with a new recording. It has not and it will not. Although political and regulatory influences have played a part (e.g. in determining technical standards), the invention and development of all the technologies used to record moving images have been commercial ventures. The investors behind them believed – correctly, in the short term – that the money which was to be made out of the media content recorded using these technologies would be generated in a relatively short time following production. So by 'permanent', the individuals and organisations which used film and videotape usually thought in terms of weeks; months at most. Simply put, they neither knew nor cared what would happen to those media after that time.

The idea of systematically preserving moving image content as a public record did not come from within the media industries, but instead from museum curators, librarians and academics who believed that its cultural value was significant, even if its long-term commercial value was negligible. The first and longest established moving image archives, therefore, were and are public sector organisations. The idea of film preservation goes back almost as far as film itself; specifically to a pamphlet published in 1898 in which a Polish photographer described film as an 'honest and infallible eye-witness' of historically important events, and worth preserving on those grounds.[3] The first film archives (in any meaningful sense of the word) were established in Europe and the United States during the 1930s, under the umbrella of organisations such as the Museum of Modern Art (USA), the British Film Institute (BFI) (a government agency established in 1933 to promote educational uses of the moving image) and the Nazi propaganda ministry. Commercial film production interests (e.g. Hollywood studios) in most cases did not attempt to preserve their output beyond the point of initial distribution, and in some cases did not start routinely doing so until well into the 1950s. There were a number of reasons for this: vaults and storage facilities were expensive to build and maintain; even more so, given the health and safety precautions necessary when storing significant quantities of nitrate in one location. Furthermore, the principal commercial markets for archive footage – television and videotape sales – did not exist until the 1950s and 1980s respectively. Before that point, many studios actually destroyed prints (the silver in black-and-white film emulsion is recyclable and has a cash value) after distribution as a safeguard against piracy (see chapter six). It has been said that between 20 and 50 per cent of all commercially produced films and television output made in the developed world have survived, depending on whose figures you believe. Whatever the actual figure is, it is indisputable that a significant proportion has been lost.[4] This is not just because producers chose not to keep it.

Chemical decomposition of film

We recall from chapter one that all pre-polyester film bases are manufactured, essentially, by dissolving cellulose (a derivative of wood pulp and/or cotton fibre) in an acid of some description and then processing the resulting brew into a flexible, transparent solid. For the first half of the twentieth century the most common method was to use nitric acid, thereby producing cellulose nitrate film. During the period 1948–92 approximately,[5] this was substituted with acetic acid, the resulting compound being cellulose triacetate. While these organic materials (i.e. ones which are produced by mixing and processing naturally occurring substances) will keep their flexible and transparent properties for as long as is needed in the immediate production, distribution and exhibition process, it has been found that in the long term, they will decompose. Eventually, both nitrate and acetate film will decompose to the point at which the images and sounds on them can no longer be recovered.

The basic chemical process which causes this decomposition is common to both nitrate and acetate, a group of compounds known as cellulose esters. These chemical combinations are extremely susceptible to hydrolysis, a process in which exposure to moisture in the atmosphere can cause the acidic content to leach out of the film base, thus releasing nitric or acetic fumes into the atmosphere. Within a sealed environment such as a film can, these acidic vapours attack what remains of the base, thereby sustaining the process. Nitrate decomposition is probably the most widely acknowledged, understood and addressed problem among the archive community. Archivists tend to prioritise preservation work on nitrate: possibly because, to borrow the words of a former curator of the UK's National Film and Television Archive, 'there is an end point ... the stuff just goes'.[6] The fact that some sort of chemical change takes place in the nitrate film base almost from the moment of manufacture predated the formal practice of film archiving by two to three decades. One industry commentator writing in 1910 described this change as 'a ripening, or seasoning process'[7] and various descriptions of the phenomenon can be found in other written sources from the early twentieth century. Very few, however, seemed concerned with ensuring the long-term preservation of moving image content: the vast majority of technicians and scientists who investigated the process of chemical change in nitrate film were mainly concerned with understanding and minimising the fire risk. Because nitrate was only manufactured on an industrial scale from the 1890s onwards, it is likely that hardly any of the film base in circulation would have started to decompose to the point of visibility for several decades afterwards. It was not until 1941, when the British archivist Harold Brown discovered a reel of nitrate which, in the space of six months, had started to 'go sticky', that a process of research began which eventually uncovered the full extent of the problem; i.e. that without preventative measures, the content of nitrate film would eventually be lost.[8]

By 1965, the decomposition process in nitrate was more or less fully understood. A report published by the International Federation of Film Archives (known by the initials of its French name, FIAF) in that year described it as 'the following sequence of physical changes: (i) the silver image undergoes a brownish discolouration and fading; (ii) the emulsion becomes sticky; (iii) there is a partial softening of the reel of film (formation of "honey"), the appearance of blisters and a pungent smell; (iv) the entire film congeals into one solid mass; (v) the film base disintegrates into a brownish powder, giving off an acrid odour. In this last stage the film has a very low ignition temperature and is highly explosive.'[9]

Tragically, the existence of the fifth stage had encouraged the destruction of nitrate both before and after systematic film archiving had come into existence. The early generation of technicians who had sought to understand the decomposition process quickly discovered that as nitrate film ages, it becomes more volatile, and that warm storage conditions can exacerbate the risk. In the unusually hot summer of 1949, for example, an abnormally high number of nitrate fires were reported to have taken place, in response to which a series of experiments was carried out to try and establish if nitrate was capable of spontaneous ignition. The abstract of the paper which reported the results makes a brutally stark recommendation:

> Cellulose nitrate film in the advanced stages of decomposition is liable to ignite spontaneously. The danger of such ignition is reduced by inspecting stored film stocks and removing and destroying all decomposing film.[10]

Of course it was not just decomposing film that was destroyed: throughout its existence as the principal film base in widespread use, destruction was the safety measure of choice for most stock which had become surplus to requirement.

When acetate superseded nitrate in the 1948–50 conversion, archivists thought they had the perfect solution to the decomposition problem. This took the form of a method which is known by archivists now as 'copy to preserve', i.e. photographically duplicating nitrate originals onto acetate stock for what they thought would be permanent preservation. To make things even worse, destruction of the nitrate originals after duplication was a routine practice, even among the more enlightened archives, for all the reasons given above (i.e. not wanting their vaults to go 'wallop' in the middle of the night). Indeed, there have been several instances of archive vaults literally going wallop: possibly the two best-known were at the Cinémathèque Française on 10 July 1959, when (again, on an abnormally hot day) it is believed that hundreds of feature films were irrevocably lost in a fire; and on 4 July 1993, when the Hendersons laboratory in South London, which specialised in archival duplication work, sufferered a similar fire which practically destroyed the building along with many irreplaceable elements. Both fires are believed to have been caused by the spontaneous ignition of nitrate.

When it was initially developed, it is not difficult to understand why archivists thought that cellulose triacetate was the answer to their prayers. At the time, nitric acid was thought to be the culprit where decomposition was concerned: so much so that the 1965 FIAF report stated emphatically that triacetate 'does not generate injurious gases and does not produce any symptoms of disintegration'.[11] Archives quickly adopted a strategy summed up by the slogan 'nitrate won't wait', of duplicating as much footage as possible as quickly as possible. An article written to launch a project intended to 'dupe' the National Film and Television Archive's remaining stocks of nitrate as late as 1987 declared that 'all nitrate material in Britain, whether in public or private hands, and irrespective of who owns the copyright, or what its historical "value" is thought to be at the present time, should be transferred to acetate stock'.[12]

In fact, triacetate was susceptible to decomposition; and as with nitrate, the scale of the problem took decades to emerge. Although triacetate is a compound based on acetic rather than nitric acid, the underlying chemical processes which take place during its manufacture are very similar.[13] Both nitrate and acetate are 'pendant chain' cellulose ester polymers, and as such are acutely sensitive to hydrolysis (reaction to water). In nitrate the result of this reaction is the decomposition process described above; in acetate it is slightly different, but just as destructive in the long term. By 1989, an updated version of the FIAF preservation manual acknowledged the existence of preservation issues associated with triacetate, noting that 'film manufacturers and research institutions have published further information about this grave problem, which they are currently investigating in greater depth'.[14]

The formal term for acetate decomposition is 'deacetylation' (loss of acetic acid from the film base). Among archivists it is more often referred to as 'vinegar syndrome', named after the characteristic smell of a deacetylating element. The initial symptom is usually shrinkage, which causes difficulties in running an affected element through any film transport mechanism (such as a printer, projector or telecine) because the perforations become smaller and the space between them is reduced. The film base will then become warped and distorted, making it difficult to produce an image with uniform focus. As the process continues the base will become brittle, and the extent of the physical deformity will eventually prevent access to the images and sounds recorded on it. In an enclosed container, such as a film can, the increasing accumulation of acidic fumes will cause the reaction to become 'autocatalytic', i.e. self-sustaining to the point at which the rate of decomposition accelerates.

It has been understood for decades that as a rough rule of thumb, storage in cool and dry conditions can inhibit the decomposition process of both nitrate and acetate. Until the early 1990s, however, it was not believed that atmospherically controlled storage could extend the service life of nitrate or acetate to the point at which this was a viable preservation technique to replace copying it as soon as possible. One significant milestone came in 1985, when the Image Permanence Institute (IPI) was established in New York. This was a research organisation, bankrolled by film and equipment manufacturers, set up in order to identify solutions to a number of problems related to the long-term preservation of film, photographic paper and electronic images. Within a decade, their scientists had established that the 'nitrate won't wait' approach of prioritising copy to preserve above all other methods might not be the most effective solution.

Preservation methods: 'copy to preserve'

Copying to preserve film is essentially the same thing as duplicating it during the production process, as described in chapter one. Ideally this will be done using a continuous contact printer, because placing the source and destination stocks in direct physical contact with each other minimises generational fading, while the full width aperture in a continuous printer will also duplicate information outside the photographic frame itself (such as camera apertures and markings which archivists can use to identify the date and location that the original stock was manufactured) which can be useful to preserve. But this method is not always possible if the original element has suffered decomposition or other physical damage, and for this reason a range of archival printing techniques have been developed for obtaining the best duplicate material possible from poor quality elements. The problems which may need to be addressed as part of the duplication process are physical defects affecting image quality (such as scratches or mould growth on the base or emulsion surface) and physical defects affecting film transport (shrinkage, brittleness or perforation damage).

A number of techniques have been evolved in order to address these problems during the duplication process. Step contact printers have proven effective in dealing with shrunk or deformed base, though sometimes at the expense of picture stability in the duplicate, because of the fractional difference in the size and position of each frame relative to each other. This in turn can be corrected to a certain extent using a step optical printer, but at the expense of contrast and definition loss introduced by the lens. 'Wet gate' printing can reduce the visibility of scratches on the original. In this method the original element is immersed in a liquid with a similar refractive index to the film base itself. This liquid has the effect of 'filling in' the indentations in the film which are visible on the screen as scratches, because as light passes through it onto the destination stock, the liquid refracts it to an almost identical extent as the film base itself.

It is now understood that destruction of original elements after duplication is potentially a very bad idea. There are two key reasons for this. Firstly, we now know that close control of environmental storage conditions can extend the service life of a nitrate or acetate element for much longer than had originally been thought (of which more below); and secondly, the duplication technology at archivists' disposal is evolving and improving all the time. For example, in 1989 FIAF advised that 'serious problems' would be encountered if any attempt were made to duplicate significantly shrunken stock using the continuous contact method.[15] In 2004, however, a continuous contact printer designed specifically for archival duplication went on the market, which is claimed by its manufacturer to be able to handle elements with up to 2 per cent shrinkage, without any visible effect on picture stability in the duplicate.[16] So, for example, a nitrate element with slight shrinkage that had been duplicated in the 1970s could now (assuming that it had not suffered any further decomposition in the meantime) be copied again, only to a much higher image quality. Furthermore, the widespread availability of polyester film base from the early 1990s onwards provides what is almost certainly a much less problematic alternative to triacetate for use in copying nitrate originals. While some doubts have been expressed about the long-term stability of the method used to bind the emulsion layer to the base, it is generally believed that, being an inorganic substance, polyester will prove to be a far more resilient medium on which to create archival preservation masters. This would, of course, be impossible if an original element had been destroyed after the initial duplication in the 1970s. But, nevertheless, destruction following duplication remained standard practice for many commercial footage libraries (who did not want the risk and expense of maintaining large quantities of volatile material) and public sector archives (which could not justify the cost both of preservation copying and storing the originals). Even the advice leaflet on nitrate handling and storage issued by Eastman Kodak advised destruction until it was revised in 2002 to suggest that only elements 'which have reached the third stage of decomposition or have no historical value' should be disposed of.[17]

Preservation methods: passive conservation

The growth of passive conservation as an archival preservation strategy is not so much a case of new technology appearing as the greater understanding of the effectiveness of an existing one. Coupled with a growing realisation that copying methods were improving over time, research carried out by the IPI and a group of scientists at Manchester Metropolitan University in Britain had, by the mid-1990s, established that temperature- and humidity-controlled storage could extend the life of both nitrate and acetate elements for a lot longer than was previously thought possible.[18] Among the key findings of these research programmes was that reducing humidity had a bigger impact than temperature in inhibiting the decomposition process, that nitrate did not necessarily decompose more quickly than acetate, and that by maintaining optimum conditions, the service life of an original element (assuming that it had not previously been stored in inappropriate conditions for any significant length of time, which of course most have) could be extended for hundreds of years rather than tens, as was previously believed.[19] In the light of this work many archives have began to move the emphasis of their preservation work away from 'copy to preserve' to a 'passive conservation' model. In 1994 an IPI scientist argued that 'it is time to broaden the scope of film preservation activities and abandon a narrow focus where copying in anticipation of future decay claims such a large share of funding and staff resources'.[20]

For most archives these resources are finite, hence the use of environmentally-controlled storage of the original to buy time for preservation copying. Unlike with videotape preservation, in which the problem of format obsolescence prevents such an approach, a combination of storage and monitoring can be used to delay, often for centuries, the moment at which an archival master element has to be duplicated in order to prevent loss of content through decomposition. To this end a number of collection management tools have been developed. For detecting deacetylation in triacetate elements the IPI supplies a testing device which is supplied either as a paper strip similar to Litmus paper (the 'acid detection strip' or 'a/d strip') or a semi-transparent plastic stud which is mounted in a hole punched in the side a film can. This will change colour according to the concentration of acetic acid vapour inside the can, thereby indicating the extent of decomposition.

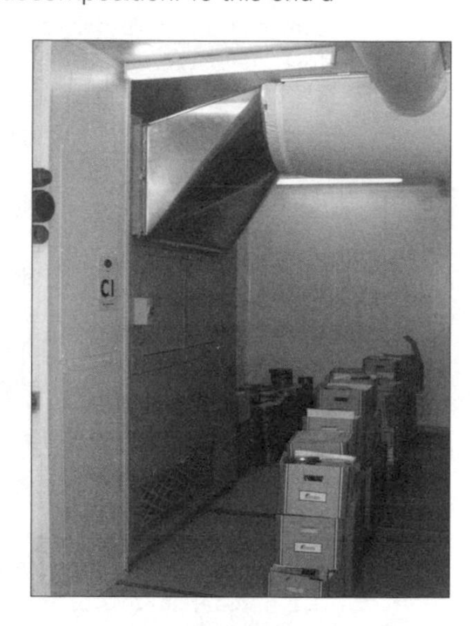

Fig. 7.1 Passive conservation – temperature and humidity control in an archival film store. Air from the vault passes through the filter at the bottom of the air handling unit. Inside the unit it is cleaned of acidic vapours and then cooled and dried before being recirculated through the manifold at the top. The fabric channels mounted on the ceiling will allow air to pass out, but not in.

The most reliable test for nitrate decomposition is slightly more complicated, as it involves using the 'accelerated aging' process. This is a technique developed by the Swedish chemist Svandte Arrhenius in 1887. It is based on the principle that if the speed of chemical change is induced primarily by heat, then by artificially increasing the level of heat exposure it is possible to simulate the effect of lower heat levels over a greater time period on a given substance. In the case of nitrate film, this is done by taking a small sample of the element being tested and placing it in a test tube containing an indicator dye which is formulated to change colour in response to nitrogen dioxide released from the film. The time this takes is proportional to the remaining useful life of the element when stored in appropriate conditions.

Through a combination of atmospheric control in storage areas and tests which can identify and predict the rate of decomposition, it is possible to inhibit and manage the process of film-base deterioration to the point at which most elements can be stored for centuries. This ensures that the only copies which need to be produced are for access purposes and removes the need to proactively make copies which may well be of lower quality than ones made closer to the moment at which they are actually needed for viewing.

Colour dye fading

This is yet another example of a decomposition problem which was only discovered thirty years or so after the affected technologies entered mainstream use. The emergence of dye-coupler colour emulsions in the 1950s (see chapter three) proved to be the catalyst which marked the beginning of the end for black-and-white as a mainstream production medium for feature films, as it was significantly cheaper and more versatile than any of the alternatives which had previously been available. The fact that all three subtractive colour dyes were combined onto a single strip of film made colour cinematography possible using normal, unmodified cameras and without the extra camera rental, film stock, lighting and laboratory costs incurred by the three-strip Technicolor process. Unlike the single-strip reversal stocks which had been used on a limited scale by amateur filmmakers in the 1930s and 1940s, it could be mass-duplicated using the negative/positive system with no more generational loss than was caused by any other form of analogue duplication. But this was only achieved by the use of some fearsomely complicated chemistry, and therein lay the problem.

These stocks contain three chemical layers which are photosensitive to each of the primary colours. The developing process activates the chemical coupler, which converts the substance from a photosensitive emulsion into a visible dye of its corresponding coupler. In the late 1960s it was established that these chemical reactions did not cease altogether after the film was processed. In fact, the dyes which form the green and blue layers on a positive element (cyan and magenta on a negative) fade gradually over an extended period of time, with the result that the overall colour balance is disrupted and the image takes on a pink hue. This phenomenon only affected Eastmancolor negatives and reversal materials derived from this technology (colour reversal intermediates were found to fade especially quickly); Technicolor dye-transfer prints were immune, because no chemical changes were induced in the substances which formed the dyes after application on the gelatine-coated film base, thereby making these elements very stable as far as colour was concerned. However, most dye transfer elements were nitrate, as were the vast majority of camera negatives (in a sad turn of fate, the film industry abandoned the three-strip camera in favour of Eastmancolor just as acetate superseded nitrate). Technicolor separation elements therefore suffer their own colour preservation issues, largely arising from the three strips shrinking at slightly different rates, thereby making it very difficult to combine them accurately. Kodachrome (apart from the first generation of stocks sold between 1935 and 1938) was also found to be extremely stable, though as a relatively high contrast reversal stock it was difficult to copy and therefore never used on the same scale as Eastmancolor. The Eastmancolor fading process was described by one writer as follows:

> The first colour elements to fade are usually the yellows and the greens. Since these are frequently the softer, less intense colours, their absence is less obvious. A change in flesh tones occurs, and the outdoor scenes are less vital ... In the extreme stages of colour fading, the picture images are gradually reduced to ugly pinkish purple shadows of the original.[21]

Even after the problem was discovered, progress in researching possible solutions was complicated by a number of high-profile campaigns which, to all intents and purposes, accused the Eastman Kodak company of being negligent in marketing colour film which was prone to fading. Perhaps the most vocal was headed by the prominent film director Martin Scorsese, who repeatedly called for action on Kodak's part in formulating new stocks which were more resistant to long-term dye fading.[22] As part of his campaign he screened faded prints at the 1980 New York Film Festival and secured the support of prominent art-house directors, including Federico Fellini, Jean-Luc Godard and François Truffaut.[23] The end result was the introduction of so-called 'low fade' coupler stocks by Eastman Kodak, starting with type 5384 35mm print stock in 1981, and subsequently by its two principal rivals (Agfa-Geveart and Fuji). But by that point archivists had been given a legacy of thirty years' worth of colour film elements which had either gone pink or were at risk of doing so; and, due largely to the public criticism they had received, film manufacturers were reluctant to co-operate in researching the problem or releasing their own data.[24]

The research which Scorsese's campaign kick-started revealed that as with cellulose ester base decomposition, colour dye fading can also be inhibited by storing elements in atmospherically controlled conditions. Unlike with triacetate base (nitrate is not a significant issue here, because Eastmancolor was not introduced until just after the nitrate/safety conversion), coupler dyes are more sensitive to temperature than humidity, though vinegar syndrome in the base can catalyse and accelerate dye fading. As a general rule, the colder the vault, the slower the fading process. Colour film stores are therefore kept at far lower temperatures than their equivalents for black-and-white or magnetic media, and some archives even go as far as to freeze colour elements in hermetically-sealed plastic shrink wrap (at that temperature, the production of off-gases from deacetylation is not considered sufficient for sealing to increase the risk) to prevent contamination from the moisture produced by artificial refrigeration systems. However, the cost of maintaining freezers is considerable, and some archivists are reluctant to freeze films because they fear that damage may be caused during the freezing and thawing processes.

If it is too late for preventative passive conservation of coupler colour elements and significant fading has already taken place, a number of 'copy to preserve' methods, both analogue and digital, have been developed which attempt to restore the colour balance of the newly processed element. The digital approach is covered in the following chapter. Some of the analogue methods can be surprisingly effective, but none are cheap or straightforward. Probably the cheapest option is to make an 'as is' telecine transfer of the affected element and then manipulate the colour balance using video post-production technology. However, the corrected version will then only exist on video and will not be available for viewing with the higher tonal range and resolution of film. As an alternative, the colour temperature of the light used in a printer can be manipulated to compensate for the disproportionate density of the three dyes, thereby producing a better balanced duplicate (on coupler stock).

Until the emergence of digital methods, the 'gold standard' in colour restoration was an adaptation of the Technicolor method. The affected element would be printed three times, through red, green and blue (or yellow, cyan and magenta, depending if the source element was a positive or a negative) filters onto panchromatic black-and-white stock. This would produce three separation negatives, similar to those exposed in the three-strip Technicolor camera. When a new colour print or video master is needed for viewing, the three negatives are exposed consecutively onto new coupler colour stock, thereby reconstituting the three colour records. Since the master colour records themselves are black-and-white, the accurate long-term preservation of colour information is therefore guaranteed, as the silver which forms the monochrome emulsion is known to be extremely stable and not prone to fading or any significant chemical change. At the time of his campaign, Scorsese promoted this method of preservation heavily, demanding that studios create separation masters immediately after completion, before dye fading necessitated any attempt at restoration.

That having been said, none of these restoration methods are foolproof. When manipulating the colour balance during duplication of an element which is already faded, no objective means exists of determining the original colour balance of the source element in order to compare it with the copy. The restored colour record, therefore, is often based simply on what 'looks right' to the laboratory technician doing the grading, or sometimes from evidence outside the film itself, such as the colour of reference landmarks photographed in the film.[25] When printing separations as preservation masters at the point of production, extreme care needs to be taken in order to ensure that accurate registration can be obtained when recombining them into a new negative. To be absolutely certain that an accurate colour record for preservation has actually been obtained, it is necessary to print a new coupler internegative from the separation masters, even though one may not be needed to produce viewing materials for many years into the future. But doing so adds even more to the already high cost of this preservation route, with the result that many studios simply have not bothered.

Preservation and restoration

From this discussion of the way archivists have addressed the colour dye fading issue, it will have become apparent that we are really talking about two separate activities. Firstly, we have *preservation*, which, broadly speaking, can be defined as the technologies and methods used to prevent damage or decomposition to audio-visual recordings from happening in the first place. Examples of preservation activity include the use of temperature- and humidity-controlled storage conditions, restricting the use of preservation elements in a way which risks scratching and pre-emptive copying to preserve (i.e. printing separation masters as soon as the production of a film is complete). *Restoration* is what archivists are forced into doing if effective preservation has not taken place, i.e. the only copies of a film which are known to survive are not believed to represent what a film would have looked and sounded like in its original state. Altering the colour temperature of the light source used in printing in order to correct dye fading in a duplicate element comes into this category, as would wet-gate printing to mask scratches. Film restoration involves doing either or both of two things:

> While 'restoration' refers to visual quality of the image, 'reconstruction' refers to a philological activity of putting the programme or narrative back to something like an 'original'.[26]

Given that the activity Paul Read and Mark-Paul Mayer term 'reconstruction' also involves making changes (as distinct from preservation, the object of which is to keep an element in the same technical condition as when an archive received it), this discussion will use the terms 'technical' and 'reconstructive' do distinguish between these two forms of restoration.

The object of technical restoration is usually to correct physical and/or visible defects in an original source element or to copy it in such a way that those defects are not reproduced. If an archive holds more than one copy of the same footage, the first stage will usually be what is termed 'technical selection'. This involves examining all the available elements and determining which one is in the best condition and should therefore be used to produce the restored version. All the relevant aspects of an element's condition will be examined and noted, including which generation it is, whether it is complete, the extent of any base decomposition, shrinkage, perforation damage, poor splices, scratching on the base or emulsion, colour dye fading and contamination by other substances (such as mould growth or projector oil). Where a large number of copies exist for a given title, the element used to make the restoration may be pieced together from sections taken from several of them, in order to form the best quality copy of the film as a whole.

A number of treatments and processes can be applied to the source element itself in order to minimise the effect of any defects when it is duplicated. Ultrasonic cleaning, in which the film is immersed in a liquid through which ultrasonic energy is conducted, can be used to remove dirt. Adhesive tape can be used to replace or reinforce damaged perforations. Dried up cement splices can be remade after scraping away the residue. The underlying rule of thumb which any responsible archivist seeks to follow when making changes to original elements directly is not to apply any treatment which is irreversible unless there is absolutely no alternative.

For example, chemical treatments which are routinely used in cinema projection boxes for cleaning release prints (e.g. Filmguard, Renovex, Ecco) would not be considered acceptable for archival restoration, as the long-effects of those chemicals are not fully understood and may be impossible to undo. The chemicals used in ultrasonic cleaning, on the other hand, have been thoroughly tested and are known to evaporate without leaving any residue. The adhesive tape used to repair perforation damage can, if necessary, be removed without trace. If the restoration process does involve assembling a master element from a number of different sources, meticulous record-keeping as to what was taken from where will enable each source to be returned to its original state if required. And if a soundtrack is rerecorded (for example, for purposes of noise reduction), the original (or a 'straight' duplicate, if the original is at risk of decomposition) will be kept for reference.

The object of reconstructive restoration is not necessarily to correct technical defects. This is carried out when the *content* of a film is not known to survive in the way that it would have been shown on its initial release or distribution, and a number of different sources are used to reassemble it. While this is a similar process to using different sources to produce a 'best' master element in the sense of image and sound quality, it is being done with a different aim in mind. Let us take one hypothetical example. Two copies of a Hollywood 'B' feature from the 1930s are known to survive. One is the edited camera negative, held by the studio's own archive. However, three scenes are missing from it because the film was re-released a few years after production. For the second release the censors demanded cuts, which were effected by simply removing footage from the original picture and track negatives before the new prints were struck directly from them. The missing sections have since been lost. The library of a major television broadcaster holds a release print of the same film, which dates from the original release and therefore includes the censored scenes. But other footage was removed, as this 78-minute film was arbitrarily cut to fit a 70-minute broadcast slot, and the deleted sections were simply thrown in the bin. In this scenario, a dupe negative of the censored scenes could be struck from the television print and the footage cut back into the camera negative, thereby enabling the complete film in its original state to be duplicated for preservation. As the censored material would be at least three generations removed from the camera negative there would be a noticeable drop in image quality as it appeared in the new version, but at least we would be able to see the film as it was originally shown.

This is a very simplistic example, but it does illustrate the principles used in reconstructing films which, for a variety of reasons, may not have survived in their original state. Perhaps the most celebrated film reconstruction ever undertaken was Kevin Brownlow's work on the five-hour epic *Napoléon* (1927, dir. Abel Gance) in the 1970s. This used literally hundreds of source elements, consisting of everything from sections of camera negative to 9.5mm prints. The end game in any restoration which involves direct technical intervention on original elements is duplication – there is no point in only passively conserving an asembled element which has undergone restoration work, as it will almost certainly not be viewable in this state.

As has been illustrated above in relation to colour dye fading, this stage can also introduce technical changes in order to correct physical defects in the original (such as optical and/or wet-gate printing).

There is an ethical dimension to restoration work which is not as big an issue when simple preservation is the object of the exercise. This relates to the question of what, exactly, constitutes 'original', and whether that is what you're aiming to restore. For example, Jacques Tati's debut feature, *Jour de fête* (1947) was filmed simultaneously using two cameras, one of which used an experimental colour process known as Thomsoncolor (a dry-screen system not dissimilar to Dufaycolor), and the other of which exposed conventional, panchromatic black-and-white stock. It proved impossible to produce any colour release prints, and therefore the film was released in black-and-white only. But the Thomsoncolor negatives survived, and in the mid-1990s a restoration project was undertaken in which they were copied onto coupler stock and the film was re-released in colour.[27] Could the restoration legitimately be described as recreating the original film, given that the colour version had never been seen when the film was initially released?

There have also been occasions when a simple lack of communication has resulted in some very contentious 'restoration'. The British director Bill Douglas's autobiographical feature *My Childhood* (1972) was shot on 16mm Eastmancolor stock, but the release prints were struck on black-and-white deliberately, in order to give the image a low-contrast feel which Douglas hoped would evoke the look of Edinburgh in the 1950s. When a new print was made for screening at a festival shortly after the filmmaker's death in 1991, no one thought to tell the lab about this, with the result that the film was screened in colour for the first time.

Even more ethically contentious are those 'restorations' which quite deliberately seek to re-release a film in a different form to that which was seen upon initial release. At least the colour negatives of *Jour de fête* were artefacts from the original production, even if it proved impossible to screen their content at the time. As an example of contentious technical restoration, *Vertigo* was never released nor intended to be released with six-channel stereo sound, but that is exactly the form in which it was 'restored' for redistribution in 1997. And how do we define 'original' where reconstruction is concerned? If a celebrated director is overruled by studio bosses, who cut or reedit his film, which is 'original' – the version approved by the director or the version that was initially seen by the public? The number of 'director's cut' videos on sale from the mid-1990s onwards would suggest to the cynically minded that, despite the potential complexity of the question, its existence is being commercially exploited with impunity. Sometimes a director himself will make changes to a film after release, as was the case with *Lawrence of Arabia* (1962, dir. David Lean), when approximately 20 minutes of footage were trimmed in the months following the initial screenings. Most of it was recovered and reinstated when the film was restored in 1989. One review questioned the decision to do this:

> When a film is cut on grounds other than censorship, there is generally good reason for it results in a tighter, faster, more compelling picture ... We are talking here of making films

as good again as they were when new. Too many other restorers are making them worse and more tedious than they have seemed for years.[28]

While to a certain extent this journalist is missing the object of the exercise (whatever you would call re-editing a film to enhance its attractiveness to a mainstream audience, it is not restoration), the definition of 'original' which an archivist uses to underpin the nature and extent of any reconstruction or technical intervention highlights the importance of what Paolo Cherchi Usai called 'the old debate on what curatorial expertise should render unto Caesar and what to God'.[29]

Preservation and restoration of videotape

In maintaining moving image content originated on (or, at any rate, which exists on) film, it will have become apparent that archivists have been given a mountain to climb, thanks mainly to the medium simply not having been designed with longevity in mind. The videotape mountain is even higher. Being an analogue medium which is delivered through the technologies of chemistry, optics, simple electronics and mechanical engineering, the images and sounds on film are relatively straightforward to recover. The intermittent mechanism and other key components of a printer can be modified to cope with shrunken or brittle stock, faded dyes can be reprinted by using prisms and filters to control the colour temperature of the light source used to print it, dirt can be removed simply by cleaning the film itself, and so on. Furthermore, the effectiveness of international standardisation has kept the number of gauges and formats in circulation comparatively low, thereby minimising the equipment and expertise needed to maintain archival film collections.

In terms of the preservation challenge it poses, videotape differs from film in two crucial respects. Firstly, the hardware needed to record and play back the images and sounds is far more complex. Secondly – and as we have seen in chapter six – the role of standardisation was never as effective in television and video technologies as it was with film. From an archivist's standpoint, the wrong precedent had been set as early as 1941, when the adoption of the NTSC broadcast standard made US broadcasting fundamentally incompatible with the British 405-line system. When Ampex introduced the first 2-inch studio VTRs in the late 1950s, separate versions had to be made for NTSC, 405-line and subsequently PAL, each of which was completely incompatible with the others. Every subsequent generation of broadcast standard videotape technology has entailed a fundamental redesign of the way in which signals are encoded on the tape, the design of the spool or cassette and the design of the VTRs. As a general rule of thumb, the standard videotape format used by broadcasters and producers changes every decade or so. When this happens, the superseded tapes become what is known as a 'legacy format', meaning that manufacturers no longer provide equipment, blank media or spare parts.

This creates real problems for archivists charged with maintaining collections of legacy format tapes. The are basically two ways they can proceed: one is to maintain legacy format VTRs as well as an inventory of spares and the expertise needed to operate and maintain them, while the other is to undertake the video equivalent of 'copy to preserve' in film, a process known as 'continuous format migration'. This is exactly as it sounds – undertaking to copy all media to a new format before or as an old one goes from current to legacy. The cost of this is significant, and as with the archives which until recently destroyed nitrate originals after duping them, a decision has to be made as to whether to keep the original after duplication. If so, this has storage space (and therefore more cost) implications. Furthermore, and again, as with film, the copying process introduces generational loss. Ultimately, some sort of format migration will be unavoidable if the content of legacy videotapes is to be preserved; and unlike nitrate film, there is no known way of putting off that day for centuries – and only just for decades. This is because of the sheer complexity of the mechanical and electronic systems inside a VTR, e.g. the helical scan head system, the tape transport mechanism and the integrated circuits needed to process the video and audio signals coming in and out of the machine. It simply would not be possible to build a 2-inch VTR with which to copy a collection of 1950s tapes from scratch in the same way that it would be possible to build a printer for copying shrunken 35mm nitrate.

Conclusion

During the second half of the twentieth century the technical processes of archival preservation and restoration of moving image content originated on film and videotape have become a key part of the way our access to moving image heritage is technologically mediated. With both film and video, these processes fall into one of two categories: *preservation*, which attempts to safeguard the technical integrity of content as it is received by the archivist, and *restoration*, which affects technical changes to it in order to recreate the look and sound of content which is believed to have existed once, but no longer does in that state.

Almost all the techniques used to do this have been evolved in arrears, so to speak. Indeed, the need for any 'techniques' at all (other than simply putting films on a shelf and forgetting about them) did not become apparent until over three decades after film had become big business and a mainstay of cultural activity, when the process and implications of nitrate decomposition started to be understood. History repeated itself when it came to developing the techniques and technologies for enabling preservation and restoration in other areas, notably in dealing with vinegar syndrome and dye fading. Perhaps the key reason for this is that, unlike the evolution of technologies for originating and financially exploiting moving images, the technologies needed to preserve them did not originate in the commercial sector. Indeed, the majority of moving image content was not deemed to have any significant monetary value after its initial release, a situation which persisted until the 1950s when the growth of television created a new market for archive film. In the public sector archives of most developed countries, moving image preservation has tended to come some way down the pecking order in allocating money and other resources, even though moving images are probably the most difficult and expensive form of archival document to preserve.

chapter eight | new moving image technologies

'"Digital" has become a trendy buzzword, redolent of the computer – the icon of our age. Calling anything digital implies praise and precision, even though the meaning of the term is rarely understood.'[1]

'Everything in this place is either too old or it doesn't work ... and you're both!' – Will Hay in *Oh, Mr. Porter!* (1937, dir. Marcel Varnel)

The 'D word' has now become an inescapable element of moving image technology. Since the early 1990s, digital videotape has been replacing its analogue predecessors in broadcast studios. The last mainstream professional analogue format, Betacam SP, was declared obsolete in 2002 when Sony announced that it would no longer manufacture VTRs in this format and would phase out spare parts availability over a seven-year timespan. In the domestic arena the Digital Versatile Disc (DVD)[2] has all but replaced VHS for the sale and rental of prerecorded media in the developed world. The digital stills camera has become one of the fastest selling consumer electronics products in the first decade of the twenty-first century, leading to a rapid decline in film-based photography by amateurs. The use of digital imaging in film production began primarily with the integration of digitally generated special effects – a technique known as computer generated imagery (CGI). This was marked spectacularly in 1992 by the release of *Terminator 2: Judgement Day* (dir. James Cameron) and *Jurassic Park* (dir. Steven Spielberg).

By the end of the decade low-budget feature films were being originated and edited entirely on digital video before being transferred to 35mm film for release. By the time of writing computer-based technologies for cinema projection were becoming available, though specific economic factors exist which will prevent their widespread use as an alternative to 35mm for some years to come. Moving image archivists have been using digital processes for restoring audio recordings since the early 1990s. From around the turn of the century the image manipulation technology which was previously only available in the CGI departments of lavishly funded Hollywood studios came down in price to the point at which archivists could start using it to scan films and the correct defects such as scratches and colour dye fading. Long-term storage of media content in digital form is proving to be a major problem, of which more below.

So, what does the 'D word' actually mean, why has this technology come – seemingly from nowhere – to establish itself so widely and what are its implications for the long-term future of recorded moving images?

Computers and moving images

With the exception of digital audio, which has been covered in chapters four and five to the extent that it is necessary to contextualise the role of related technologies, all the other technologies discussed in this book record and reproduce moving images by using what is termed *analogue* (from the Greek *analogos*, meaning equivalent or proportionate) methods. The terms 'analogue' and 'digital' can to all intents and purposes be treated as antonyms. An easy way to imagine how analogue recording works is to think of it as an analogy: images or sounds are represented, or analogised, as continuously variable physical quantities which can be read and written by a machine which converts them to and from visible images and audible sounds. These variable physical quantities can take a number of different forms. For example, the pattern of silver salts on photographic film is an analogue of the pattern of light which existed at the time and in the place a photograph was taken; or the pattern of magnetised iron oxide on an audio tape is an analogue of the changes in air pressure in a given space at a given time, as they were detected by a microphone. In the case of still photographic images, the only machine needed to read this recording is the human eye. In others, notably videotape, some very sophisticated mechanical and electronic engineering is needed to produce the hardware which can reproduce the recording as meaningful images and sounds.

Digital media does not record a direct representation of a continuous process of change. Instead, it represents that process as information, or *data* (from the Latin 'datum', meaning 'something given'). That information takes the form of numbers, hence the word 'digital'. Computers are used to encode images and sounds into digital data, and then reproduce them by converting the data back into an analogue signal which can be displayed on a screen or played through a loudspeaker. Doing this may at first sight seem illogical because it would appear to introduce an extra, unnecessary process. But representing recorded images and sounds digitally has one crucial advantage from which several others flow: digital data can be copied with 100 per cent accuracy, whereas an analogue recording cannot.

Copying an analogue recording of any description involves losing some of its accuracy, or quality. Even if you duplicate a film by contact printing it, the individual grains of dye in the source element could never be placed in precise alignment with their counterparts on the unexposed duplicate stock. Therefore, a small amount of distortion will be introduced into the copy. If a film is printed optically, even more distortion is introduced, because the light which is shone through the source element will be refracted by a lens before reaching the destination stock. To copy an analogue magnetic audiotape it has to be played back using one machine, and the electrical signal which results fed to the recording head in another. The amplification and signal processing electronics between the two tapes will inevitably degrade the signal by introducing unwanted modulation (noise) during the copying process.

In the digital domain none of this applies. Because the image or sound recording is represented as a series of numbers, it is only necessary to copy those numbers (the data) accurately – something which can be checked by comparing the source and destination elements after the copying is complete – in order to produce a 'clone' or perfect reproduction of the original, totally free of any imperfections. For an industry which relies almost entirely on the underpinning technology of being able to mass-produce copies of an original recording, this is obviously a significant advantage. The introduction of visual effects (not just obvious and spectacular ones such as dinosaurs, but routine effects such as fades to black, which are used repeatedly in virtually every feature film) previously worked by applying the effect optically during the copying process. So, for example, a scene could be filmed in which an actor performs against a plain background. This could be superimposed with footage taken at a given location in duplication to a new element which combines the two images. But as the combined image is a generation removed from either original, it is of lower quality. Working digitally, these images could be manipulated in a computer and the result output to film again without any loss of resolution, contrast or colour depth. This is just one example of the actual and potential advantages of using digital media rather than analogue.

A computer is essentially a very large, very powerful calculator. It stores and performs calculations on data electronically. This is achieved by exploiting the fact that an electrical circuit has two states – open or closed. For example, the light switch in your living room is either on or off. We can then give these states a number – 0 for off, 1 for on. Each one of these states is the basic, smallest unit of digital data possible, known as a *bit*. It is then possible to combine sequences of bits to form larger numbers. So for example, sequence of two bits would offer four possible states – 00 (1), 01 (2), 10 (3) and 11 (4). The grouping of bits used by almost all modern computers as the working unit of data is a sequence of 8 (256 possible states), known as a *byte*. Possibly the commonest data units represented as bytes in a computer are letters of the alphabet – for example, a capital 'A' is 01000001 (which is 65 if expressed as a decimal integer). A computer stores and performs calculations on bytes by containing literally billions of tiny electronic circuits which its processor can open or close as required, thereby representing the data. Given that the byte is so small as to be meaningless as an expression of 'real life' data storage, the memory (data capacity) of a computer is usually expressed in multiples: a *kilobyte* consists of 1,024 bytes (2^{10}) and is referred to by the abbreviation 'k' or 'kb'; a *megabyte* is 1,048,576 bytes, or 1,024k (2^{20}), with the abbreviation of 'mb'; a *gigabyte* is 1,024mb (2^{30}) with the abbreviation of 'gb'; and a terabyte is 1,024gb (2^{40}) and is so big that it is not used frequently enough for a common abbreviation to have become established. Moving image data usually falls into the megabyte or gigabyte category – for example, a feature film on DVD will typically occupy between 4 and 8gb.

The first machines which could store data and carry out mathematical calculations on it were mechanical. Arguably the earliest design of any significance was that of the 'Difference Engine', produced by the British mathematician Charles Babbage in 1822. Although it was too mechanically complex to actually be built at the time, one was built by the Science Museum in London in 1991 from Babbage's original plans, and the three-ton computer was found to work exactly as he had predicted. The Difference Engine stored data and programs (a set of instructions determining what calculations a computer carries out and in what order)[3] on punched cards and performed calculations using clockwork mechanisms. Though its processing power was infinitesimal compared to anything electronic, it is estimated that Babbage's computer could execute simple calculations three to five times faster than the human brain.[4]

Mechanical devices for simple and repetitive data processing functions became established on an industrial scale in the latter half of the nineteenth century. A notable pioneer in this field was the American engineer Herman Hollerith (1860–1929), who, from 1879 onwards, developed machines for counting and sorting punched cards electromechanically, and counting each instance of a data item encoded on them. Because the only data processing carried out by these machines was to accumulate a record of each time a given pattern of holes passed through a mechanism – they did not perform any actual calculations – they could not be termed computers in the strict sense of the word. But they established the precedent for automated data processing on an industrial scale, one which would eventually provide the groundwork for the computers which would be capable of dealing with the volumes of data needed to represent images and sounds. The company which would eventually become International Business Machines (IBM) was founded on the back of Hollerith's inventions in 1911.

The next significant step was the emergence of machines that could perform full-scale data processing operations (i.e. ones based on carrying out mathematical functions) electronically, i.e. without the need for mechanical moving parts. The technology used to achieve this was the same one with which Lee de Forest was able to electronically amplify audio signals: the thermionic valve. They acted as the basic unit of storage for each 'byte', and because they stored data in the form of electrical energy rather than through a mechanical, solid-state process, enabled calculations to be performed far more quickly than was the case previously. The first generation of valve-driven computers, which were developed in the 1940s for code breaking and other military applications during World War Two, could hardly be described as 'micro', though: one such example, the American ENIAC (Electronic Numerical Integrator and Calculator), built by the US Army to calculate the trajectory of artillery shells, weighed 25 tons and consumed over 200 kilowatts of power (enough to light a small town). Its processing power was roughly equivalent to that of the first hand-held calculators, sold in the 1970s.

The story of computing from that point on was one of decreasing physical size and increasing processing power. The transistor, invented in 1947, did the same job as the valve but was about a hundredth of the size and consumed an insignificant amount of power. The 'integrated circuit', more commonly known as the 'silicon chip' or 'microchip', was first produced in 1958, and in 1971 the first microprocessor – a device which combines most of a computer's processing functions onto a single chip – went on sale. Within three decades the processing power available from 25 tons of power-guzzling valves had been replaced by a miniature system of circuits little bigger than a postage stamp. Furthermore, the speed and complexity of the calculations performed by microchips has been rapidly increasing ever since. The term 'Moore's Law' is commonly used to describe this phenomenon, which refers to a prediction made by Gordon Moore, the founder of the world's largest microchip manufacturer, Intel. In an essay written in 1965, Moore stated his belief that the number of transistors on an integrated circuit (and therefore its processing power) could be doubled, approximately every eighteen months (an estimate he later revised to two years).[5] So far, this prediction has proven to be roughly accurate.

The product with which the microprocessor has become primarily associated is the personal computer (PC), which had its origins in the late 1970s and began to be sold on a significant scale from the early 1980s. It was 'personal' in the sense that the processor, memory, offline storage (such as floppy and hard disc drives) and user interfaces (i.e. keyboard, screen and later, mouse) were all combined into a single unit, without the processing power being shared between a number of users. It should be borne in mind, however, that these are not the only devices to contain microprocessors, which are now an integral part of practically every item of hardware used to record and reproduce moving images.

In crude terms, it needed roughly twenty years of Moore's Law taking effect before the processing power of an average microprocessor was sufficient to handle audio encoded as digital data with a quality equivalent to that of an LP record or an FM radio transmission, and then a further fifteen until it was able to manipulate digital video at a resolution which is comparable to a broadcast PAL or NTSC transmission shown on a consumer television set. The following quote is taken from a British national newspaper article which appeared in 1994, describing the use of 'multimedia' computer-based video images in teaching:

> The picture is grainy, jerky and usually runs in a tiny window in the middle of your computer screen, to minimise the amount of memory and disc space needed to store and display the moving images.[6]

Within a decade, video manipulation and editing would be possible using home computers of average power and memory capacity. In the rest of this chapter we shall consider some of the ways in which computing has shaped the evolution of moving image technologies at the start of the twenty-first century.

Sound

Audio was the first element of moving image technology to be computerised, with location and studio sound for film and video being recorded digitally by the mid-

1980s in most of the developed world. Put crudely, this is because the volume of data needed to represent an audio waveform and the complexity of sums involved in converting it to and from sound and data are significantly lower than with video or film – so much lower, in fact, that 10–15 years of Moore's Law separates the systematic use of computer technology for audio and moving images.

At some stage, all media content which is encoded as digital data has to undergo what is called *analogue to digital conversion*. This is the process whereby the representations of light and air pressure which are detected by the eye and the ear are turned into the ones and zeros which can be stored and processed by a computer. In the case of audio, the underlying technique for achieving this is known as *sampling*. As will be recalled from chapter one, a 'moving' image consists of multiple still photographs, exposed and projected in rapid succession. To computerise this impression of movement, it is only necessary to create digital versions of each still frame, and then to determine the speed and duration at which they will be displayed. The passage of time, therefore, is neatly divided into a series of discrete segments. The same does not hold true for a recorded soundtrack. In the analogue domain, this consists of a single pattern of modulation which varies infinitely across the surface area of the film, disc or tape on which it is recorded. By placing the medium in contact with the playback device at the same speed is the one which recorded it, the passage of time can be made identical in both recording or playback.

With digital audio it is not as simple as that. Numercial data is not infinitely variable – it is a set volume of ones and zeros. A zero does not gradually become a one over a millimetre or so of magnetic tape: that tape will encode a zero followed by a one, in just the same way that our light switch is either on or off. Therefore a way is needed of determining how that data can be made to represent changes in air pressure (which is detected by the human ear as sound) over time. This is where sampling comes in. The technique has its origins in an influential research paper written by Harold Nyquist, a Swedish physicist who emigrated to the US in 1928 and ended up as a researcher for the Bell Telephone Laboratories.[7] As with Lee de Forest and electronic amplification, Nyquist was primarily interested in the potential application of digital audio in long-distance telephone links. He determined that the passage of time had to be carved up into a series of discrete segments (comparable to frames in a film or each interlaced scan in a cathode ray tube). The *sampling rate* determines how long the audio information in each 'sample' is reproduced for, or, in other words, how many samples are recorded and reproduced within a set period of time. Within each segment, it is necessary to record data which represents the frequencies of the sounds to be reproduced. Because most audio recordings consist of many different frequencies being audible at once, the sampling rate directly affects the frequency range which can be recorded and reproduced, and consequently the perceived quality of the audio to the human ear. Basically, the higher the sampling rate, the better the sound. However, not only does the sound quality increase with the sampling rate; so does the memory, storage space and computer processing power needed to record and reproduce it digitally.

At the time of Nyquist's research, the computing power did not exist to put any of his ideas into practice. Even the valve-driven behemoths such as the ENIAC were still over a decade away. But, as with Babbage and the Difference Engine, further experiments during the mid-1950s vindicated his theories, and the first commercial applications for digital audio were indeed in telecommunications. But because the memory and processing requirements increased in proportion to the perceived quality of the recorded audio, digital sound for radio, recorded music, film and television did not become feasible until microprocessors were available in the mid-1970s. The key developments in digital audio during the 1970s and 1980s are covered in chapter four, though it is worth noting here that sound capture and editing using personal computers did not become a reality until the mid-1990s.

While the studio recorders which used modified UMatic and Beta videocassette mechanisms and the CD audio players which were sold to consumers since 1982 did contain microprocessors, the speed at which they needed to carry out analogue to digital conversion and vice-versa was relatively low. This was because the first generation of digital audio recording technology used an uncompressed signal. *Compression* is an essential technique for digitally encoding many types of video and audio. In essence, it reduces the volume of data needed to encode a given amount of media content to a given quality, and does so in one of two ways. *Lossy* compression works by discarding (or 'losing') data relating to parts of the audio spectrum which it is believed most human ears cannot detect (extremely high or low frequencies, mainly), leaving only the audible parts left. *Lossless* compression, on the other hand, preserves the entire encoded spectrum, but uses complex mathematical equations, known as algorithms, in the software that carries out the encoding in order to compress it into a smaller data storage space. The main benefit of compression is obvious: you can store more sound using less data. The key drawback is that more processing power is needed in the analogue-digital-analogue conversion process than with an uncompressed data stream. Furthermore, if heavy lossy compression is used, the sound quality will suffer. For example, many of the compression protocols now in use would enable Beethoven's ninth symphony to fit on a 1.44mb floppy disc, if you were prepared to listen to it with the sound quality of a telephone handset. In contrast, an uncompressed recording of it on a compact disc (which were designed for a playing time of 74 minutes, it is rumoured because Beethoven's ninth symphony was the Sony founder Akio Morita's favourite piece of music), in stereo, and with a 16-bit sampling frequency of 44,100 per second, would occupy close to 650mb.

In the first generation of digital audio technology no compression was used, the reason being that on tape-based systems and CDs, storage space was not at a premium and, unlike with consumer digital video a decade later, there was no economic barrier to providing sufficient processing power to decode 'raw' digital audio without the helping hand of asymmetric compression. Furthermore, the electronics industry was anxious to prove that digital audio could match the quality of the most expensive analogue systems then in use. This was one of the reasons why the compact disc was designed for a storage capacity of 650mb – almost 500 times the capacity of any removable data storage medium in widespread use at the time of its launch – despite the formidable engineering challenge that this imposed.

The convergence of digital audio and the personal computer took longer. Moore's Law gradually boosted PC processing power throughout the 1980s, but storage space remained at a premium. The data storage methods used by dedicated digital audio hardware in the early 1980s imposed limitations which made them incompatible with PCs. Either they used magnetic tape, access to which was strictly linear (i.e. it was impossible to access any part of the data stream instantaneously or at random), or the optical compact disc, which until the late 1990s was a 'read only' medium which could only be mass-duplicated in a factory.

The key device which enables a personal computer to store and retrieve large amounts of data quickly and at random is the hard disc. The first ones (sometimes known as Winchester discs after a type of rifle, because their inventor, Ken Haughten, was a keen amateur hunter) were sold by IBM in 1973, and consisted of a sealed unit containing one or more rigid steel platters coated with magnetic oxide, and mounted on a rotating spindle. The reading and writing head was on a pivoting assembly suspended above the rotating platter, which was divided into 'tracks' and 'sectors' on which blocks of data were stored. Being a magnetic medium the data could be erased and rewritten without practical limit, and being a disc, random access was possible. The problem was that until the early 1990s, hard discs were very expensive, limited in capacity and not very reliable. 'Head crashes', in which the mechanism which positioned the head above its required track and sector failed, causing it to make physical contact with the disc surface and thereby destroying it, were common occurrences (and even with today's hard discs, they are not unheard of). By the late 1980s, a top-of-the-range PC costing four figures might contain a hard disc with a capacity of 20mb; in 1995, 500mb was a typical figure – still less than the data capacity of a single CD.

The ability to work with uncompressed audio data on a personal computer did not become firmly established until the late 1990s (in another illustration of the exponential effect of Moore's Law, this was only around 3–5 years before the same could be said of digital video). Three factors had the greatest influence here: the ongoing effect of Moore on processor power, steady but significant increases in typical hard disc capacities and the emergence of the recordable compact disc. Soon analogue audio recording and post-production had almost totally disappeared from the professional film and broadcast industries.

Television and video

A video signal is digitised by dividing the frame into individual dots, or 'pixels'. The number of pixels per frame determines the *resolution* (the amount of detail in the image, and consequently the volume of data needed to represent each frame) of the image, and this varies according to the format in use and the television system being encoded. For example, the system used for encoding the video stream on a conventional DVD uses a resolution of 720 (horizontal) x 576 (vertical) pixels for PAL

(625 lines of analogue horizontal resolution), or 720 x 480 for NTSC (525 lines). Data recording the luminance and chrominance of each pixel is stored on a predetermined scale, the detail of which varies in a similar way to the sampling rate in digital audio. For display on a conventional television the digital to analogue converter in a VTR will convert this grid of pixels into conventional, 525- or 625-line signal, at the appropriate scanning rate and interlaced if need be.

Once microprocessors which were fast enough to carry out these functions became available, it made obvious sense for videotape recording to go digital. One of the reasons that videotape had superseded 16mm film for the majority of origination in the television industry during the 1970s and 1980s was the lower cost and greater flexibility in editing. With film it was often necessary to make a separate print (the 'cutting copy'), cut the camera negative to match it only when all the editing decisions were finalised, in order to minimise damage to the original, and only then strike the transmission print. This was a very time consuming and expensive process. In circumstances when it was logistically impossible (such as editing news footage), editing the original camera element directly was necessary and this often caused substantial, visible damage (such as scratches and dirt) in transmission. Archivists have had to expend considerable time and energy in restoring television material from the 1960s, 1970s and 1980s which has been mutilated in this way.[8] With videotape, the way in which content is edited is to selectively copy footage from several original camera source tapes, assembled onto a single edited master. Sometimes, for example, if special effects or complex titles, visual effects or sound dubbing is required, this can mean going through several generations of tape, thereby lowering the picture quality through generational signal loss – if the video format is analogue. Digital video offered the potential to eliminate this problem, thereby enabling the final edited master of each production to 'clone' the quality of the original material.

This was the next major area of moving image technology to go digital after audio, and did so following a similar pattern: professional studio use first, then digital television and video marketed directly to consumers. In the broadcasting industry of the 1980s, the analogue Betacam SP format superseded UMatic and retained a dominant market share throughout the decade. The first commercially marketed digital videotape format for televison production, the Ampex 'D2' system, was launched in 1986 but failed to achieve significant sales. The D2 VTRs only had analogue inputs and outputs: all the analogue-digital-analogue conversion took place inside the machine. This negated the principal advantage of going digital, as the process of decoding and then re-encoding the signal when copying between two machines for editing introduced the same generational loss as with analogue systems. Although it was uncompressed, the D2 format only encoded a composite video signal in order to minimise the volume of data and processing power needed.

The format which signalled the systematic introduction of digital video into the broadcast industry was Sony's Digital Betacam, launched in 1993. On a pragmatic level, it used the same design of cassette and tape transport mechanism as Betacam SP, thereby meaning that the new VTRs would be 'backwards compatible', i.e. they would also play the old analogue tapes, of which many broadcasters possessed libraries of tens of thousands. Like its analogue predecessor, 'Digibeta' (as it soon came to be known) encoded a *component* signal, which split the video information up into three signals. This is a further refinement of the NTSC principle of broadcasting luminance and chrominance information separately, in which the luminance is transmitted as before but the colour information is divided into two extra signals. Thanks to being a component system, Betacam SP had acquired a reputation for being comparatively resistant to generational signal loss even in the analogue version. Although the digital version incorporated a 2:1 compression ratio in order to keep the physical tapes compatible, it quickly became established as the gold standard in broadcast video technology.

The demonstrable image quality of Digibeta was, however, reflected in the price of VTRs and media which, throughout the format's life cycle, have remained at levels where it is only really available to major broadcasters and production companies. It was therefore inevitable that a group of other tape-based formats would emerge to fill the gap left by what had become 'semi professional' analogue video formats, notably UMatic and S-VHS. Perhaps the most widespread were Sony's DVCAM (launched 1995), which used a cassette design similar to the DAT audiotape, and Panasonic's DVC Pro (1996). These have much higher compression ratios than Digibeta – typically between 5:1 and 10:1, depending on the specific format, and therefore do not match its image quality or durability. But the far lower price of hardware and blank media marked the beginning of digital videotape becoming available to a wider user base, with the new generation of formats catering to a range of markets, ranging from the amateur 'home movie' makers who would have used Super 8 a generation earlier to producers of promotional and training videos.

A process of 'going digital' was, therefore, well underway by the mid-1990s. But none of this would have been apparent to viewers in the home at this point, who were still watching their 525-line NTSC or 625-line PAL analogue CRT television sets, which received an analogue composite signal either from the airwaves or a VHS VCR. The image compression technology available thus far had enabled professional and semi-professional digital video to become a reality, but had not bought prices down or reliability up enough for it to be mass-marketed to consumers. The technology which enabled it to make that jump has the somewhat uninspiring title of MPEG-2.

MPEG stands for Motion Picture Experts' Coding Group, a committee of computer scientists founded by Leonardo Chiariglione under the umbrella of the International Standards Organisation in January 1988. Its aim was to establish software standards for encoding and decoding digital video which could be used in a range of consumer and industrial applications, from the small displays in hand-held videogames to digital cinema projection on full-sized screens.[9] In the following years it produced five main sets of standards (MPEG-1, -2, -3, -4 and -7), of which MPEG-2 has become the method of choice for the two large-scale consumer digital video applications: digital terrestrial television transmission, and the DVD. MPEG-2 proved especially suitable for this role because of two unique aspects of the lossy compression technique it uses. Firstly, it is what is termed an 'asymmetrical' system:

> In audio and video compression, where the encoder is more complex than the decoder, the system is said to be assymetrical. MPEG works in this way. The encoder needs to be algorithmic or adaptive whereas the decoder is 'dumb' and carries out fixed actions. This is advantageous in applications such as broadcasting where the number of expensive and complex encoders is small, but the number of simple and inexpensive decoders is large.[10]

From this explanation we can deduce that the key to putting new technology in the hands of the consumer is that it has to be as cheap and simple as possible. The computing power needed to decode the video data on a Digibeta tape, for example, could never be produced at a price which consumers would be able or willing to pay. MPEG-2 solves this problem by not only encoding the moving images themselves, but also small amounts of instructional code which tells the player's decoder how to interpret the image data. To a certain extent, it could be argued that some of the software is built into the media.

The broadcast standard tape formats which preceded it, by contrast, simply contained the image data in a predetermined format, which the VTR had to decode using entirely its own software. Compared to the 'dumb' decompression of MPEG this has a tendency to sap processing power, especially if the compression in use is lossless (because in this case it has to extract the uncompressed data first, before decoding it into an analogue picture). To understand how this technique works let us consider a hypothetical example. A camera photographs a static landscape shot, half of which depicts clear blue sky. A broadcast standard tape format would probably encode all the different blue pixels in the frame individually, even if the difference in luminance and chrominance is so slight as to be invisible to the naked eye. Even if the next shot is a panning close-up on fast-moving action, each frame would generate the same volume of data as the landscape shot which preceded it, and the decoding process would require the same computing power. This is because it would decode and convert to analogue every pixel individually, regardless of the relative similarity or difference between their respective luminance and chrominance properties.

MPEG-2 goes about it a different way. In our landscape shot, instead of the data stream individually representing tens of thousands of blue pixels, it might contain just one, followed by an instruction which says 'repeat this for the next X pixels'. In this way thousands of bits of data storage can be dispensed with instantly, and the microprocessor doing the decoding only has to perform one operation rather than thousands. In the bottom half of the frame, where more detail is needed, more of the pixel characteristics can be defined individually. This is, of course, what defines MPEG-2 as being a 'lossy' compression system – in this example we are 'losing' the detail from the sky, even if the varying shades of blue are so slight that most people would not notice them. MPEG-2 carries this principle even further, by analysing the relationship between whole frames during the encoding process. The video stream is divided into a sequence of 'intra' (I) and 'predicted' (P) frames (the exact composition of the sequence is flexible and can be determined when encoding). If, for example, a sudden flash of lightening were to appear against the sky in our hypothetical landscape shot, then instead of encoding two entire frames, the second could be substituted for an instruction which says 'repeat frame A as frame B, but in frame B shade these pixels as so where the lightening bolt appears'. Once again, this reduces both the volume of data needed to represent the video stream, and the number-crunching required to decode and display the picture.

Both of these abilities were essential requirements for the first consumer medium which utilised MPEG-2, the DVD. News of the imminent launch of the format came in 1995:

> 'Sony of Japan and Philips of the Netherlands plan to mount an assault on Hollywood to persuade US movie studios to make films and other entertainment products on their new digital video disc system. Toshiba and Matsushita have an alternative, non-compatible system. Although both camps have tried rapprochement, they seem on course for a format war in the digital disc market just as over the first generation of video cassettes when Sony struggled to establish its Betamax format over Matsushita's ultimately successful VHS system.'[11]

Despite its infamous precedent, the anticipated format war did not materialise, and the first pre-recorded Hollywood films on DVD went on sale in Japan in the winter of 1996, and in March 1997 in the United States. Unlike the case of CD audio, processing power *was* at a premium this time, and so was data capacity. Not using any compression at all simply was not an option – uncompressed NTSC video, for example, would gobble up 26mb *per frame*! By greatly increasing the writing density of a DVD (which, by taking advantage of a decade's development in electromechanical technology, was made possible by using a lower wavelength laser) and sandwiching two data storage layers, each of 4.7gb, onto each disc which could be read by adjusting the frequency of the laser pickup head, it was possible to squeeze 8.5gb onto a carrier with the same physical dimensions as a 1982 CD. To encode a two-hour feature film into this capacity and with a picture quality that would look acceptable on an average domestic television set, a compression ratio in the order of 20:1 was needed. It was to achieve this that the asymmetric design and dynamic, lossy compression features of MPEG-2 were exploited. Due mainly to the economies of scale gained by the fact that the disc transport mechanism in many consumer players was a slightly modified version of those already in mass production for audio and data CD drives, market saturation happened very quickly. Sales in the US almost tripled in the three years to 2003, and at the time of writing this format has effectively superseded VHS as the principal format for sales of pre-recorded media.[12]

Digital terrestrial television has been the other main consumer application for MPEG-2. Since digitisation became a commercial reality in videotape recording during the late 1980s, engineers started looking for ways of transmitting the pictures digitally, replacing the VHF and UHF analogue radio signal used to carry television images since that method was developed by EMI and RCA in the 1930s. The reason is simple. Only a very small part of the total electromagnetic spectrum (all possible wavelengths of electromagnetic energy)[13] can be used for the wireless transmission of electromagnetic energy which can be made to represent images and sounds, known as radio waves. Within the radio waves spectrum, the number of wavelengths which can be used to carry individual signals is finite, and each analogue PAL or NTSC television broadcast consumes a disproportionately large amount of this total 'bandwidth'. As John Watkinson explains:

> There is only one electromagnetic spectrum, and pressure from other services such as cellular telephones makes efficient use of bandwidth mandatory. Analogue television broadcasting is an old technology and makes very inefficient use of bandwidth. Its replacement by a compressed digital transmission will be inevitable, for the practical reason that the bandwidth is needed elsewhere.[14]

As with the scheduled broadcasts in the two years before the outbreak of World War Two, Britain has led the way with digital television broadcasting. The first steps came in the early 1980s when the BBC's engineers developed the NICAM (Near Instantaneously Compacted and Expanded Audio Multiplex) protocol. This was a means of broadcasting two channels of digital audio alongside the existing PAL analogue video transmission. Its first public broadcast was of a classical music concert in July 1986, and it has continued to be used with all terrestrial analogue broadcasts in the UK ever since (at the time of writing practically all televisions and consumer VCRs sold in the UK include a NICAM audio receiver). Although NICAM has also been adopted in some Scandinavian countries, the system was not modified for NTSC use or used on any significant scale outside Europe, where a variety of analogue stereo audio methods for television use emerged during the late 1970s and early 1980s.

At this point MPEG-2 was still some years away, but the issue of digital television was now firmly on the agenda. In September 1993 the Digital Video Broadcasting (DVB) group was formed as a consortium consisting of broadcasters and electronics manufacturers, principally in Europe. Over the following five years it developed and tested three transmission standards, all of them based on MPEG-2: DVB-T (terrestrial), DVB-S (satellite) and DVB-C (cable). In addition to the MPEG-2 video and audio streams, DVB also included data transmission standards for related services, such as the so-called 'interactive' content which can be broadcast alongside each television channel, e.g. subtitles, alternative audio tracks or additional textual information about the programme being broadcast.

After two years of experiments and development of infrastructure, the first regular DVB service began in the UK on 15 November 1998. Sweden followed suit in April 1999, and over the next two years most of Western Europe and Australia did likewise. At the time of writing, the US still uses NTSC analogue for terrestrial broadcasting, though the French owned cable operator Canal+ had offered DVB-C transmissions since 1997. Thanks to the very low bandwidth needed to broadcast the MPEG-2 programme stream it is possible to transmit tens of digital channels alongside the existing analogue ones using the DVB-T system – between thirty and forty in the case of Britain, depending on the transmitter.

This of course leads into the question as to when the original purpose of going digital will be realised, by switching off the analogue signal. History shows us that the greater the market saturation of a consumer technology, the greater the resistance is to the onset of its obsolescence. Videotape formats tend to be superseded comparatively quickly (every decade or so, roughly speaking) in the broadcast and professional sectors because these users derive direct economic benefits from investing in the new technology – for example in the greater flexibility by not having to worry about generational loss when editing video content digitally. The same is not true of media technology sold directly to consumers, who tend to be reluctant to invest unless there is either a demonstrable functionality benefit or they are given no choice. When the BBC announced its competition between the Baird and EMI transmission systems in 1935, the Baird 30-line standard which the winner was intended to supersede could be killed off quickly and uncontroversially, because only a tiny number of 30-line receivers had been sold. The Goldmark/CBS colour system was abandoned after small-scale experimental broadcasts for the same reason.

With consumer technology, however, the situation is very different. The 405-line to PAL conversion in Britain, which began in July 1967, offers a typical example of how technological change tends to happen. At first, the majority of the population were not willing to invest in a new television – the slightly sharper picture simply was not worth what the new sets cost (and the price of colour was stratospheric). Furthermore, 'set-top boxes' were sold to convert the new BBC2 channel, which was only broadcast in PAL, to 405-line so that it could be received on existing sets. As a result the two systems had to be broadcast side-by-side until January 1985, and therefore the conversion took over 17 years. Another example is the commercial lifetime of VHS, which has lasted for almost thirty years and it is still the principal consumer format for off-air recording. Despite JVC's attempt to 'do a Betacam' by introducing a digital version of VHS which was backwards compatible with analogue tapes in 1995, the hardware has hardly sold at all, except to a few enthusiasts.

The set-top 'digibox',which includes a digital to analogue converter and connects by wire to an existing PAL set, has also been the primary method through which DVB has been introduced, and, as with the DVD, the rate of market penetration has certainly been significantly higher than with any comparable analogue technology. The UK regulatory body for public broadcasting, Ofcom, estimated that as of 30 June 2004, 55 per cent of households had access to digital broadcasting, either in terrestrial, satellite or cable form.[15] Given that in most Western countries, television broadcasting also has a public service dimension (it has even been argued that having access to television should be considered a basic human right), any change in the broadcast standard which will render expensive capital items in the homes of low-earners obsolete inevitably becomes as much of a political issue as a technological one, as has the regulation of public broadcasting throughout its history. A report published by the UK government in October 2004 concluded that 'digital television [has to] be near-universally available, generally affordable and have been taken up by the majority of consumers before the analogue signal can be switched off'.[16] Its authors set a target of 95 per cent of the population to be receiving digital television 'on their main television set' before the analogue plug is finally pulled[17] and hinted at the potential political fallout that would result from imposing 'the penalty costs of forced conversion'.[18] It is likely, therefore, that analogue television broadcasting will still be with us for some time to come, and furthermore that another potential benefit of DVB will take a long time to roll out. At the time of writing, only the Faroe Islands and an experimental zone in Berlin consisting of one or two suburbs had switched off analogue television. It should also be remembered that while the 'digibox' remains the receiving technology of choice, converting the MPEG-2 stream to an analogue PAL signal and then passing it by wire to a standard television results in a far lower quality of picture than displaying the digital pixels directly on a CRT or TFT, as is the case on the monitor of a personal computer.

Piracy

As has been noted in chapter six, the illegal copying and/or commercial exploitation of media content has been a major concern pretty much ever since moving image technology has been used on a commercial basis. The industry argued that the advent of consumer VCRs in the late 1970s turned this into an endemic problem, i.e. that the extent of piracy was systematic and that it was resulting in significant loss of revenue. The response of copyright owners has traditionally taken either or both of two forms: legal action and technological barriers. The latter have usually worked on the twin principles of making it sufficiently difficult to copy prerecorded media content as to stop everyone from doing it except organised criminals and a small minority of consumers with above-average technical skills, and a preference for distributing content on media which are inherently resistant to unauthorised use (e.g. the VHS format, which suffers massive generational loss when copied).

The advent of digital media raised the stakes significantly. It will be recalled that the key advantage of digitising video and audio is that it can then be 'cloned' – copied without any loss of signal quality. If legitimate users of the technology can do this, then so can pirates. When the first mass-marketed form of digital media, the audio CD, went on the market, the music industry initially did not perceive the move to digital as being a problem. The discs could only be mastered and duplicated in purpose-built factories, and although purchasers could make analogue copies on magnetic tape, they had been doing that with LP records for decades previously and the industry had still survived. The emergence of the recordable CD (CD-R) in the late 1990s changed all that.

Both CDs and DVDs encode data as minute indentations in the surface of a reflective disc. In the factory-pressed CD or DVD, the discs are produced from a 'glass master' etched by a laser and then coated with a transparent, reflective coating on a production line. In 1989 the Japanese electronics manufacturer Taiyo Yuden produced a stand-alone CD recorder that 'burnt' the indentations into a dye which was already present on the pre-manufactured disc. The machine cost $149,000 and even the blank discs cost three figures. Its main application was to burn a single copy of a completed CD in order to check it for errors before the expense of producing a glass master was incurred. The price of this technology dropped gradually during the early 1990s, but nowhere near the point at which consumers could make cloned copies of commercially sold audio CDs for less than their cover price. All that changed in 1995, when Philips produced the first computer-based CD writer which sold below the $1,000 mark. Set-top audio CD recorders soon followed, and today the ability to copy CDs and DVDs is a routine feature of virtually every personal computer sold in the Western world. Furthermore, the blank media typically retail for less than $1 a piece. Recordable DVD technology was marketed to the public from the launch of the Panasonic DVR-A03 in June 2001, and since then its cost and availability has followed a similar trajectory.

The 'Video CD' (VCD) was the first optical disc video format to be systematically exploited by pirates. It was a very low resolution system, based on the MPEG-1 protocol, which could encode up to 70 minutes of video on a conventional CD. It was used extensively as a consumer format in the Middle East, China and Asia during the late 1990s, but failed to gain a market foothold in the West. This was due mainly to the poor picture quality and the fact that Hollywood refused to release media on Video CD because by that stage, the DVD was already in an advanced stage of development. Due to the unprecedentedly low cost of hardware and blank media, it proved to be the ideal medium for pirates in poorer, developing countries: so much so that it was estimated that in China in 1998, pirate VCDs outsold legitimate copies of feature films by a ratio of 35 to 1.[19]

With the digital piracy genie well and truly out of its bottle, the industry had to resort, once again, to its two traditional lines of attack, legal and technical. At the time of writing almost all commercially published DVDs for retail sale incorporate at least two copy protection methods: Macrovision (see chapter six) to prevent unauthorised duplication to VHS, and the 'Content Scrambling System' (CSS) to prevent unauthorised duplication of the digital data. CSS works by exploiting a so-called 'hidden area' of the DVD near the inner rim, which can be written onto on a factory-pressed disc but not a home-recordable one. The hidden area includes an encryption key, which enables the software in a DVD player to decode the encrypted video and audio data. Simply cloning the data from the main area of the disc onto a blank will produce an unplayable copy, because the CSS encryption key will not have been copied along with it. Unfortunately, the designers of this system failed to take account of the fact that if the hidden area can be read by a set-top player, it can also be read by illegally used decryption software running on a personal computer. At the time of writing such software is said to be easily available from the Internet, and will enable the pirate to decrypt the video and audio data before burning the 'en clair' version to a recordable DVD.

Although the same rationale applies to CSS as with other technological attempts to restrict piracy (i.e. that only organised criminals and computer geeks will have the motivation and the technical knowledge to circumvent it), the media industry of the early twenty-first century increasingly felt that it was fighting a losing battle. The bottom line is that if human brains can dream up encryption systems, other human brains can crack them: as one film industry technical specialist recently told me, 'as soon as we build a ten-foot wall, they build an eleven-foot ladder'. But in the digital domain the stakes are higher, so the laws are getting tougher. This has resulted in the emergence of new legislation designed specifically to combat copyright fraud perpetrated using digital means, both in the US and Europe. In 1998 the United States Government passed the Digital Millennium Copyright Act, while the European Union Copyright Directive of 2001 is currently being implemented by EU member states. Both laws specifically make it an offence to circumvent anti-piracy measures incorporated into commercially-sold computer software or digital audiovisual media, and the latter even makes it an offence to use equipment or software designed for this purpose, even if no copyright is actually infringed by doing so. In 2004 following intense lobbying by Hollywood and the music industry, the US government also went one step further. It proposed to introduce legislation that would have the effect of nullifying the Betamax decision of 1981 (see chapter six), in effect outlawing the possession any equipment which *could* be used for piracy, even if its owners did not use it for this purpose. When civil rights campaigners pointed out that this would turn all VCR owners into criminals, the government backed down – but the fact that it was ever seriously considered in the first place gives a powerful demonstration of just how worried copyright owners had become over the issue of digital piracy.

The Internet

> The Information Highway will transform itself, even more than it is at present, into the Information Toll Road.[20]

This issue of piracy leads directly onto the other key area of computing technology that is used to store and deliver audiovisual media in digital form – the Internet. The Internet is, in effect, a global telecommunications network for computers. The function it performs is no more and no less than providing the ability to transport large quantities of digital data over long distances, including across physical (such as oceans) and international boundaries, in a very short time and at relatively low cost.

The Internet's origins lie in the Cold War, specifically in a project which was ostensibly a communications network for the mainframes used for research in university computing departments in the 1970s, the real purpose of which was a military communications network designed to enable the government to continue functioning in the aftermath of a nuclear attack. This was ARPAnet, named after the US Government's Advanced Research Projects Agency, which was established by the Eisenhower administration in 1957. The network began operation in 1969, and over the next two decades grew to encompass a number of other long-distance computer communications systems which had been established independently. This was achieved through a combination of dedicated data communications lines, existing telephone lines, radio and satellite links.

The process which enabled this infrastructure to grow and transform into the Internet we know now encompasses both political and technological factors. The first key political one was a gradual move from military and governmental control of the interlinked networks which comprised the system as a whole, to civilian and private sector management. Two key milestones on this path were in 1979, when commercial use of the Internet was permitted for the first time, and in 1995 when the US government effectively privatised the network infrastructure within its own borders. Flowing from these political decisions is the economic structure which finances the Internet. Being a largely autonomous collection of interlinked computers, the user (be that an individual or organisation) pays a flat rate to use the network as a whole rather than a price calculated on distance, as is the case with other mass communications networks such as the postal and telephone services. Users who wish to receive data which is made available through the Internet and users who wish to make their data available to others buy access from an Internet Service Provider (ISP), which acts as a gateway to the wider network. In the case of the former, the charge is usually levied by the 'online' time a consumer's computer is connected to the ISP's. In the case of the latter, the data provider's computer, known as a 'server', is permanently connected to the ISP's, who generally charges by bandwidth (i.e. the volume of data flowing in or out). The ISPs in turn purchase access to the telecommuications 'backbones' which make up the global Internet and amortise that cost in what they pass on to their customers. The important thing to note about this economic model is that charges for data transfer are *not* levied according to where that data has come from or is being sent to, as with the telephone and postal services. To the end user, therefore, it is no more expensive to communicate with the other side of the world than the next street. Given that legal attempts to combat piracy depend on legislatures that are, by and large, incapable of working across international boundaries, an infrastructure which enables audiovisual media not only to be copied but digitally 'cloned' across international boundaries quickly and easily is obviously a big concern for rights holders.[21]

The other factors which grew the Internet into a medium of mass communication were technological. To start with there was the effect of three decades of Moore's Law, which in this case did not only apply to the processing power of individual computers, but also to the speed at which data could be reliably modulated into an analogue signal for transmission along a conventional telephone line. This meant that by the mid-1990s, the average personal computer in someone's home or office was powerful enough to carry out a range of consumer applications which would potentially benefit from the use of a long-distance data communications network, and the Internet had become fast and accessible enough to deliver that use effectively. The other important factor was the old chestnut of standardisation. Just as Dickson's 35mm film format became a truly global medium because the core technical variables were standardised, so did the Internet because the same thing happened to the software used to pass data through the Internet. These standards include the Transmission Control Protocol (TCP), which emerged during the late 1970s, the File Transfer Protocol (FTP), Hypertext Transfer Procotol (HTTP) and Hypertext Mark-Up Language (HTML), which form the basis of the World Wide Web (WWW), the system of transferring text and graphics 'pages' through which most end users access the Internet.

There was a significant repositioning of technological goalposts when Internet-based audiovisual media began to emerge, relative to consumer hardware which was designed specifically and only for recording and reproducing it. The device through which the media was delivered was the personal computer, a machine which is designed to have other applications besides recording and playing audio and video. In particular, its principal data storage medium was the hard disc, which until the late 1990s was a lot less reliable and could not offer the same capacity as removable media such as CDs, DVDs and digital videotape. And unlike with offline media, bandwidth was very severely restricted. Until the early 2000s, the highest speed at which a personal computer could connect to the Internet equated to approximately 5kb per second – at that rate, the volume of data equivalent to a feature film on DVD would literally take days to download. With the advent of the Digital Subscriber Line (DSL), colloquially known as 'broadband', in the early years of the twenty-first century, this has increased tenfold to around 50kb per second, but even at that speed our feature film would still take 10–12 hours.

The first audiovisual medium to be transmitted by Internet on a significant scale was audio, for the same reasons that it led the way with offline media. The breakthrough came with the emergence of the 'MP3' format in the late 1990s, which was a very heavily lossy compression format that just about enabled 2–3 minutes of CD quality audio to be squeezed into under a megabyte. Being an asymmetric system the encoding process used more computing power, but it was still well within the capabilities of an average PC. At around the turn of the century large amounts of content, mainly popular music, had been illegally encoded into this format and made available for public download on websites throughout the world. The industry responded through a combination of legal action against the owners of the servers holding the illegal media content in the country where they were situated, and the establishment of officially-sanctioned sites which legitimately sold MP3 files for download. Even so, music industry representatives were still repeatedly claiming at the time of writing that illegal downloading was hitting their turnover to the order of tens of per cent.

Use of the Internet for video broadcasting (and for that matter, piracy) is today still not as big an issue as it is with audio, because the infrastructure of the Internet itself is simply not yet capable of moving the volume of data needed fast enough to deliver an equivalent quality picture to a PAL or NTSC television broadcast in anything like real time, even with the heavy compression of MPEG-2 and its competing protocols which are now starting to emerge. Assuming that Moore's Law applies to the growth of Internet capacity as it always has done, this will start to be an issue around the late 2000s, when we can expect to see the same profile of legitimate and illegitimate Internet use with video as we have with audio. While digitised pirate films have been transferred over the Internet on a small scale, these are so heavily compressed and with such a small image that even the 'grainy, jerky picture which runs in a tiny window of your computer screen' reported by the *Guardian* journalist in 1994 would seem like a complimentary description. Another issue is that, while MPEG-2 is bound to be with us for a long time due to its use in dedicated DVB and

DVD hardware, it is by no means the only video encoding method available for PC use. There are several in widespread use, with two key competitors to MPEG being the proprietary systems developed by the two major producers of PC operating system software, Microsoft's Windows Media format and Apple's Quicktime. Both offer demonstrable advantages over MPEG for certain specific applications, and before Internet video broadcasting becomes a commercial reality, some serious standardisation will have to take place. At the moment, therefore, consumers without a conscience are far more likely to buy pirate DVDs from their local market stall than to download movies from illegal websites.

The one significant use of digital video delivered specifically via the Internet to date has been to distribute political propaganda which for a number of reasons would not be broadcast by terrestrial television, because the latter is politically regulated on a national basis. Probably the best known example is the use of the Internet for this purpose by Islamic terrorists. It has included the distribution of lectures by the terrorist leader Osama bin Laden following the destruction by his organisation Al-Qaeda of the World Trade Center in New York on 11 September 2001, and videotaped footage of the murders by beheading of a number of Western hostages in the aftermath of the Iraq War. The VCD is also believed to have been used on a significant scale for distributing terrorist and extremist political propaganda before the Internet became capable of that function, and for similar reasons – the discs are very cheap and easy to produce, and are small enough to be illicitly shipped without detection.[22] The fact that the videos themselves are of such low resolution that they do not even approach the definition of VHS is not the issue. The point is that these Islamic terrorists are specifically exploiting a unique technological attribute of the Internet: its ability to pass digital media across international boundaries without national governments being able to do much about it. The nature and extent to which the Internet will come under legal regulation is an interesting topic for speculation, but so far the only countries where this has happened have been those in which all ISPs are either government-owned or heavily regulated, China and Iran being two key examples. In any Western democracy, barring access to any given content would require every ISP (i.e. public point of access to the Internet) in a given country to co-operate by blocking access to the servers hosting that content by its customers, and even if they did so an individual would still have the option of establishing a telephone link to a foreign ISP and accessing the banned content that way (albeit very slowly and at a significant cost).

One striking and almost farcical example can be found in the form of a story published in 2003 by an Italian newspaper, which alleged that the Prince of Wales is bisexual and had had a gay affair with his butler. A court injunction was swiftly obtained and the grandees in Buckingham Palace doubtless thought that that was that. Far from quelling any interest, their action probably ensured that the allegation gained more credence than if they had just ignored it. The British print and broadcast media referred extensively to the allegation and its original source without explicitly stating what the allegation actually was, thereby inviting anyone who spoke Italian to head for the nearest PC and download the whole story in a matter of seconds.

Prince Charles consequently discovered that a court order obtained in London had no jurisdiction over an Internet server situated in Rome.

In the early years of the twentieth century, therefore, the significance of the Internet in moving image technology is principally political; though its eventual use as a mass communications medium for digital video in similar ways to that of broadcasting and offline media is almost certainly imminent.

Film

Film comes next down the list of moving image technologies to be affected by the digital revolution in chronological terms, despite having been the first one to be invented. Once again, the reason is Moore's Law. Because the first and primary application film was designed for was to enable the projection of a moving image on a surface area of up to hundreds of square feet, the amount of detail (termed *definition* or *resolution*) in the emulsion of a 35mm film element is many times that of videotape or broadcast television. To put this in perspective, a digital representation of a PAL television frame using the MPEG-2 system is 720 x 576 pixels (about 25mb per frame uncompressed), but a digital representation of a 35mm film frame with the emulsion densities being produced at the turn of the twenty-first century is around 5,000 pixels square (40gb per frame). An uncompressed feature film stored to the resolution of a 35mm camera negative would require tens of terabytes of hard disc space. At the time of writing, the total replacement of film by computer technology is even further in the distance than full resolution Internet video.

It will be recalled from chapters one and two that the production of film-based moving images has three stages: origination in the camera, the post-production stages (such as editing, the introduction of special effects and duplication) and projection in the cinema (and/or telecine transfer for television and video use). The first of these stages to introduce computer technology was post-production, specifically in the emergence of CGI, hence the significance of *Jurassic Park* and *Terminator 2*.

As with the role of telecine technology in providing the interface between film and video-based media, the use of digital intermediate stages in the production of feature films for cinema exhibition requires a way of converting the images on exposed and processed film into digital data, and then, after the footage has been edited and manipulated digitally, outputting the result back onto film so that it can then be duplicated and distributed as any other cut camera negative would be. The first of these processes involves the use of a scanner, sometimes termed a *datacine* in order to distinguish it from a telecine, the first of which were produced by Eastman Kodak using the trade name 'Cineon' in the early 1990s. The difference is that a telecine outputs an analogue signal which conforms to a specific broadcast television standard (usually PAL or NTSC), dividing the picture up into horizontal lines. A datacine contains a CCD array which scans the picture as a grid of individual pixels and simply outputs the data in that form, without dividing it up into lines or interlacing it. The advantage of this method is that with sufficient computing power to process the resulting data, the output of a single scan can be used to produce every sort of element from a VHS tape to a high resolution data stream for digital cinema projection:

> This concept started to become feasible in 1996 when Philips introduced the Spirit DataCine film scanner. It was the first telecine that, in addition to standard and high-definition video output, provided data output with direct film density representation. Its application to long-form film transfer work, e.g. full-length feature films, was made possible by an increase in transfer rate of one to two orders of magnitude over previous data output-type devices.[23]

The volume of data produced for each frame, or the resolution, of a given scanner is commonly expressed in multiples of thousands of pixels squared, e.g. '2K' usually describes an image offor example, 2,048 x 1,744 pixels, assuming the Academy ratio.[24] At the time of writing, datacine scanners for professional film industry use are being produced by four principal manufacturers: Kodak/Cineon, Oxberry, Philips/Spirit and Quantel, with resolutions ranging from 2K to 6K. It is estimated that the emulsions of most 35mm camera negative stocks have a grain density roughly equivalent to 4–5k.

Once the source footage has been scanned it can be edited and its visual properties manipulated using a wide range of software tools which have been developed for the purpose. When the use of digital post-production was in its infancy it was usually the case that the majority of editing was done conventionally (i.e. using a 16mm or 35mm cutting copy), with only those sections of camera negative which combined CGI and photographically-originated images being processed digitally, before being output back to film and then spliced into the assembled negative. Throughout the 1990s and early 2000s the role of computer technology in post-production has been steadily expanding. The use of analogue video copies made from a telecine transfer for editing dates back to the mid-1980s, when this method started to become substantially cheaper than producing a cutting copy on film. By the late 1990s editors were increasingly making use of PC-based editing software, which at the end of the process would output an *edit decision list* (EDL). An EDL contains precise instructions that enable the as yet untouched camera negative to be *conformed*, or cut in order to exactly match the edited sequence in the computer, before duplication using the conventional route. One of the many advantages this method offers, especially in low-budget production, is to minimise handling of the camera negative, thereby reducing the chances of contamination or damage.

Outputting the processed image data back to film is a key requirement of digital intermediate work, and will remain so unless and until film totally disappears from all stages of production, post-production, distribution and exhibition. Film recorders expose each individual pixel onto a 35mm fine-grain internegative or interpositive element in one of two ways. Until around the turn of the century, CRTs which expose red, green and blue (or yellow, cyan or magenta, depending on whether the destination film stock is positive or negative) colour records sequentially and through filters (it will be recalled from chapter six that the luminance generated by a CRT does not in itself have any colour) were the norm. During the late 1990s laser-operated film recorders started to be developed, though they remained prohibitively expensive for most operations until the early 2000s. As this description of Kodak's Lightning recorder suggests, lasers are a far more accurate method of burning image data onto film:

> The Lightning Recorder's technology is unique in that it uses red, green and blue lasers to expose negative film. The three lasers write directly to each colour layer of the intermediate stock. This combination produces images of unparalleled sharpness and colour saturation. Cinesite [the post-production company from whose website this text is quoted] outputs to 5244 intermediate stock, which is virtually grainless (ASA 3) and has the ability to record the grain pattern from faster stocks so that the digital transformation is unnoticeable in the final edit.[25]

At the time of writing, studio cameras which originate images as digital data are not being used on any significant scale for major feature-film production, although they are starting to replace 16mm and Super 16 for low-budget feature-film and high-budget television drama and documentary production. The 'Digital High Definition' (known as 'HD') format was developed by Sony in the late 1990s, the key breakthrough being the launch of the HDC-F900 camera in 2000. This incorporated the '24-p' format meaning that, as in a datacine, the camera's CCD exposed progressive, discrete frames at the rate of 24 per second, in order to maximise compatibility with film. HD's most vocal advocate in the film world to date has probably been the science fiction director George Lucas, who shot one of his *Star Wars* series of films almost entirely using the following generation of HD camera, the HDC-F950, in 2003. In the main, though, feature studios have preferred to stick with film: new generations of stock remain compatible with their existing cameras, the perceived image quality of HD is not believed by most people to match the best that film can offer and as yet there is no demonstrable economic advantage in abandoning film for studio origination in cases where extensive special effects are not needed.

Cinema exhibition

> Any adoption of a new methodology or a new technology must take into account two issues: (i) does the new way decrease the cost of getting the film into distribution, or (ii) is there a different benefit, such as increased creativity? Change for the sake of change in the film world is not common practice.[26]

Of all the areas of moving image technology in which computers are playing an increasing role, its use for projection in cinemas has arguably generated the most controversy within industry circles, and its future is possibly the most difficult to predict.

It is worth bearing in mind that the electronic projection of television images onto large screens is nothing new. Probably the earliest successfully demonstrated and commercially used technology for this purpose was the Eidophor (from the

Greek *Eidos* – 'image', and *phor* – 'conveyor'), developed by the Swiss scientist Fritz Fischer, first demonstrated in 1943 and remaining on the market in one form or another until 2000. The light source of the Eidophor was the same as in a conventional cinema projector – carbon arc at first, with later models being lit by xenon arc bulbs. The imaging device consisted of a thin layer of oil, known as the 'Eidophor liquid', coated on a reflective surface, the opacity of which could be modulated by an electron beam similar to the ones used in CRT television receivers. In effect, the coated surface acted as the equivalent of the gate in a film projector: by projecting the arc light through it and focusing it with a lens, it was possible to project a television image onto a large screen.[27]

The Eidophor was used on a small scale for cinema exhibition in Europe and North America, usually for live feeds of special events such as sports and election coverage. Its core market, though, was in non-theatrical venues such as university lecture theatres. Other television projection technologies emerged during the 1970s and 1980s which gradually superseded the Eidophor. The use of high-power CRTs for large-screen projection was pioneered by, among others, the Belgian-American Radio Corporation (Barco): these designs featured three separate tubes, one for each of the primary colours, each of which had to be aligned with the others to produce a properly focused image in projection. In the early 1990s LCD arrays that could withstand the heat of a projection lamp were developed, resulting in the production of small, portable projectors that were suitable for displaying video or the display of a PC monitor in settings such as classrooms and business meetings.

Until this point no one had seriously considered the use of electronic projection as a replacement for 35mm and 70mm film in cinemas. The only moving images that Eidophors, CRT and LCD-based projectors had been able to display thus far were those supplied either by a videotape or a live broadcast – in other words, electronic projection was limited to the definition of PAL or NTSC did not even come close to the image quality offered by film. Two factors changed that: the effect of Moore's Law, which, by the early 2000s, had given us the computers and data storage space capable of handling the volume of data needed for digital moving images to approach the resolution of film, and the invention of the Digital Light Processing (DLP) imaging device.

The DLP technique was initially developed by Larry Hornbeck of the American electronics giant Texas Instruments in 1987. The guts of the system is the 'digital mircomirror device', a microscopic mirror mounted on an adjustable hinge. When a light source is placed at right angles to the device, the angle of the micromirror determines whether or not it will refract the light through a lens in front of it. Each DLP chip contains an array of micromirrors of a similar density to the imaging CCDs used in HD cameras and datacines. A DLP cinema projector will contain three chips, one for each of the primary colours. Today, 2K DLP cinema projectors are being marketed, with the launch of 4K models said to be imminent.

DLP has been aggressively promoted by the electronics industry as a replacement for release prints on film. Theoretically it provides the final piece in the digital imaging jigsaw where feature films are concerned: the combination of HD cameras,

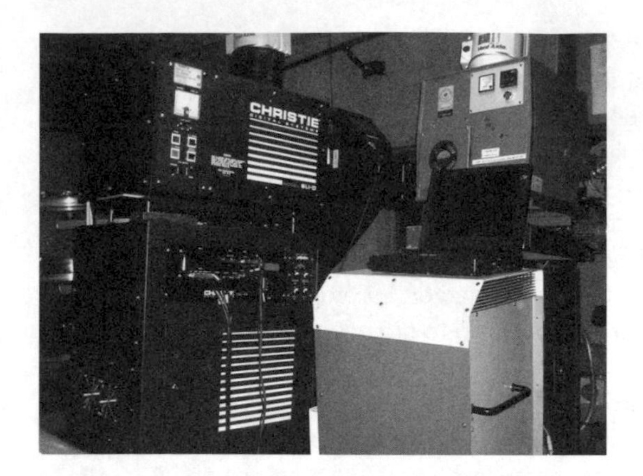

Fig. 8.1: A 1.3k DLP projector installed alongside a conventional 35mm film projector in a modern cinema.

computer-based post-production and DLP cinema projection is being promoted as finally enabling moving images to be produced and shown with a quality that was previously only available using film, only without any film. The detractors of digital imaging point out that resolution is not the only relevant criterion in determining the quality of a projected image. Colour depth – the number of separate shades it is possible to colour each pixel – is also an issue, with film's supporters pointing out that with a photochemical emulsion this is infinite, whereas in the digital domain it is fixed and carries a data storage penalty the higher you go. The inevitable response to this is that Moore's Law will solve that problem sooner or later.

There is another significant issue involved, though; one which will not necessarily be solved by Moore's Law. This is the economic structure through which distribution and exhibition on film has traditionally been supported. A research study on the economic potential of digital cinema projection carried out by Wall Street financial researchers in 2002 noted that 'the hype surrounding the technology was overwhelming' when they first examined the sector in 2000, but concluded that 'structural issues in the industry were likely to serve as a major impediment to adoption'.[28] Those structural issues still exist, and if anything represent an even greater impediment than they did in 2002. A key characteristic of 35mm is that the imaging device is contained within the film itself, in the form of the emulsion. The research and development which goes into producing each successive generation of camera negative, intermediate and release print stocks results in a product which delivers a demonstrably 'better' image than its predecessor, but which remains compatible with existing cinema projectors. The projectors themselves are essentially very simple, mechanical devices. A 35mm projector built in the 1940s will, if properly maintained, be able to display the benefit of a new generation of release print stock sold in the twenty-first century. At present a typical new 35mm projector mechanism costs around $30,000. A state-of-the-art DLP projector costs over $150,000.

With DLP the technology is in the hardware. Not only is the projector a far bigger investment, but the resolution and colour-depth possible from a given imaging device is fixed and cannot be upgraded. In this respect Moore's Law may even be a negative factor rather than a positive one – whereas with film, a 35mm projector can reasonably be assumed to have a service life of several decades, the rapid growth in computing power may well mean that a year or two after a cinema invests in DLP, its rival venue acquires a newer model of projector, producing a better image and rendering theirs obsolete and economically uncompetitive.

Furthermore, history shows that new technologies which have required a substantial investment at the exhibition end usually fail. The industry wanted exhibitors to install widescreen in the early 1930s, but exhibitors decided that it was not worth it. The same happened when Twentieth Century Fox tried to package magnetic sound with CinemaScope in the 1950s and when Eastman Kodak tried to launch a digital sound system that was not backwards compatible, CDS, in the early 1990s. It is hardly surprising that the electronics industry is lobbying Hollywood hard over DLP, stressing the potential savings to be made by now avoiding the cost of striking large inventories of 35mm prints, which in the case of a big blockbuster can run into millions of dollars. But cinema exhibitors are not about to spend the huge amounts of money needed even for the cinemas in North America and Europe to convert on a meaningful scale, let alone the rest of the world, without a substantial proportion of the studios' savings being passed on to them. So far, no sign of that happening is apparent.

Other obstacles remain, chiefly that of standardisation. Hollywood representatives have for some time taken the line that 4k should be considered a bare minimum for any systematic rollout, in order to show-off high-budget production values to the best advantage. Independent filmmakers and the 'art house' sector are calling for standardisation at 2k, as this would substantially lower the cost of encoding the image data from whatever source it was originated on to the format used by the cinema DLP systems. As things stand, 35mm film is a global, universally agreed standard, and no-one in the industry wants the nightmare of it being replaced by a myriad of incompatible formats and systems, which is precisely what has happened with videotape. Furthermore, film industry representatives are increasingly voicing fears over the potential risk of piracy, especially if the method of delivering the image data to cinemas is some form of online data transmission, which could theoretically be intercepted. Were this to happen the stakes would be even higher still than with DVD piracy, because criminals could potentially get their hands on clonable, cinema-quality copies of films weeks or months before they were scheduled to be broadcast or released on consumer media. Unlike most electronic media, 35mm film is inherently very safe as far as piracy prevention goes: a release print of a feature film is bulky, very heavy, kept in buildings which can be made reasonably secure, requires specialist technical skills to handle, requires expensive, specialist telecine equipment to duplicate and being of much higher contrast than intermediate elements, does not telecine very well.

The result of all these factors is that cinema exhibition is about the one area of moving image technology in which the industry actually seems to be withdrawing from the digital route. A projection engineer I spoke to recently told me that 'the industry doesn't want it [digital projection] and isn't willing to pay for it'. As things stand, the only people who are persisting with digital projection in its current form are public sector arts organisations, of which several in European countries are advocating the use of DLP for enabling low-budget, specialist and archival re-release films to be shown in 'art house' venues when the cost of striking a 35mm print would be a significant barrier. They are operating in the usually mistaken belief that DLP will, for this market, bring the overall cost of distribution and exhibition down while pushing the range and technical quality of material shown up. One particular example is the British Government's film agency, the UK Film Council (UKFC). In 2003 it announced its 'Digital Screen Network' initiative, a £15 million fund intended to equip up to 250 independent cinemas with DLP projectors. The statements which appeared on UKFC's website, characterised by emotive rhetoric such as 'freeing cinema from the tyranny of 35mm', during the following year demonstrated quite obviously that whoever had thought up the idea had very little technical knowledge; had not addressed fundamental questions of compatibility, standardisation, the cost of encoding and distributing digital media to venues, the cost of maintenance and rapid equipment obsolescence and much else besides; had not sought the advice of any real technical experts (or if they did, had ignored it); and had failed to do the simple sums which would have revealed that UKFC's distribution and exhibition objectives could have been met far more cheaply and efficiently using 35mm. That such a scheme ever got past the drawing board is, in the opinion of this author, almost a criminally negligent waste of taxpayers' money, and a powerful demonstration of the spin and propaganda which surrounds the 'D word' where moving image technology is concerned.

It is impossible to predict if and when cinema projection will eventually go digital, because the barriers to conversion lie more in the economic and political domain than the technological. The speed at which the quality of digital imaging improves is going to have to start outpacing that of film. Up to now it has not; indeed it could even be argued that Moore's Law applies to the evolution of film emulsions as much as with computer-based imaging. There will also need to be a shift in the economics of film distribution so that the investment in digital projectors can be amortised more evenly across the different sectors of the industry than would be the case if cinemas just went out and bought them. Once again, there is no sign of any meaningful progress on this front yet.

Archival restoration and preservation

The 'D word' essentially means two things for archivists, one of them a godsend, the other a nightmare. The godsend is in the use of digital intermediate technology to carry out film restoration functions which could previously only be dealt with by photochemical methods. Until the late 1990s, for example, scratches and dirt could only be removed by invasive chemical treatments applied to original elements and/or expensive photochemical duplication using techniques such as wet-gate printing. Elements with faded colour dyes could also only be restored by photochemical duplication, specifically by manipulating the light source used to expose the destination stock.

These and other defects can now be corrected using what is essentially the same technology that gave us the special effects in *Jurassic Park* and *Terminator 2*. An original element can be scanned, the image manipulated using a computer and the result output to whatever format is needed for access purposes – potentially anything from a low resolution MPEG file to a new, laser-burnt 35mm intermediate. The importance of retaining original elements (or as near a generation to them as can be acquired) for long-term preservation in appropriate atmospheric conditions is thus heightened, as this enables them to be rescanned as necessary as and when Moore's Law gives us higher-quality scanning and image-manipulation technology. At the time of writing the use of digital technology for restoration work is just coming within the financial reach of well-financed private libraries and large, national public-sector archives, but is not yet cheap enough to be a routine tool used by smaller and more specialist organisations. But in this case it should just be a case of sitting back and waiting for Moore to deliver the goods.

In fact, the issue with which archivists will increasingly have to grapple in relation to digital technology is probably more ethical than technical. In 1990, the joke in *Gremlins 2* in which a cable TV station broadcasts '*Casablanca* – now in full colour and with a happy ending!' was effectively science fiction. In 2004, colourising *Casablanca* to broadcast standard is now well within the capabilities of an average home PC, given the requisite software. This opens up the possibility – indeed, some would say the *probability* – that films will find themselves being 'restored' by individuals and organisations which either do not have the contextual and historical knowledge needed to evolve a coherent model of the 'original' state they are aiming to restore (a hypothetical example would be digitally removing the Dufaycolor réseau pattern in the belief that it is a defect, even though it is in fact a characteristic of the original film) or are working more to a commercial agenda than a cultural one (e.g. digitally removing the Dufaycolor réseau pattern in footage being broadcast in a documentary, in the full knowledge that doing so destroys the technical integrity of the footage but deliberately, on the grounds that uneducated viewers would believe it to be a technical fault). Some PC video editing software in widespread use today (such as Final Cut Pro or Adobe Premiere) actually has the facility to *add* digitally simulated image characteristics such as scratches, flat contrast and a pink, faded hue – presumably in order to suggest to viewers that footage is 'old'.[29] Once this digital genie fully emerges from its bottle, therefore, archivists will have to develop effective and considered strategies for protecting the technical integrity of the footage in their care, once representations of it enter the digital domain.

If digital restoration is the archivists' godsend, attempting to preserve 'born digital' moving images is their nightmare. The term 'born digital' refers to media content which has been originated digitally and has never existed in an analogue form considered of equivalent quality. Examples would include a feature film shot on HD, edited digitally and only ever projected using DLP without any film transfer ever taking place, television news footage originated on Digibeta or DVCAM, or home movies shot on Digital 8 or Mini DV. It will be recalled from chapter seven that the only known reliable method of preserving footage originated on videotape in the long term is 'continuous format migration', i.e. copying from one tape format to another as the former approaches obsolescence. When starting with an analogue medium this is bad enough, but with digital two new problems also have to be addressed. In an analogue videotape, partial signal loss will result in a visible defect (a common

Fig. 8.2 The challenge of preserving 'born digital' in the form of a British newspaper advertisement for consumer DVD recorders, published in the summer of 2003. Many archivists would regard the claim that the data recorded on this medium will be preserved 'forever' as naïve at best, and a cynical manipulation of their customers at worst – especially as the disc shown in the picture is of a format (RW) which is specifically designed to be erased and rerecorded! The implicit and probably unintentional acknowledgement by a major manufacturer of VCRs that their tape format of choice is effectively useless for long-term preservation ('What I used to tape...') is nevertheless remarkable.

example is the momentary loss of picture sync known as a 'dropout'), but the rest of the image will still be displayed. But if, for example, a significant amount of digital data is lost due to chemical decomposition of the magnetic oxide on a tape, one has potentially lost a much larger part of the moving image sequence. So-called error correction protocols – software which uses surviving data to 'guess' the lost part – will work up to a set volume of data loss, but beyond that the whole picture will be lost. In crude terms, an analogue videotape suffering chemical decomposition will play with an imperfect picture on the screen, but a digital one with an equivalent level of damage will play nothing at all. Another issue is that data compression reduces the tolerance of error correction – so, for example, if you lose ten seconds' worth of video on an uncompressed tape, the same volume of data on a compressed tape might represent 20 or even 30 seconds. In practice, archivists have no choice but to try and pre-empt format obsolescence, and to continually examine collections of digital tapes in their care in an attempt to pre-empt physical damage or decomposition.

Compression also imposes another problem, one which actually negates the 'clonability' advantage of digital media. Virtually all the digital video formats in widespread use implement compression of one sort or another. In most cases the only way to decompress them is through digital to analogue conversion. For example, no Digibeta VTR currently on the market will output the datastream bit for bit as it is encoded on the tape. So format migration can only be accomplished by converting from digital to analogue and then back to digital using a different compression system, or, if one is lucky, 'transcoding' the data from one system to another in the digital domain. Either way will introduce generational loss (the latter marginally less so), just as analogue copying methods will.

The volume of 'born digital' video content being generated in a range of market sectors is now simply so large that it will prove impossible to preserve even a significant proportion of it in the long term. This situation is bought about by the fact that no offline data storage medium currently available has been shown to preserve data integrity in long-term storage for anything approaching that of analogue film, and furthermore no new solutions have emerged (as with the temperature and humidity controlled storage of film, for example) to alleviate this situation. Magnetic tapes shed oxide and suffer from dropouts leading to data loss, and even if they were to refrain from doing so the format obsolescence problem remains. Optical discs – CDs and DVDs – are to all intents and purposes an unknown quantity, but overwhelming anecdotal evidence suggests that recordable DVDs burnt shortly after the format's launch in 2001, handled carefully and stored in cool, dry conditions, have already undergone chemical changes which have rendered them unreadable. In a generation's time we may well find ourselves in the strange position whereby film records from the twentieth century have survived intact, but our digital moving image record of the early twenty-first has largely disappeared.

Conclusion

The 'D word' in essence signifies the use of computers to originate, manipulate, distribute and display moving images and sounds. The theoretical possibilities this offered were realised and developed in the early twentieth century, but the data processing power needed to make them a reality did not become widespread until the 1990s. At the time of writing – an indicative though unintentional catchphrase of this chapter – they had made a profound impact on some forms of moving image technology (for example, videotape recording had almost entirely 'gone digital' save for the continuing use of VHS as a consumer medium) but less so in others (principally cinema exhibition). The only reliable tool we have for predicting what happens next is the ubiquitous Moore's Law – and even then, as demonstrated by the commercial failure of digital cinema projection thus far, it is often not the only factor in play.

notes

introduction

1 This model was originally proposed by Laura Mulvey in 'Visual Pleasure and Narrative Cinema' in *Screen*, vol. 16, no. 3 (1975), pp. 6–18.
2 Sokal's essay, 'Transgressing the Boundaries: Toward a Transformative Hermeneutics of Quantum Gravity', can be found in *Social Text*, vol. 46, no. 47 (Spring-Summer 1996), pp. 217–52. For more on the controversy its publication caused, see Alan Sokal & Jean Bricmont, *Intellectual Impostures: Postmodern Philosophers' Abuse of Science*, London, Profile Books (1998).
3 Alan Sokal, 'Throwing a Spanner in the Text', *The Times Higher Education Supplement*, 7 June 1996, p. 19.
4 Barry Salt, *Film Style and Technology* (2nd ed.), London: Starword, 1992.

chapter one

1 *The Observer*, 3 March 2002 (See http://www.observer.co.uk/uk_news/story/0,6903,661093, 00.html for text of complete article: quotation taken from access on 11 January 2002).
2 John Bailey, 'The Future of Cinema: Evolution or Revolution?', *SMPTE Journal*, vol. 108, no. 9 (September 1999), p. 630.
3 Joseph & Barbara Anderson, 'Motion Perception in Motion Pictures', in Teresa de Lauretis & Stephen Heath (eds), *The Cinematic Apparatus* (Basingstoke: Macmillan, 1980), p. 87. See also Stephen Herbert, 'Persistence of Vision', essay published online at http://www.grand-illusions.com/percept.htm.
4 Herbert, ibid.
5 For a more detailed account of the evolution of pre-film technologies in the nineteenth century, see Brian Coe, *The History of Movie Photography* (Westfield, NJ: Eastview Editions, 1981), chapters 1 and 2; Bernard Happé, *Basic Motion Picture Technology* (London: Focal Press, 1971), chapter 1; and Laurent Mannoni, *The Great Art of Light and Shadow* (trans. Richard Crangle, Exeter: Exeter University Press, 2000), passim.
6 Other variables can affect this, notably the camera's aperture setting. These are discussed in chapter two.
7 Josef Maria Eder, *History of Photography* (New York: Dover Publications, 1972), p.700.
8 Mannoni, op. cit, p. 253.
9 US Patent no. 420,130.
10 For a full description of the camera, see E. Kilburn Scott, 'The Career of L. A. A. le Prince', *Journal of the SMPE*, vol. 17 (1931), reproduced in Raymond Fielding (ed.), *A Technological History of Motion Pictures and Television* (Berkeley: University of California Press, 1967), p. 78.

11 Christopher Rawlence, *The Missing Reel* (London: Collins, 1990), p. 27.

12 For more on the Kinesigraph, see Stephen Herbert, *Industry, Liberty and a Vision* (Hastings: Projection Box, 1998), passim.

13 C. F. Jenkins, *Animated Pictures* (Washington D.C.: H. L. McQueen, 1898), pp. 26–44.

14 E. C. Worden, *Nitrocellulose Industries* (New York: D. van Nostrand Co., 1911), vol. 1, pp. 22–4.

15 For details of this research and development see Earl Thiesen, 'The History of Nitrocellulose as a Film Base', *Journal of the SMPE*, vol. 20 (1933), pp. 259–62.

16 Ibid., p. 261.

17 Rawlence, op. cit., p. 246.

18 W. K. L. Dickson, 'A Brief History of the Kinetograph, the Kinetoscope and the Kineto-phonograph', *Journal of the SMPE,* vol. 21 (1933), reprinted in Fielding, op. cit., pp. 11–12.

19 Ibid., p. 12.

20 David Bordwell, Janet Staiger and Kristin Thompson, *The Classical Hollywood Cinema: Film Style and Mode of Production to 1960* (London: Routledge, 1985), pp. 276–8.

21 J. H. McNabb, 'Film Printing', *Transactions of the SMPE*, no. 14 (1922), pp. 41–2.

22 Thanks to this practice, the shape and pitch of film perforations varied widely and can thus provide archivists with important clues as to the provenance of very early film elements.

23 Dickson, op. cit., reprinted in Fielding, p. 12.

24 Barry Salt, *Film Style and Technology: History and Analysis* (2nd edn., London: Starword, 1992), p. 36.

25 Cecil Hepworth, *Came the Dawn: Memories of a Film Pioneer* (London: Phoenix House, 1951), p. 67.

26 C. A. Kenneth Mees, 'History of Professional Black-and-White Motion-Picture Film', *Journal of the SMPE,* vol. 63 (1954), reprinted in Fielding, op. cit., p. 125.

27 Ibid.

28 Bordwell, Staiger and Thompson, op. cit., p. 294.

29 Colin Harding and Simon Popple, *In the Kingdom of Shadows: A Companion to Early Cinema* (London: Cygnus Arts, 1996), p. 43.

30 For more information see T. M. Crozier, *Report to the Right Honourable Secretary of State for Scotland on the Circumstances Attending the Loss of Life at the Glen Cinema, Paisley, on the 31st December, 1929* (London: HMSO, 1930) and the 2004 essay by Brian McGuire at http://www.glencinema.org.uk/history.php (accessed by the author on 8 February 2005). A television documentary about the incident, *The Glen Cinema Disaster*, was produced in 2004 and broadcast by BBC1 in Scotland on 1 February 2005.

31 Carlos Bustamente, 'The Bolex Motion Picture Camera', in John Fullerton and Astrid Soderbergh-Widding (eds) *Moving Images: From Edison to Webcam* (Eastleigh: John Libbey, 2000), p. 61.

32 Brian Winston, *Technologies of Seeing: Photography, Cinematography and Television* (London: British Film Institute, 1996), p. 61.

33 Quoted in Glenn E. Matthews and Raife G. Tarkington, 'Early History of Amateur Motion-Picture Film', *Journal of the SMPE*, vol. 64 (March 1955), reprinted in Fielding, op. cit., p. 130.

34 Charles R. Fordyce, 'Improved Safety Motion Picture Film Support', *Journal of the SMPE*, vol. 51, no. 4 (October 1948), p. 331.

35 *International Photographer*, vol. 18, no. 3 (April 1945), p. 22.

36 See Leo Enticknap, 'The Film Industry's Conversion from Nitrate to Acetate-based stocks in the late 1940s: A Discussion of the Reasons and Consequences' in Roger Smither & Catherine Surowiec (eds) *This Film is Dangerous: A Celebration of Nitrate Film* (Brussels: FIAF, 2002), pp. 202–12.

37 Fuji Photo Film Co. Ltd., *Fifty Years of Fuji Photo Film* (1984), p. 33.

38 D. R. Whit, D. J. Glass, E. Meschter and W. R. Holm, 'Polyester Photographic Film Base', *Journal of the SMPE*, vol. 64 (January 1965), pp. 674–8.

39 George J. van Schil, 'The Use of Polyester Film Base in the Motion Picture Industry', *SMPTE Journal*, vol. 89, no. 2 (February 1980), p. 109.

40 Eastman Kodak press release, see http://www.kodak.com/country/US/en/corp/pressReleases/

pr19960509-01.shtml (access by author on 3 February 2003).

41 Henry Wilhelm and Carol Brower, *The Permanence and Care of Colour Photographs: Traditional and Digital Color Prints, Color Negatives, Slides and Motion Pictures* (Grinnell, IO: Preservation Publishing Co., 1993), p. 303.

42 For more on Ilford, see R.J. Hercock and G.A. Jones, *Silver By the Ton: A History of Ilford Ltd.* (Maidenhead, 1979), passim.

chapter two

1 Bernard E. Jones, *The Cinematograph Book* (London: Cassell, 1915), p. 9.

2 For more on the Kinetograph, see W. K. L. Dickson, 'A Brief History of the Kinetograph, the Kinetoscope and the Kineto-phonograph', *Journal of the SMPE*, vol. 21 (December 1933), reprinted in Raymond Fielding (ed.) *A Technological History of Motion Pictures and Television* (Berkeley, CA: University of California Press, 1967), pp. 9–16.

3 Barry Salt, *Film Style and Technology: History and Analysis* (2nd edn., London: Starword, 1992), p. 32.

4 Jones, op. cit., p.18.

5 Ibid., p. 45.

6 H. Mario Raimondo Souto, *The Technique of the Motion Picture Camera* (4th edn., London: Focal Press, 1982), p.20.

7 Ibid.

8 Jack Fay Robinson, *Bell and Howell Company – A 75 Year History* (Chicago: Bell and Howell, 1982), p. 16.

9 Salt, op. cit., p. 21.

10 David Bordwell, Janet Staiger and Kristin Thompson, *The Classical Hollywood Cinema: Film Style and Mode of Production to 1960* (London: Routledge, 1985), pp. 26–9.

11 Brian Winston, *Technologies of Seeing: Photography, Cinematography and Television* (London: British Film Institute, 1996), p. 86.

12 I am grateful to Barry Salt for this information.

13 Salt, op. cit., p. 207.

14 Winston, op. cit., p. 85.

15 Salt, op. cit., p. 261.

16 *International Photographer*, vol. 2, no. 1 (February 1930), p.19.

17 John Belton, 'The Origins of 35mm Film as a Standard', *SMPTE Journal*, (August 1990), p. 656.

18 Paul C. Spehr, 'Unaltered to Date: Developing 35mm Film' in John Fullerton and Astrid Soderbergh-Widding (eds) *Moving Images: From Edison to Web Cam* (Eastleigh: John Libbey, 2000), pp.10–17.

19 Michael Chanan, *The Dream that Kicks: The Prehistory and Early Years of Cinema in Britain*. 2nd edn. (London: Routledge, 1996), pp. 193–6.

20 Jeanne Thomas Allen, 'The Industrial Context of Film Technology: Standardisation and Patents', in Teresa de Lauretis and Stephen Heath (eds) *The Cinematic Apparatus* (Basingstoke: Macmillan, 1980), p.34.

21 Brian Coe, *The History of Movie Photography* (Westfield, NJ: Eastview Editions, 1981), p.88.

22 Quoted in Bordwell, Staiger and Thompson, op. cit., p. 105; my emphasis.

23 'Motion Picture Standards Adopted in Committee of the Whole Society', *Trans. SMPE*, 16–17 July 1917.

24 This has not always been the case: before the arrival of sound some technical manuals and publications expressed speeds in feet per minute.

25 Kevin Brownlow, 'Silent Films – What Was the Right Speed?', *Sight and Sound*, vol. 49, no. 3 (1980), p. 165.

26 Richard Rowland, 'The Speed of Projection of Film', *Transactions of the SMPE*, no. 33 (1927), p. 79.

27 Scott Eyman, *The Speed of Sound: Hollywood and the Talkie Revolution, 1926–1930* (New York: Simon & Schuster, 1997), p. 112.
28 Quoted in Brownlow, op. cit., p. 166.
29 A clapperboard is a device which enables separate picture and sound recordings to be combined, in synchronisation, onto a single element in printing. The board is operated in front of the camera at the beginning of a shot. When the picture and sound negatives have been processed, the picture frame at which the two wooden surfaces converge and the point on the optical soundtrack at which the 'clap' is visible can be positioned in synchronisation with each other and the two printed together to produce a combined print.
30 Quoted in Eyman, op. cit., p. 222.
31 Sergei Eisenstein, 'The Dynamic Square', in Jey Leyda (ed.) *Film Essays and a Lecture* (Princeton: Princeton University Press, 1982), pp. 48–65.
32 John Belton, *Widescreen Cinema* (Cambridge, MA: Harvard University Press, 1992), p. 44.
33 Ibid., pp. 71–4.
34 As the focus setting for a projection lens did not change during the zoom (i.e. the distance between a cinema projector and screen never changes, whereas the distance between a camera and subject will do if either the camera or subject moves during a shot), the restrictions in early zoom lens technology did not inhibit its use for projection.
35 Belton, op. cit., p. 46. At first glance Belton's statement would seem to be nonsensical, because by replacing the projection lens with one of a different focal length it is possible to magnify any frame area to fit any screen size (within certain limits). However, part of the thinking may have been to minimise the cost of additional projection equipment (i.e. not having to buy additional lenses), in which case the physical dimensions of many auditoria would have dictated that the four-perf pulldown be retained.
36 A detailed account of the technicalities of shooting in Grandeur by *The Big Trail*'s cinematographer, Arthur Edeson, can be found in *American Cinematographer*, September 1930.
37 Quoted in James Cameron and John F. Rider, *Sound Pictures and Trouble Shooters' Manual* (Manhattan Beach, 1930), p. 668.
38 Piers Brendon, *The Dark Valley: A Panorama of the 1930s* (London: Jonathan Cape, 2000), p. 61.
39 Cameron and Rider, op. cit., p. 668.
40 'Securing a Wide Screen Picture', *Journal of the SMPE* (January 1931), p. 83ff.
41 Eastman Kodak Company, *How to Make Good Home Movies* (Rochester: Eastman Kodak Company, 1958).
42 Patricia Zimmerman, *Reel Families: A Social History of Amateur Film* (Indianapolis: Indiana University Press, 1995), p. 113.
43 The two fictional features were *How the West was Won* (1962, dir. Henry Hathaway *et al.*) and *The Wonderful World of the Brothers Grimm* (1962, dir. Henry Levin and George Pal).
44 Belton, op. cit., p. 106.
45 The technology has its origins in stereoscopic still photographs in the nineteenth century: see Coe, op. cit., pp. 156–60. Cinemas at that time would routinely have been equipped with at least two 35mm projectors, as 'long play' projection of an entire feature programme from a single spool or platter was not widespread until the late 1960s. Most, however, would only have had a spool capacity limited to 2,000 feet (approximately 20 minutes) per projector, meaning that these 3-D presentations could only have been presented with regular breaks in the screening without significant investment in extra equipment. See chapter 5 for more on this issue.
46 Eventually a system was devised enabling black-and-white 3D exhibition from a single projector, in which the left and right eye images were printed onto a single strip of colour print stock through red and green filters respectively. Red and green tinted spectacles (as distinct from the polarised ones worn with the two-projector system) then allowed the viewer's brain to identify and position the two.
47 As with *The Big Trail*, TCF hedged its bets with *The Robe* by shooting each scene simultaneously using two cameras, one in CinemaScope and the other using the conventional 35mm Academy

Ratio. The latter version was shown in cinemas which had not converted to CinemaScope. It was also produced as an insurance policy against CinemaScope following Fox Grandeur in failing to establish itself as a mainstream exhibition technology. For a detailed explanation of the various reduction printing routes from VistaVision, see Robert E. Carr and R. M. Haynes, Wide Screen Movies (Jefferson, NC: McFarland. 1988), pp. 144–8.

48 Geoffrey Macnab, *J. Arthur Rank and the British Film Industry* (London: Routledge, 1993), p. 211.

49 Walter Lassally, 'Alice in Format Land', *Image Technology*, vol. 79, no. 10 (1997), p. 14.

50 Robinson, op. cit., p. 45; emphasis in original.

51 Glenn E. Matthews and Raife G. Tarkington, 'Early History of Amateur Motion-Picture Film', *Journal of the SMPTE*, vol. 64 (March 1955), reprinted in Fielding, op. cit., pp. 130–1.

52 'The Pathéscope Home Cinematograph', *British Journal of Photography*, vol. 60 (March 1913), pp. 216–17.

53 Owen Wheeler, *Amateur Cinematography* (London: Pitman, 1929), p. 21.

54 Alan Kattelle, *Home Movies: A History of the American Industry, 1897–1979* (Nashua, NH: Transition Publishing, 2000), p. 81.

55 Ibid., p. 96.

56 For examples, see Philip Grosset, *Making 8mm Movies* (London: Fountain Press, 1959), pp. 179–89.

57 *International Photographer*, vol. 18, no. 3 (April 1945), p. 22.

58 David Cleveland, 'Don't Try This at Home', in Roger Smither and Catherine Surowiec (eds) *This Film is Dangerous: A Celebration of Nitrate Film* (Brussels: FIAF, 2002), p. 197.

59 For more on the initial development of the system, see *Forbes Global Business and Finance*, 1 June 1998, p. 92.

60 Carr and Haynes, op. cit., p. 185.

61 'Credits roll on the huge-screen revolution as Imax cinema closes', *The Independent on Sunday*, 4 January 2004, p. 9.

chapter three

1 Brian Coe, *The History of Movie Photography* (Westfield, NJ: Eastview Editions, 1981), p. 112.

2 Richard Abel, *The Ciné Comes to Town: French Cinema, 1896–1914* (Berkeley, CA, UNiversity of California Press, 1994), p. 13.

3 Barry Salt, *Film Style and Technology: History and Analysis* (2nd edn., London: Starword, 1992), p. 44.

4 R. T. Ryan, *A Study of the Technology of Color Motion Picture Processes Developed in the United States* (Unpublished PhD thesis, University of Southern California, 1977), p. 127.

5 David S. Hulfish, *Cyclopaedia of Motion Picture Work*, 3rd edn. (New York: American Technical Society, 1915), p. 37.

6 Ryan, op. cit., p. 16.

7 Eastman Kodak Co., *Chronology of Motion Picture Films, 1889–1939*, http://www.kodak.com/US/en/motion/about/chrono2.shtml, accessed by the author on 22 October 2002.

8 *Today's Cinema*, 7 January 1933; quoted in Linda Wood (ed.) *British Films, 1927–39* (London: British Film Institute, 1986), p. 27.

9 From the Greek *orthos*, meaning 'straight' or 'upright'. The phrase was probably originally used to denote a more 'straight' (as in accurate) form of photographic reproduction than the blue-only emulsions which preceded it.

10 From the Greek *pan*, meaning 'all' or 'everything', i.e. all three colours which form the visible spectrum.

11 Luke McKernan, 'The Brighton School and the Quest for Natural Colour', in Vanessa Toulmin & Simon Popple (eds) *Visual Delights 2: Audience and Reception* (Eastleigh: John Libbey, 2005), pp. 205–18.

12 G. A. Smith, 'Animated Photographs in Natural Colours', *Journal of the Royal Society of Arts*, vol.

57, no. 2925 (11 December 1908), pp. 70–6; quoted Ibid., p. 207.

13 McKernan gives the figure of 30fps (p. 8); Salt 32 (p. 79) and Coe 32 (p. 117).

14 Terry Ramsaye, *A Million and One Nights* (New York: Simon and Schuster, 1926), p. 562.

15 Nicola Mazzanti, 'Rasing the Colours (Restoring Kinemacolor)' in Roger Smither and Catherine Surowiec (eds) *This Film is Dangerous: A Celebration of Nitrate Film* (Brussels: FIAF, 2002), pp. 123–5.

16 Coe, op. cit., p. 121.

17 Not to be confused with the subsequent use of the trade mark 'Kodacolor' from the 1960s onwards as a tripack colour film for use in still cameras.

18 C. E. K. Mees, 'Amateur Cinematography and the Kodacolor Process', *Journal of the Franklin Institute* (January 1929), pp. 1–17.

19 For a fuller technical description of Dufaycolor, see 'The Dufaycolor Process' – promotional leaflet published by Spicer-Dufay Ltd.; undated but believed by the author to be c.1935 (copy available in the BFI Library).

20 Simon Brown, 'Dufaycolor – The Spectacle of Reality and British National Cinema', http://www.bftv.ac.uk/projects/dufaycolor.htm , p. 6.

21 Herbert T. Kalmus, 'Technicolor Adventures in Cinemaland', *Journal of the SMPE*, vol. 36, no. 6 (December 1938), p. 565.

22 A number of two-colour Technicolor features have survived and been preserved. Some were available on VHS and/or DVD at the time of writing. For example, the UCLA Film and Television Archive's restoration of *The Toll of the Sea* (US 1922, dir. Chester Franklin), is available on the second edition of the US National Film Preservation Foundation's *Treasures From American Film Archives* DVD, published in May 2005 (see www.filmpreservation.org/treasures/dvd.html), while *Legong: Dance of the Virgins* (US 1935, dir. Henri de la Falaise) was published by Milestone in November 2004 (http://www.milestonefilms.com/movie.php/legong/).

23 Salt, op. cit., p. 198.

24 Duncan Petrie, *The British Cinematographer* (London: British Film Institute, 1996), p. 40.

25 Watford is a suburb north of London.

26 Rachael Low, *Film Making in 1930s Britain* (London: George Allen and Unwin, 1985), p. 105.

27 Technicolor standard contract: copy in 'Technicolor' subject file in BFI library.

28 François Truffaut, *Hitchcock by Truffaut* (2nd edn., London: Paladin, 1986), p. 265.

29 In fact, in the paper in which Leopold Mannes and Leopold Godowsky first detailed the Kodachrome process, it was described as a technology aimed specifically at amateurs – see Leopold Mannes and Leopold Godowsky Jr, 'The Kodachrome Process for Amateur Cinematography in Natural Colour', *Journal of the SMPE*, vol. 25 (1935), pp. 248–57.

30 Quoted in Alan Kattelle, *Home Movies: A History of the American Industry, 1897–1979* (Nashua, NH: Transition Publishing, 2000), p. 186.

31 Pat Jackson, *A Retake Please* (Liverpool, 1989), p. 161. *XIV Olympiad: The Glory of Sport* (UK 1948, dir. L. Castleton Knight) also used Monopack negative to film the winter sports sequences.

32 Eric Rentschler, *The Ministry of Illusion: Nazi Cinema and its Afterlife* (Cambridge, MA: Harvard University Press, 1996), p. 110.

33 Henry Wilhelm and Carol Brower, *The Permanence and Care of Color Photographs: Traditional and Digital Color Prints, Color Negatives, Slides and Motion Pictures.* (Grinnell, IO: Preservation Publishing Co.), p. 21.

34 Kattelle, op. cit., p. 186.

35 Leo Enticknap, 'The Film Industry's Conversion from Nitrate to Safety Film in the Late 1940s: A Discussion of the Reasons and Consequences', in Roger Smither and Catherine Surowiec (eds) *This Film is Dangerous: A Celebration of Nitrate Film* (Brussels: FIAF 2002), pp. 202–12.

36 Dudley Andrew, 'The Post-War Struggle for Colour', in Teresa de Lauretis and Stephen Heath (eds) *The Cinematic Apparatus* (Basingstoke: Macmillan, 1981), pp. 70–1.

37 *50 Years of Fuji Photo Film*, Fuji Photo Film Ltd. (1984), p. 33.

38 F. P. Gloyns, 'Processing Over Fifty Years: The Work of the Film Laboratories', *The BKSTS Journal*,

vol. 63, no. 1 (January 1981), p. 35.

39 I am grateful to David Pierce for this information.

40 *Variety*, 6 April 1988, p. 44.

41 See also Michael Dempsey, 'Colorization', Film Quarterly, vol. 40, no. 2 (Winter 1986-87), pp. 2-3.

42 Kerry Segrave, *Movies at Home: How Hollywood Came to Television*. (Jefferson, NC and London: McFarland, 1999), p. 156.

chapter four

1 The British music-hall star Teddy Elben, appearing in an experimental sound film in 1926 ('Teddy Elben and his Irish Jewzaleers play *When the Yiddisher Band Played an Irish Tune*', De Forest Phonofilms – source: National Film and Television Archive, London).

2 Jean-Paul Kauffman, *The Dark Room at Longwood* (London: Harvill Panther, 1999), p. 21.

3 John Coleman, 'Deaf in the Afternoon', *Financial Times*, 2 June 1978, p. 4.

4 Scott Eyman, *The Speed of Sound* (New York: Simon & Schuster, 1997), p. 343.

5 Laura Mulvey, 'Now You Has Jazz', *Sight and Sound*, vol. 9, no. 5 (May 1999), p. 18.

6 Quoted in Brian Coe, *The History of Movie Photography* (Westfield, NJ: Eastview Editions, 1981), p. 93.

7 W. K. L. Dickson, 'A Brief History of the Kinetograph, the Kinetoscope and the Kineto-Phonograph', *Journal of the SMPE*, vol. 21 (December 1933), reprinted in Raymond Fielding (ed.) *A Technological History of Motion Pictures and Television* (Berkeley, CA: University of California Press, 1967), pp. 12–13.

8 According to a widely repeated folklore, the words Edison first spoke into his phonograph: see, for example, Coe, op. cit., p. 92.

9 David Morton, *Off the Record: The Technology and Culture of Sound Recording in America* (New Brunswick, NJ: Rutgers University Press), 2000, pp. 4–5.

10 Walter Read and Leah Brodbeck Stenzel Burt, *From Tinfoil to Stereo: The Acoustic Years of the Recording Industry, 1877–1929* (Gainsville, 1994), pp. 96-97.

11 For more on this see Rick Altman, 'The Silence of the Silents', *Musical Quarterly*, vol. 80, no. 4 (1996), pp. 648–718.

12 Richard Crangle, 'Next Slide Please: The Lantern Lecture in Britain, 1890–1910' in Richard Abel and Rick Altman (eds) *The Sounds of Early Cinema* (Bloomington, IN: Indiana University Press, 2001), p. 46.

13 Noël Burch, *Life to Those Shadows* (London: British Film Institute, 1990), p. 107.

14 Though interestingly, an early example of an office Dictaphone in use can be seen in an equally early example of the feature-length narrative film: *Traffic in Souls* (1913, dir. George Loane Tucker).

15 For a detailed discussion of one specific example, including printed excerpts from the score itself, see Neil Brand, 'Distant Trumpets: The Score to *The Flag Lieutenant* and Music of the British Silent Cinema', in Andrew Higson (ed.) *Young and Innocent? The Cinema in Britain, 1896–1930*. (Exeter: University of Exeter Press, 2002), pp. 208–24.

16 Coe, op. cit., pp. 91–2.

17 Stephen Bottomore, 'The Story of Percy Peashaker: Debates about Sound Effects in the Early Cinema', in Richard Abel and Rick Altman (eds) *The Sounds of Early Cinema* (Bloomington, IN: Indiana University Press, 2001), pp. 129–31.

18 Akira Kurosawa, *Something Like an Autobiography* (New York: Alfred A. Knopf, 1982), p. 74.

19 Emmanuelle Toulet, 'Le Cinéma à l'Exposition Universelle de 1900' in *La Revue d'Histoire moderne et contemporaine*, vol. 33 (1986), pp. 186–7.

20 A claim made by Messter in the documentary *Als man anfing zu filmen* (*The Beginnings of Filming*, 1934, dir. Nicholas Kauffman). See also Messter's autobiography, *Mein weg mit dem Film* (Berlin Schöneberg: Max Hesse Verlag, 1936).

21 John Scotland, *The Talkies* (London: Crosby, Lockwood & Son, 1930); foreword by Cecil Hepworth, p. 19.

22 Ibid.

23 Eyman, op. cit., pp. 27–8; Coe, op. cit., p. 96.

24 Edward W. Kellogg, 'The History of Sound Motion Pictures, Part 1', in *Journal of the SMPTE*, vol. 64 (1955), reproduced in Raymond Fielding (ed.) *A Technological History of Motion Pictures and Television* (Berkeley, CA: University of California Press, 1967), p. 174.

25 Eyman, op. cit., p. 27.

26 Kellogg, op. cit., pp. 174–5.

27 David Bordwell, Janet Staiger and Kristin Thompson, *The Classical Hollywood Cinema: Film Style and Mode of Production to 1960* (London: Routledge, 1985), p.195.

28 Ibid., p. 234.

29 Rick Altman, 'The Sound of Sound: A Brief History of the Reproduction of Sound in Movie Theatres', http://www.geocities.com/Hollywood/Academy/4394/altman.html, undated (hereafter 2003), accessed 27 November 2003.

30 John Eargle, *The Microphone Book* (Boston: Focal Press, 2001), p. 1.

31 Timothy D. Taylor, 'Music and the Rise of Radio in 1920s America', in *Historical Journal of Film, Radio and Television*, vol. 22, no. 4 (2002), p. 429.

32 Morton, op. cit., p. 50.

33 Jerrold Northrop Moore, *Sound Revolutions: A Biography of Fred Gaisberg* (London: Sanctuary, 1999), pp. 243–4.

34 Morton, op. cit., pp. 13–47.

35 Kellogg, op. cit., p.176.

36 Early historical accounts can be found in Benjamin Hampton, *History of the American Film Industry* (New York: Covici Friede, 1931), Lewis Jacobs, *The Rise of the American Film* (New York: Teachers' College Press, 1939); Kellogg, op. cit.; more recent research in the area is typified by Douglas Gomery, 'Warner Bros. and Sound', in *Screen*, vol. 17, no. 1 (1976), pp. 40–53 and Donald Crafton, *The Talkies* (Berkeley, CA: University of California Press, 1997); and some more polemical writing on the subject is represented by Eyman, op. cit. and Mulvey, op. cit.

37 Harold B. Franklin, *Sound Motion Pictures* (New York: Doubleday, Doran & Co, 1929), p.79.

38 In the last year of Vitaphone's life as a primary production format for studio recording, Warner Bros. developed a 'brutally complicated' means of editing Vitaphone sound, which worked by selectively copying a mix of several source discs onto a single copy – similar in principle to the techniques for linear videotape editing used in the television industry from the 1970s until the late 1990s – see Eyman, op. cit., pp. 203–5. As any form of editing an analogue recording which necessitates copying it degrades the signal quality in each subsequent generation, this method was used in both instances simply due to the absence of any viable alternative.

39 Edward Bernds, *Mr. Bernds Goes to Hollywood: My Life and Career in Sound Recording at Columbia with Frank Capra and Others.* (Lanham, MD and London: Scarecrow Press, 1999), p. 68.

40 James R. Cameron, *Motion Pictures With Sound* (Manhattan Beach, NY: Cameron Publishing Co., 1929), p. 249.

41 Earl Sponable, 'Historical Development of Sound Films', *Journal of the SMPE*, no. 48 (April 1947), p. 281.

42 Coe, op. cit., p. 104.

43 Bernds, op. cit., p. 68.

44 Thomas H. Cripps, *Hollywood's High Noon* (Baltimore: Johns Hopkins University Press, 1997), p. 101.

45 Rudolph Miehling, *Sound Projection* (New York: Mancall, 1929), p. 1.

46 Sponable, op. cit., pp. 287–90.

47 Freda Frieberg, *The Transition to Sound in Japan*, available online at http://www.filmsound.org/film-sound-history/sound (accessed 27 November 2003).

48 Eyman, op. cit., p. 360.

49 Douglas Gomery, 'Writing the History of the American Film Industry: Warner Bros. and Sound', *Screen*, vol. 17, no. 1 (Winter 1976), p. 53.

50 Salt, op. cit., p. 211.
51 Linda Wood, *British Films, 1927–39* (London: British Film Institute, 1986), p. 18.
52 Ernest Betts, 'Why 'Talkies' are Unsound', *Close Up*, vol. IV, no. 4 (April 1929), reprinted in James Donald, Laura Marcus and Anne Friedberg (eds) (1998) *Close Up, 1927–1933: Cinema and Modernism*. (London: Cassell, 1998), pp. 89–90.
53 Kristin Thompson, *Exporting Entertainment: America in the World Film Market, 1907–1934* (London: British Film Institute, 1985), p. 148.
54 For an extensive and detailed comparison of the two versions, see Charles Barr, '*Blackmail*: Silent and Sound', *Sight and Sound*, vol. 52, no. 2 (Spring 1983), pp. 122–6.
55 Don Fairservice, *Film Editing: History, Theory and Practice* (Manchester: Manchester University Press, 2001), p. 232.
56 Crafton, op. cit., pp. 228–9.
57 Edward W. Kellogg, 'The ABC of Photographic Sound Recording', *Journal of the SMPE*, vol. 44, no. 3 (March 1945), p. 172.
58 Martin Sawyer, 'The Sound of Nitrate' in Roger Smither and Catherine Surowiec (eds) *This Film is Dangerous: A Celebration of Nitrate Film* (Brussels: FIAF, 2002), p. 138.
59 Kellogg (1945), op. cit., p. 158.
60 Ibid., pp. 161–3.
61 Charles Felstead, 'What Radio has Meant to Talking Movies', *Radio News* (April 1931), reproduced at http://www.antiqueradios.com/movies.shtml (accessed 27 November 2003).
62 Brian Winston, *Media, Technology and Society* (London: Routledge, 1998), p. 62.
63 John Nathan, *Sony: The Private Life* (London: HarperCollins, 2000), p. 29.
64 John Borwick, *Sound Recording Practice* (Oxford: Oxford University Press, 1976), p. 117.
65 Ibid., p. 380.
66 'Stereo' is a term deriving from the Greek *stereos*, meaning 'solid'. In the context of domestic audio it is commonly used to describe two soundtracks played from separate speakers within a given space but simultaneously (sometimes a straight copy of the signal from two microphones used to make the original recording, but more usually mixed down from a greater number of channels), in order to give the listener a more accurate sense of spatiality than would be the case with a single, monaural channel. In the film, television and multimedia industries, however, it is routinely used as a generic adjective to describe any sound reproduction system carrying multiple channels – up to eight in the case of some systems.
67 Coe, op. cit., p. 111.
68 Robert Alexander, *The Inventor of Stereo* (London: Focal Press, 1999), pp. 80–92.
69 Morton, op. cit., p. 38.
70 John Belton, *Widescreen Cinema* (Cambridge, MA: Harvard University Press, 1992), p. 207.
71 Altman (2003), op. cit.
72 The term 'cassette', from the French 'shell', refers to a plastic moulded casing which houses both the feed and take-up spool for a length of magnetic tape. It can be inserted into a recorder more easily than open-reel tape and with no risk of misthreading, hence the popularity of this device in consumer audio-visual technologies and professional ones which need to be operated by staff without special technical training.
73 http://www.dolby.com/company/chronology1970_1979.html (accessed 14 July 2004).
74 Morton, op. cit., p.183; Nathan, op. cit., p. 260.
75 Mulvey, op. cit., p. 16.

chapter five

1 'McGinty at the Living Pictures' – a popular Music Hall song from the early twentieth century. At the time of writing, an Edison cylinder recording of it, performed by Edward S. Favor in 1907, can be heard at www.tinfoil.com.
2 W. K. L. Dickson, 'A Brief History of the Kinetograph, the Kinetoscope and the Kineto-Phonograph',

Journal of the SMPE, vol. 1 (December 1933), reprinted in Raymond Fielding (ed.) *A Technological History of Motion Pictures and Television* (Berkeley, CA: University of California Press, 1967), p. 13.

3 For a more detailed description of the Kinetoscope, see Brian Coe, *The History of Movie Photography* (Westfield, NJ: Eastview Editions, 1981), pp. 64–5.

4 Ibid., p. 195.

5 Reproduced from James R. Cameron, *Motion Picture Projection*, 3rd edn. (New York: Technical Book Co., 1922), p. 481.

6 Herbert Tümmel, *Deutsche Laufbildprojektoren: Ein Katalog*. Berlin: Stiftung Deutsches Kinemathek, 1986), p. 131.

7 Don G. Malkames, 'Early Projector Mechanisms', *Journal of the SMPE*, vol. 66 (October 1957), reprinted in Fielding, op. cit., p. 101.

8 For a detailed account of the incident, see H. Mark Gosser, 'The *Bazar de la Charité* Fire: The Reality, the Aftermath and the Telling', *Film History*, vol. 10, no. 1 (1998), pp. 70–89.

9 For more on nitrate fires, see Roger Smither, 'Calendar of Film Fires' in Roger Smither and Catherine Surowiec (eds) *This Film is Dangerous: A Celebration of Nitrate Film* (Brussels: FIAF, 2002), pp. 429–53.

10 *The Showman*, 5 January 1901, iv.

11 I. L. Thatcher, *From Magic Lantern to Stereophonic Sound: The Story of Motiograph, Inc.* – Motiograph advertising brochure based on an article originally published in *Box Office*, 8 May 1954 (copy in BFI library).

12 Tümmel, op. cit., pp. 29–37.

13 David Bordwell, Janet Staiger and Kristin Thompson, *The Classical Hollywood Cinema: Film Style and Mode of Production to 1960* (London: Routledge, 1985), pp. 294–7.

14 Charlotte Herzog, *The Motion Picture Theater and Film Exhibition, 1896-1932* (unpublished PhD thesis, Northwestern University, 1980), pp. 156-157.

15 Lewis M. Townsend, 'Problems of a Projectionist', *Transactions of the SMPE*, no. 27 (January 1927), pp. 79-88; reprinted at http://www.cinemaweb.com/silentfilm/bookshelf/3_ettow3.htm (accessed 13 April 2004).

16 Cameron, op. cit., p. 195.

17 *British Journal of Photography*, 1 December 1897, p. 937.

18 Photograph from sales brochure, author's collection. The text refers to a previous model launched in 1921, suggesting that this one went on sale significantly later.

19 W. S. Ibbetson, *The Kinema Operator's Handbook: Theory and Practice*. London: E. and F. N. Spon, 1921), p. 47.

20 Cameron, op. cit., p.27.

21 Victor Milner, 'Speed of Projection', *American Cinematographer*, July 1923, p. 6; reprinted at http://www.cinemaweb.com/silentfilm/bookshelf/19_ac_23.htm (accessed 13 April 2004).

22 K. Owens, 'Joy Riders of the Theatre', *Photoplay Magazine*, November 1916, p. 72; reprinted at http://www.cinemaweb.com/silentfilm/bookshelf/18_pp_5.htm (accessed 13 April 2004).

23 F. H. Richardson, 'Theoretical vs. Practical as Applied to Standardisation and Some of the Things to be Considered as Proper Subjects for Standardisation', *Transactions of the SMPE*, no. 6 (1918), pp. 33–5; reprinted at http://www.cinemaweb.com/silentfilm/bookshelf/18_18_12.htm (accessed 13 April 2004).

24 Ibid.

25 F. H. Richardson, 'The Various Effects of Over-Speeding Projection', *Transactions of the SMPE*, no. 10 (1920); reprinted at http://www.cinemaweb.com/silentfilm/bookshelf/18_20_14.htm (accessed 13 April 2004).

26 'Projectionists Say Film Run at High Speed May Cause Fire', *Moving Picture World*, 27 January 1923, p. 322; reprinted at http://www.cinemaweb.com/silentfilm/bookshelf/18_pro_6.htm (accessed 13 April 2004).

27 'Report of the Standards and Nomenclature Committee, September 1927', *Transactions of*

the SMPTE, no. 36 (September 1927), pp. 443-449; reprinted at http://www.cinemaweb.com/silentfilm/bookshelf/18_27_19.htm (accessed 13 April 2004).

28 Harold B. Franklin, *Sound Motion Pictures* (New York: Doubleday, Doran & Co, 1929), passim.

29 Ibid., pp. 78–9.

30 *Today's Cinema*, 28 December 1931, p. 8.

31 Herzog, op. cit., p. 139.

32 *Sound Films in Schools – The Report of an Experiment Undertaken Jointly by Certain Education Authorities and by the National Union of Teachers in the Schools of Middlesex'*, London, The Schoolmaster (1931), p. 9.

33 *Kinematograph Weekly*, 1 December 1955, p. 8.

34 John Belton, *Widescreen Cinema* (Cambridge, MA: Harvard University Press, 1992), p. 108.

35 Ibid., p. 152.

36 Don E. Kloepfel (ed.) *Moving Picture Projection and Theater Presentation Manual* (New York: Society of Motion Picture & Television Engineers, 1969), p. 66.

37 *Financial Times*, 23 September 1957, p. 9.

38 *Screen Trade*, June 2003, pp. 46–7.

39 The following statistical information applies to the UK only, but it paints an indicative picture of the trend across Europe and North America. Source: The British Film Institute's *Film and Television Handbook*, 1998 edition, pp. 30–42.

40 A customer survey carried out by Granada Theatres in 1946, the peak year of admissions, revealed than less than half visited a given cinema specifically in order to see a given film. Other important factors included its proximity to their home, how comfortable the seats were and whether or not it had air conditioning.

41 *Daily Film Renter*, 30 April 1945, p. 4.

42 Barry Fox, 'Squeezed Sound for the Silver Screen', *New Scientist*, 13 July 1991, p. 26. A 70mm variant of the CDS system was also developed.

43 Jim White, 'Never Mind the Movie – Just Feel That Sound', *Daily Telegraph*, 19 January 2004, p. 18.

chapter six

1 Jeff Kisseloff, *The Box: An Oral History of Television, 1920–61* (New York: Penguin, 1995), p. 3.

2 Albert Abramson, *Zworykin: Pioneer of Television* (Urbana: University of Illinois Press, 1995), xiii.

3 Neil Robson, 'Living Pictures Out of Space', *Historical Journal of Film Radio and Television*, vol. 24, no. 2 (June 2004), p. 224.

4 Quoted on www.xtvworld.com/tv/bbc/baird_vs_bbc.htm, accessed 21 August 2004.

5 For an account of how the NTSC's function in this regard was established in the US, see William F. Boddy, 'Launching Television: RCA, the FCC and the Battle for Frequency Allocations, 1940–47', *Historical Journal of Film, Radio and Television*, vol. 9, no. 1 (1989), pp. 45–57.

6 Robson, op. cit., p. 228.

7 David Morton, *Off the Record: The Technology and Culture of Sound Recording in America* (New Brunswick, NJ: Rutgers University Press, 2000), pp. 48–73.

8 For a detailed account of Farnsworth's research, see David E. Fisher and Marshall John Fisher, *Tube: The Invention of Television* (Orlando, 1996), pp. 135–52.

9 Transcript of the opening broadcast quoted in Robert Alexander, *The Inventor of Stereo: The Life of Alan Dower Blumlein* (London: Focal Press, 1999), p. 198.

10 Robson, pp. 223–4.

11 For a detailed account of this event, see Ron Becker, 'Hear and See Radio in the World of Tomorrow: RCA and the Presentation of Television at the World's Fair, 1939–40', *Historical Journal of Film, Radio and Television*, vol. 21, no. 4 (October 2001), pp. 361–78.

12 For more on Nazi television, see William Urrichio (ed.) *The History of German Television, 1935–44*, special edition of *Historical Journal of Film, Radio and Television*, vol. 10, no. 2 (1990).

13 Frank Coven (ed.) *The Daily Mail Television Handbook* (London: Daily Mail Publications, 1950), p. 5.
14 Boddy, op. cit., passim.
15 Fisher and Fisher, op. cit., p. 364.
16 Reproduced from Coven, op. cit., p. 52.
17 Asa Briggs, *The History of Broadcasting in the United Kingdom – Competition, 1955–1974* (Oxford: Oxford University Press, 1995), p. 849.
18 John Nathan, *Sony: The Private Life* (London: HarperCollins, 1999), p. 42.
19 For more on the development of the Trinitron tube, see ibid., pp. 42–8.
20 Note: pixels are also the individual units which form a digital image, whether derived from a CCD or not. In this instance, the electrical energy which flows out sequentially from each pixel in a CCD can be recorded either as an analogue waveform or digital data.
21 Chris Forrester, *The Business of Digital Television* (Oxford, 2000), p. 16.
22 Briggs, op. cit., pp. 850–4.
23 Paul Dambacher, *Digital Broadcasting* (London: Institution of Electrical and Electronic Engineers, 1996), p. 155.
24 'Colour Television for the World' (RCA brochure, 1964; copy in British Film Institute library), p. 5.
25 Richard H. Kallenberger and George D. Cvjetnicanin, *Film into Video: A Guide to Merging the Technologies* (Boston and London: Focal Press, 1994), p. 2.
26 Kerry Segrave, *Movies at Home: How Hollywood Came to Television* (Jefferson, NC and London: McFarland, 1999), p. 132.
27 For a detailed account of the evolution of telerecording technology in the 1930s and 1940s, see Albert Abramson, 'A Short History of Television Recording', *Journal of the SMPTE*, vol. 64 (February 1955); reprinted in Raymond Fielding (ed.) *A Technological History of Motion Pictures and Television* (Berkeley, CA: University of California Press, 1967), pp. 250–4.
28 Edward Pawley, *BBC Engineering, 1922–1972* (London: BBC Publications, 1972), pp. 491–2.
29 Anthony Kamm and Malcolm Baird, *John Logie Baird – A Life* (Edinburgh: National Museums of Scotland Publishing, 2002), pp. 82–3.
30 Details of the reconstruction work can be found at www.tvdawn.com/tv1strx.htm (accessed 4 October 2004).
31 For more on VERA see Pauley, op. cit., pp. 490–6; and Peter Axon, 'VERA – An Experimental Broadcast VTR' in David K. Kirk (ed.) *Twenty Five Years of Video Tape Recording* (Bracknell: 3M Corporation, 1981), pp. 11–19.
32 The word is comprised of the initials of the company's founder (Alexander M. Poniatoff) and 'ex' for 'excellence'.
33 Pawley, op. cit., p. 494.
34 Asa Briggs, *The History of Broadcasting in the United Kingdom – Competition, 1955–1974* (Oxford: Oxford University Press, 1995), p. 837.
35 Pawley, op. cit., p. 494.
36 David F. P. Machon, 'The Shape of Things to Come?', *BKSTS Journal*, vol. 63, no. 12 (1981), p. 739.
37 Nathan, op. cit., p. 106.
38 For more on CED, see Margaret Graham, *RCA and the Videodisc: The Business of Research* (Cambridge: Cambridge University Press, 1986), passim.
39 This system was originally named by JVC (Japanese Victor Corporation) engineers as 'Vertical Helical Scan'. 'Victor Helical Scan' was also used in some technical trade publications from the 1970s. When VHS eventually went on the market it was renamed as the more consumer friendly 'Video Home System'.
40 *Daily Mail*, 30 June 1969 (copy in 'video' cuttings file, BFI library; page number missing).
41 Segrave, op. cit., p. 192.
42 Brian Norris, 'Video Report', *Sight and Sound*, vol. 52, no. 2 (1983), p. 106.
43 Brian Winston, *Media, Technology and Society* (London: Routledge, 1998), p. 127.
44 Obododimma Oha, 'The Visual Rhetoric of the Ambivalent City in Nigerian Video Films', in Mark

Shiel and Tony Fitzmaurice (eds) *Cinema and the City* (Oxford: Oxford University Press, 2001), p. 196.

45 Paul Kearns, *The Legal Concept of Art* (Oxford: Hart Publishing, 1998), pp. 61–85.

46 André Gaudreault, 'The Infringement of Copyright Laws and its Effects, 1900–06', *Framework*, no. 29 (1985), passim.

47 Kerry Segrave, *Piracy in the Motion Picture Industry* (Jefferson, NC: McFarland, 2003), p. 178.

48 Ibid., p. 179.

49 *Cinematograph Exhibitors' Association (CEA) Newsletter*, no. 178 (February 1981), p. 9.

50 Winston, op. cit., p. 128.

chapter seven

1 Jean-Paul Kauffman, *The Dark Room at Longwood* (London: The Harvill Press, 1999), p. 49.

2 John Reed, 'Nitrate? Bah! Humbug! A Personal View from an Archive Heretic', in Roger Smither and Catherine Surowiec (eds) *This Film is Dangerous: A Celebration of Nitrate Film* (Brussels: FIAF, 2002), p. 225.

3 Boleslaw Matuszewski, *Un nouvelle source de l'histoire* (Paris, 1898), quoted in Penelope Houston, *Keepers of the Frame: The Film Archives* (London: British Film Institute, 1994), p. 10.

4 For a fuller discussion of this issue, see David Pierce, 'The Legion of the Condemned: Why American Silent Films Perished', *Film History*, vol. 9, no. 1 (1997), passim.

5 1992 was the year in which the bulk of cinema release printing switched from triacetate to polyester.

6 David Francis, interviewed in the television documentary *Keepers of the Frame* (US 1999, dir. Mark McLaughlin).

7 Stephen Bottomore, 'A Fallen Star: Problems and Practices in Early Film Preservation', in Smither and Surowiec, op. cit., p. 186.

8 Houston, op. cit., p. 38.

9 Herbert Volkmann, *Film Preservation: A Report of the International Federation of Film Archives* (London: National Film Archive, 1965), p. 6.

10 James W. Cummings *et al.*, 'Spontaneous Ignition of Decomposing Cellulose Nitrate Film', *Journal of the SMPTE*, vol. 54 (March 1950), p. 268.

11 Volkmann, op. cit., p. 9.

12 'Nitrate Project 2000: A Race Against Time', *Historical Journal of Film, Radio and Television*, vol. 7, no. 3 (1987), p. 329.

13 N. S. Allen, M. Edge, J. S. Appleyard, T. Jewitt and C. V. Horie, 'Initiation of the Degredation of Cellulose Triacetate Base Motion Picture Film', *Journal of Photographic Science*, vol. 38 (1990), passim.

14 Henning Schou, *Preservation of Moving Images and Sound* (Brussels: FIAF, 1989), p. 7.

15 Ibid., p. 46.

16 See www.rtico.com/bhp/shrink.html (accessed by the author on 18 October 2004).

17 www.kodak.com/global/en/corp/environment/kes/pubs/pdfs/H182.pdf (revised version accessed by the author on 18 October 2004), p. 2.

18 For a summary of the key findings, see P. Z. Adelstein, C. L. Graham and L. E. West, 'Stability of Cellulose Ester Base Photographic Film', *SMPTE Journal*, vol. 101 (May 1992), pp. 336–53.

19 For a detailed discussion see Henry Wilhelm and Carol Brower, *The Permanence and Care of Color Photographs* (Grinnell, IO: Preservation Publishing Co., 1993), pp. 320–1.

20 James Reilly, 'Keeping Cool and Dry: A New Emphasis in Film Preservation', published online at www.lcweb.loc.gov/film/storage.html (accessed by the author on 23 April 2004).

21 Paul C. Spehr, 'Fading: The Color Film Crisis', *American Film*, vol. 5, no. 2 (1979), p. 57.

22 'Colour Problem', *Sight and Sound*, vol. 50, no. 1 (1980/81), pp. 12–13; Wilhelm and Brower, op. cit., pp. 307–9; 343–4.

23 Anthony Slide, *Nitrate Won't Wait: A History of Film Preservation in the United States* (Jefferson,

NC: McFarland, 1992), p. 107.

24 'Kodak Abandons Secrecy Policy', Wilhelm and Brower, op. cit., p. 309.

25 For example, in the 1997 restoration of *Vertigo* (1957, dir. Alfred Hitchcock), samples of paint from a car used in the original production served as a reference point. For more on the restoration, see Leo Enticknap, 'Some Bald Assertion by an Ignorant and Badly Educated Frenchman', *The Moving Image*, vol. 4, no. 1 (2004), passim.

26 Paul Read and Mark-Paul Mayer, *Restoration of Motion Picture Film* (Oxford: Butterworth-Heinemann, 2000), p. 69.

27 For a full account of the restoration, see François Ede, *Jour de fête, ou, Le couleur retrouvée* (Paris: Cahiers du Cinema Livres, 1995), passim.

28 Alan Stanbrook, 'Lawrence fails to stay the distance', *The Daily Telegraph*, 1 March 1989, p. 14.

29 Paolo Cherchi Usai, *The Death of Cinema: History, Cultural Memory and the Digital Dark Age* (London: British Film Institute, 2001), p. 121.

chapter eight

1 Hans Fantel, 'Digital TV: Clear Outlines, Sharp Detail, True Colour', *The New York Times*, 10 March 1983, p. 30.

2 It is often assumed, incorrectly, that this abbreviation stands for 'Digital Video Disc'. The format was intended as a universal data storage medium, not just for video, hence the adjective 'versatile'.

3 'Program' meaning a computer program is always spelt the American way, even in UK English.

4 For more on Babbage, see Doron Swade, *The Cogwheel Brain: Charles Babbage and the Quest to Build the First Computer* (London: Abacus, 2001), passim.

5 Gordon E. Moore, 'Cramming more components onto integrated circuits', *Electronics*, vol. 38, no. 8 (19 April 1965), republished at ftp://download.intel.com/research/silicon/moorespaper.pdf (accessed by the author on 26 October 2004).

6 Nick Evans, 'From TV to Terminal: Video on the Move', *The Guardian*, 8 March 1994, p. 8.

7 Harold Nyquist, 'Certain Topics in Telegraph Transmission Theory', *Transactions of the AIEE*, vol. 47 (1928), pp. 617–44.

8 For more on 16mm film post-production for television in the pre-videotape period, see John Burder, *16mm Film Cutting* (London: Focal Press, 1976), passim.

9 For more on the work of the MPEG, see http://www.sims.berkeley.edu/courses/is224/s99/GroupG/report1.html (accessed by the author on 10 July 2003).

10 John Watkinson, *MPEG-2* (London: Focal Press, 1999), p. 1.

11 *Financial Times*, 15 August 1995, p. 4.

12 8,498,545 unit sales to year end 2000; 21,994,389 to year end 2003. Source: US Consumer Electronics Association data, reproduced at http://www.thedigitalbits.com/articles/cemadvdsales.html (accesssed by the author on 28 October 2004).

13 The entire spectrum consists of electric power, radio waves, microwaves, infra-red, visible light, ultra-violet light, x-rays and gamma rays (in that order).

14 Watkinson, op. cit., p. 5.

15 http://www.ofcom.org.uk/media_office/latest_news/20040917a_nr (accessed by the author on 28 October 2004).

16 Department for Culture, Media and Sport, *Persuasion or Compulsion? Consumers and Analogue Switch-Off* (London: HMSO, 2004), p. 4. Document available online at http://www.digitaltelevision.gov.uk/pdf_documents/publications/Consumer_Expert_Group_report.pdf (accessed by the author on 28 October 2004).

17 Ibid., p. 31.

18 Ibid., p. 39.

19 Kerry Segrave, *Piracy in the Motion Picture Industry* (Jefferson, NC: McFarland, 2003), p. 175.

20 Brian Winston, *Media, Technology and Society* (London: Routledge, 1998), p. 336.

21 For more on the history of the Internet, see ibid., pp. 321–36.

22 Darrell William Davis, 'Compact Generations: VCD Markets in Asia', *Historical Journal of Film, Radio and Television*, vol. 23, no. 2 (2003), p. 166.

23 W. Hunter *et al.*, 'Digital Film Mastering', *SMPTE Journal*, vol. 108, no. 12 (1999), p. 859.

24 Sometimes the grid can be asymmetric, e.g. a scanner which produces a frame 3K across x 2K down.

25 http://www.cinesite.com/lo/scanrec/filmrec.html (accessed by the author on 4 February 2003).

26 Thomas A. Ohanian and Michael E. Phillips, *Digital Filmmaking: The Changing Art and Craft of Motion Pictures* (Boston: Focal Press, 1996), pp. 22–3.

27 For more on the Eidophor see Kira Kitsopanidou, 'The Widescreen Revolution and Twentieth Century Fox's Eidophor in the 1950s', *Film History*, vol. 15, no. 1 (2003), pp. 32–56.

28 Gibboney Huske and Rick Vallières, *Digital Cinema, Episode II* (London: Credit Suisse First Boston, 2002), p. 2.

29 A specific case study in restoration ethics can be found in Leo Enticknap, 'Some Bald Assertion by an Ignorant and Badly Educated Frenchman', *The Moving Image*, vol. 4, no. 1 (2004), pp. 130–41.

chronology

1671	photography	Chérubin d'Orléans publishes *La Dioptique Oculaire*, which describes the operation of a basic camera obscura.
1765	moving images	Chevalier d'Arcy gives one of the earliest systematic descriptions of the illusion of continuous movement.
1802	photography	Thomas Wedgwood produces what is possibly the first photosensitive emulsion, a leather canvas saturated with silver salts.
1835	photography	Louis Daguerre carries out successful experiments to increase the sensitivity, or 'speed', of photographic emulsions to light.
1839	photography	Henry Fox-Talbot demonstrates a method for copying photographic images.
1833	moving images	Phenakistiscope first demonstrated.
1841	sound	Augustin Louis-Cauchy produces the sampling theorem, which is the basis of digital audio recording.
1845–6	film	Discovery of cellulose nitrate.
1861	colour	James Clerk Maxwell demonstrates that all visible colour can be represented as mixtures of the three primary colours: red, green and blue.
1871	photography	Richard Leach Maddox invents the gelatine bromide process, increasing the speed of photographic emulsions to the point at which moving image sequences produced photographically is technically possible.
1873	sound	The discovery by Willoughby Smith of the photoelectric properties of the element selenium provides the underlying technology for optical sound reproduction.
1876	film	Wordsworth Donisthorpe produces the Kinesigraph, a camera which exposes glass-plate photographic negatives at the rate of 8 frames per second.
1877	sound	Thomas Edison makes the first known audio recording, on a cylinder coated with tin foil.
1884	television	Paul Nipkow patents his television scanning disc, which is the core technology used in the mechanical television systems of the 1920s and 1930s.
1885	film	Eastman paper roll film goes on sale to the public.
1887	sound	Emile Berliner produces the first audio discs (gramophone records).
1888	film	Louis Augustin Le Prince builds a working movie camera which uses paper roll film.
1889	film	W. K. L. Dickson orders the first cellulose nitrate film for moving image use.
	sound	W. K. L. Dickson claims to have successfully synchronised a phonograph to a film technology, thereby converging the technologies of moving

		images and audio recording.
1891	exhibition	The Edison Kinetoscope is patented.
1892	moving images	Émile Reynaud's *Théâtre Optique* first shown in Paris.
1895	exhibition	Louis and August Lumière present what is believed to be the first commercial moving image film exhibition in Paris, using a working projector with an intermittent mechanism.
1896	film	The first purpose-built film printers, which used the step contact method, are marketed.
	colour	'An Eastern Dance', believed to be the first hand-coloured film element, is shown in London on 8 April.
	exhibition	Woodville Latham invents the 'Latham loop', used in the film path of all intermittent motion projectors since; the earliest recorded nitrate fire in a cinema takes place, in London on 10 June.
1898	archiving	Boleslaw Matsuszewski publishes the earliest known article calling for the systematic preservation of moving image media.
1899	photography	Invention of the reversal method.
	colour	Lee and Turner additive three-colour system demonstrated, unsuccessfully.
1904	film	W. C. Parkin patents a method of inhibiting the flammability of nitrate film.
	sound	John Ambrose Fleming invents the vacuum diode valve (tube).
1906	colour	Pathécolor (stencil colouring) system introduced.
	sound	Lee de Forest demonstrates the 'Audion tube', the thermionic valve which would later form the basis of electrical audio amplification.
1907	film	The Bell and Howell company is formed in Chicago.
1908	colour	Kinemacolor first demonstrated in public on 1 May; principle of lenticular process first described by Robert Berthon.
	film	Motion Picture Patents Company founded on 18 December.
1909	film	Cellulose acetate (safety) film base is first sold by Eastman Kodak.
	exhibition	Over 250 people are killed in a nitrate fire at a cinema in Acapulco, Mexico – the highest recorded number of deaths in a single incident.
1911	television	A. A. Campbell Swinton provides the first detailed technical account of a cathode ray tube (all-electronic) television display system.
1912	film	The Bell and Howell 2709 35mm studio camera goes on sale.
	colour	The first pre-tinted release print film stock is sold, by Gevaert in Belgium; Gaumont Chronochrome system first used in France.
1913	film	Panchromatic negative film marketed by Eastman Kodak.
1915	colour	The Technicolor company is formed.
1916	colour	A process was developed by Max Handschiegl for mechanising the application of colour dye to processed release prints.
1917	moving images	Society of Motion Picture Engineers (SMPE) was formed, and publishes a 'professional standard' for the 35mm film format; Motion Picture Patents Company disbanded.
1918	film	The first automated film splicer is marketed by Bell and Howell.
1921	sound	Charles Hoxie demonstrates the first variable area optical sound camera.
1922	film	Pathé introduces the 9.5mm film format, initially as a medium for viewing reduction prints of commercial films in the home. The first cameras were sold in the following year.
1923	film	Eastman Kodak introduces the 16mm format, marketed specifically for amateur use.
	sound	De Forest Phonofilms formed in New York, and demonstrates the 'AEO-light', which enables the recording of variable density optical sound.
1925	sound	Electrical amplification for use in audio recording and playback introduced

		by the record industry.
1926	film	Introduction of the first purpose-designed duplicating (intermediate) film stocks.
1928	colour	The Kodacolor lenticular process is first marketed in 16mm reversal form to amateur filmmakers.
	sound	Harold Nyquist publishes *Certain Topics in Telegraph Transmission Theory*, which proves Louis Cauchy's sampling theorem, thus proving that digital audio recording is possible.
1929	film	The launch of the Fox Grandeur format marks the first commercial use of 70mm film. Like several other widescreen formats marketed around this time, it was abandoned soon afterwards.
	sound	The 'Blattnerphone' magnetic tape recorder is demonstrated in Germany.
1932	sound	The installation of sound equipment in almost all North American European cinemas is complete; sound on-disc is abandoned as a release format.
	film	The standard 8mm format is introduced by Eastman Kodak.
	film	The Taylor-Hobson 'Varo', the first zoom lens, goes on sale.
1933	film	Bell and Howell introduces the first 16mm sound projectors.
1934	film	The Mitchell BNC 35mm studio camera goes on sale.
	film	Fuji Photo film Co. formed in Tokyo.
	sound	The German chemical manufacturer BASF starts producing magnetic tape, initially on a cellulose diacetate base.
1935	colour	The first feature to be shot in three-strip Technicolor, *Becky Sharp*, is released; the Dufaycolor (dry-screen) colour process goes on sale in 16mm & 9.5mm reversal form to amateur filmmakers in Europe; the Kodachrome (coupler) colour system goes on sale in reversal form for 35mm still and 16mm moving image form in the US.
	sound	Alan Blumlein produces the first successfully demonstrated method for recording stereo sound synchronised to film.
1936	television	1 November: The BBC begins a three-month period of trial transmissions using the Baird (mechanical) and EMI (electronic) systems, the first regular programme of scheduled public television broadcasting.
1937	film	The Arriflex, the first film camera to incorporate a reflex viewfinder, is launched in Leipzig.
1938	film	Eastman Kodak introduces acetate propionate film stock, a far more durable safety base than any previously available.
	sound	German state radio begins using the 'Magnetophon' as a time-shifting device for broadcasts.
1939	colour	A negative-positive version of the Agfacolor coupler system is launched, which is used for a number of prominent Nazi propaganda features during World War Two.
1940	sound	*Fantasia*, believed to be the first feature film with stereo sound, is shown in the US.
1941	television	1 July: commercial broadcasting begins in the US.
	archiving	Harold Brown discovers the process of nitrate decomposition.
1947	sound	The first magnetic tape recorders are sold to consumers in the US.
1948	film	Charles R. Fordyce publishes *Improved Safety Motion Picture Film Support*, announcing the launch of cellulose triacetate film base by Eastman Kodak and the phasing out of nitrate.
	colour	Ansco, the US arm of Agfa, introduces the first negative/positive coupler stocks to be commercially used outside Germany and the USSR.
1950	moving images	SMPE renamed the Society of Motion Picture *and Television* Engineers (SMPTE).

	film	Eastman Kodak ceases the manufacture of cellulose nitrate base.
	colour	Eastman Kodak launches Eastmancolor, the first mass-produced coupler system for commercial moving image production.
1952	film	Commercial launch of Cinerama, on 30 September.
	film	The first 3D feature, *Bwana Devil*, is shown in Los Angeles on 26 November.
1953	film	The first CinemaScope feature, *The Robe*, is premiered in New York on 16 September.
1954	film	The first VistaVision feature, *White Christmas*, is premiered in New York on 27 April.
	film	The Panavision company is founded by Robert Gottschalk.
1954	exhibition	Zeiss-Ikon Corporation demonstrates short-arc xenon lamps at a Berlin trade show, a technology which was essential to the growth of multiplex cinemas in the 1980s.
1955	film	The first Todd-AO (70mm) feature, *Oklahoma!*, is premiered on 10 October.
1957	digital	The US Government establishes the Advanced Research Projects Agency (ARPA), which would build the data communications network that eventually grows into what is now known as the Internet.
1961	film	VistaVision is discontinued as a primary production medium following the release of *One Eyed Jacks*.
1962	sound	Philips and Sony launch the 'Compact Cassette' audiotape format, a user-friendly medium in which both the feed and take-up spool are housed within a single, self-contained mechanism. This principle would soon be applied to videotape.
1965	film	The Super 8 format is introduced by Eastman Kodak.
1967	film	The Imax process is demonstrated for the first time.
	television	The Phase Alternate Line (PAL) transmission system is first demonstrated at the Berlin Radio Exhibition.
1969	digital	ARPAnet, the first dedicated long-distance data communications network, commences operation.
1971	film	The Super 16 format is introduced.
1976	video	The Video Home System (VHS) videotape format is launched by JVC.
1979	digital	Commercial use of what is now the Internet is permitted for the first time, and the first Internet Service Provider company, CompuServe, is formed.
1981	archiving	Eastman Kodak introduces 'low fade' coupler colour stocks following a publicity campaign to highlight the problem of dye fading organised by Martin Scorsese during the previous two years.
1982	sound	Compact disc launched by Philips: the world's first digital consumer audio medium.
1985	archiving	The Image Permanence Institute is established in New York.
1986	colour	The television magnate Ted Turner signs a deal to 'colourise' feature films for broadcast.
	digital	NICAM broadcasting begins in the UK in July, adding two-channel digital audio to the existing analogue television transmission. This marked the first scheduled broadcast of any digital part of a television transmission.
1987	sound	Sony launches the 'Digital Audio Tape' (DAT) format, which quickly supersedes ¼-inch analogue magnetic tape as a location audio recording medium for film and television.
	digital	Invention of the DLP chip.
1988	digital	The MPEG digital video standards organisation is founded.
1992	sound	Three competing systems for encoding audio in the cinema – Dolby Digital, DTS and SDDS are introduced.

1992	exhibition	US cinema owners recommend the use of polyester stock for 35mm release prints.
1993	digital	The Digital Betacam (Digibeta) format is launched by Sony; the Digital Video Broadcasting consortium is formed in Europe.
1995	digital	The US Government hands over control of its Internet infrastructure to the private sector telecommunications industry.
1996	digital	The first DVD video recordings go on sale in Japan; Philips launches the Sprit DataCine film scanner.
1998	digital	The first regular digital terrestrial broadcast television service begins in the UK on 15 November.
2000	digital	Sony introduces the HDC-F900 Digital High Definition studio camera.
2001	digital	Recordable DVD technology goes on sale to consumers.
2004	film	The British film stock manufacturer Ilford goes into receivership on 24 August, blaming the growth of digital imaging for falling sales.

glossary

N.B. Words printed in italic in the definitions are cross-references to other entries.

Acetate – see *cellulose acetate*.

Acoustic – audio recording and reproduction which is achieved purely by registering or creating changes in ambient air pressure which can be detected by the human ear, i.e. no electronic amplification is used.

Additive – in colour film technology, the process of 'adding' combinations of three primary colours to a black base in order to form a picture, such as by projecting through filters or applying combinations of colour dye to the film base, as with (for example) *Dufaycolor*.

Accelerated aging – a test used in the preservation of archival audiovisual media, in which artificial levels of heat are used to simulate the passage of time in order to predict the shelf life of film or videotape before decomposition prevents the images and sounds on them from being recovered.

Amplifier – a device which increases the strength of an electrical *audio* signal to the level needed for recording or reproduction.

Analogue – representing or broadcasting images and sound as an infinitely variable modulated waveform, which can be recorded on magnetic oxide or a photosensitive emulsion.

Arrhenius test – see *accelerated aging*.

Aspect ratio – the proportion of height to width of a film *frame* or television picture, usually expressed in the format height:width, such as 1:1.85 for the widescreen ratio most commonly used for modern cinema films.

Audio – from the Latin, literally meaning 'I hear': a generic term which refers to the technology for and practice of creating, editing, copying and reproducing sound recordings.

Bandwidth – The volume of electronic signal (either *analogue* or *digital*) which can be transmitted through a channel of communications at any one time.

Base – the flexible, transparent support onto which a photosensitive *emulsion* is coated to produce moving image film.

BBC – see *British Broadcasting Corporation*.

Beta – a series of video cassette formats, which use the same design of cassette shell but with different ways of recording on the tape inside, developed by Sony and used since 1976: Betamax was intended for consumer use but rapidly lost market share to the more successful VHS, and had virtually disappeared by the mid-1980s; Betacam was a professional format intended as a rival to UMatic for *ENG* use, and introduced in 1981; it in turn was superseded by Betacam SP in 1986, which recorded a component signal; Digital Betacam followed in 1993, and is today generally considered the highest quality videotape format in routine use by broadcast studios. Neither of these formats are compatible with each other, for example while you could physically insert a Betamax tape into a Betacam SP machine, it would not be able to reproduce the recorded signal.

Blow-up – a film *element* produced by *optical printing* on a larger *gauge* than the source, such as a 16mm print made from a 35mm source.

Born digital – audiovisual media which is originated directly as digital data, i.e. not copied from an analogue source (such as film which has been scanned or an analogue videotape which has been migrated to a digital format).

British Broadcasting Corporation (BBC) – an organisation founded in 1926 which, until the 1950s, held a monopoly for radio and television broadcasting in the UK (although it was independent from the government and therefore is not, strictly speaking, a state broadcaster). Beginning in 1936, the BBC provided the first scheduled public television broadcasts in the world.

Camcorder – a device which combines the function of a *television camera* (usually with a *CCD* as the imaging device) and *video cassette recorder* (VCR) in a single unit.

Camera – a device which records a visual image by exposing light to a photosensitive medium, e.g. *film* or a *charged couple device.*

Capacitance Electronic Disc (CED) – a short-lived video recording format developed by RCA, which was marketed between March 1981 and June 1986, in which a modified phonograph stylus measured electrical resistance between it and grooves in a record.

Cassette – a moulded plastic frame housing magnetic tape, which contains both the feed and take-up spools. Openings in the cassette housing enable the tape to be threaded automatically upon insertion into an audio or video tape recorder, thus eliminating the risk of misthreading and ensuring that the operator never has to handle the tape itself.

Cathode Ray Tube (CRT) – the principal means of recording television images before it was superseded by *charge coupled devices* in the 1980s, and a means of displaying electronic moving images, invented in the 1940s and still in widespread use at the time of writing.

CCD – see *charged couple device*

CED – see *Capacitance Electronic Disc*

Cellulose – a carbohydrate which forms most of the biomass in wood pulp, used as the base for *cellulose acetate* and *cellulose nitrate* film.

Cellulose acetate – a film *base* produced by dissolving cellulose in acetic acid.

Cellulose nitrate – a film *base* produced by dissolving cellulose in nitric acid, which was used by the world's major film industries between 1889 and 1950.

Cement splice – a method of joining nitrate and acetate film, in which a chemical compound (such as amyl acetate) is used to dissolve a thin layer of film base on two facing surfaces, which are then pressed together to form an adhesive seal.

CGI – see *computer generated imagery.*

Channel – in sound recording, each separate 'track' intended to be played simultaneously to form a complete recording.

Charge Coupled Device (CCD) – the imaging device which replaced *cathode ray tubes* in the 1980s and 1990s as the principal means of recording moving images electronically, such as in a television camera, *camcorder* or *telecine* device.

Chrome – the Greek word for 'colour', which is used in a number of technical terms and product trademarks related to film-based colour systems, such as *panchromatic* or Kodachrome.

Chromogenic – in colour photography, the process of converting a photosensitive *emulsion* into a visible dye during *developing.*

Chronochrome™ – the first successfully demonstrated three-colour photographic colour process for moving images, invented by Léon Gaumont and first demonstrated in 1912 (Greek – *chronos* = 'time').

Cinema Digital Sound (CDS)™ – the first successfully demonstrated *digital* optical sound process for 35mm and 70mm *release prints*, developed by Kodak and first shown in 1991.

CinemaScope™ – a system developed by Twentieth Century Fox in 1953 for producing a *widescreen* image on film by means of an anamorphic lens to compress the picture along the horizontal plane.

Cinerama™ – a *widescreen* process developed in 1952 in which three mechanically interlocked film cameras and projectors are used to expose the image and project it onto a deeply curved screen.

Colour Reversal Intermediate (CRI) – a *reversal* film *element* of the same orientation (i.e. negative or positive) of the one it is printed from, used to reduce the number of generations of intermediate element needed between a camera original and *release print*.

Colour temperature – a means of describing the colour balance of a given light source within the visible spectrum, usually expressed in Kelvin.

Combined – a single film *element* which contains both an image and a sound recording intended to be reproduced in synchronisation with each other.

Compact disc – an optical data carrier introduced in 1982, 5 inches in diameter, which can store up to 700mb of data. Its most widespread applications are in the consumer *audio* market and for distributing computer software. It has a number of uses in moving image technology, notably as an audio carrier for the *DTS* cinema sound system and in the Video CD medium.

Complimentary colours – the *subtractive negatives* of the three *primary colours*, i.e. magenta, cyan and yellow.

Component – in video recording, a method whereby the video information is split into three separate signals. The first carries luminance (brightness), and is equivalent to a black-and-white picture. The other two are 'colour difference' signals: the 'B-Y' signal shows the level of blue relative to luminance, and the 'R-Y' does likewise with red. There is no need to record the green level, as this can be calculated from the other two primary colours. Component video is the highest quality method of encoding video images in use at the time of writing.

Composite – in television broadcasting and video recording, a method whereby the video information is all combined into a single signal, phase-shifted to distinguish the luminance (brightness) information from the chrominance (colour). It was originally devised by engineers working for RCA in order to broadcast a single transmission which could be received by both black-and-white and colour television sets. Composite video is the lowest quality method of encoding video images in use at the time of writing.

Compression – in digital audio and imaging, the means of reducing the amount of data needed to encode media content. The software which achieves this does so in one of two ways: *lossy* compression, which discards parts of the signal which are unlikely to be detected by the human eye or ear, and *lossless* compression which uses complex mathematical algorithms in order to condense the whole signal intact.

Computer – an electronic device which performs mathematical calculations on data, including data which represents information, images or sounds.

Computer Generated Imagery (CGI) – moving images produced entirely by computers, as distinct from images originated photographically and then manipulated using a computer. CGI is used primarily for generating special effects which are then integrated with photographically originated images (e.g. the dinosaurs in *Jurassic Park*).

Contact printing – a method of copying film in which the *element* containing the source image and the unexposed stock onto which it is being copied are placed in physical contact with each other and exposed to light.

Content Scrambling System (CSS) – a form of digital data encryption intended to prevent pre-recorded DVD video discs from being illegally copied.

Continuous format migration – a method of preserving the contents of magnetic and digital media by regular copying to a newer format in order to pre-empt the original format's obsolescence.

Continuous printing – a method of *contact* or *optical* printing in which the source and destination *elements* are passed over a light source at a constant speed. This is much faster than *step printing* but is only possible when the source element is in good condition, for example, making large numbers of cinema *release prints*.

'Copy to Preserve' – attempting to preserve moving image media in an archive by duplicating the content of elements which is considered at risk of being lost (for example, through decomposition or format obsolescence), onto a medium which can be sustained in the longer term.

Coupler – in colour film technology, a photosensitive *emulsion* which is converted into a visible dye during *processing*.

CRT – see *Cathode Ray Tube*.

CSS – see *Content Scrambling System*.

Deacetylation – see *vinegar syndrome*.

Developing – the first stage in *processing* film, which changes the *latent image* into one which is visible to the naked eye.

Digital – representing text, images and sound as numerical data which can be stored and manipulated using a computer.

Digital cinema – the process of originating, storing, manipulating and displaying moving images digitally for the main purpose of cinema exhibition. The key difference between digital cinema and *digital video* is that the latter represents images digitally which conform to the scanning requirements of a television standard (such as *PAL* or *NTSC*). The images used in digital cinema are usually of a much higher resolution than would be needed for this application, and the encoding standards do not need to support *interlacing*.

Digital High Definition (HD) – moving image cameras which use a *charge coupled device* to originate a progressive moving image sequence at 24 frames per second in the form of digital data, which are designed to replace film in studio feature production.

Digital Light Processing (DLP) – the means of displaying a moving image used in some video and digital cinema projectors.

Digital Theater Systems (DTS) – a company which markets a system, first sold in 1992, for *digital* sound reproduction in the cinema, in which a timecode on the film is used to synchronise the playback of audio from a separate *compact disc*.

Digital Versatile Disc (DVD) – an optical data carrier introduced in 1998, 5 inches in diameter, which can store up to 4.7gb of data on each of two physical layers (some discs have only one). At the time of writing its most common application is as a consumer video medium, using content encoded in the MPEG-2 format.

Digital video – (i) a conventional *video* recording in which the signal is encoded as digital data, as in the Digital Betacam format; (ii) 'DV', the proprietary name of a digital videotape format launched by Sony in 1995; (iii, colloquial) the practice of and technology for editing, manipulating and displaying digital moving images, and (iv) for encoding analogue *video* footage as digital data, using a personal computer.

Digital Video Broadcasting – the name of an international organisation established in 1993 to standardise and promote methods of digital television transmission based on the MPEG-2 standard. The first scheduled terrestrial broadcasts took place in 1998. By 2004 over 30 countries were broadcasting digital television using this system.

DLP – see *Digital Light Processing*.

Dolby – a company (founded by Ray Dolby) which develops and markets products for recording and reproducing sound in cinema, television and video.

Dry Hire – the practice of an external supplier renting equipment which is then operated and/or maintained by its customers' own technical staff. This business model has been used to regulate the use of technology quite extensively in the film and television industries and for a number of reasons. One of the better-known examples is Panavision. This manufacturer will not sell its cameras and lenses outright, preferring the dry hire model in order to ensure that only the latest generation of its technology is in use by film studios.

Dry Screen – see *Lenticular*.

DTS – see *Digital Theater Systems*.

Dufaycolor™ – an *additive* film colour process first marketed in 1934, in which angled patterns of red, green and blue dye, known as 'réseau', produce the coloured image in projection. It was used mainly in the UK for amateur filmmaking until superseded by *tripack* coupler stocks in the early 1950s.

Dye transfer – see *imbibition.*

DV – see *digital video.*

DVB – see *digital video broadcasting.*

DVB-C – the DVB standard for digital cable television, first used commercially in 1997 in the United States.

DVB-S – the DVB standard for digital satellite television.

DVB-T – the DVB standard for digital terrestrial television, first used for scheduled broadcasting in the UK on 15 November 1998.

DVD – see *digital versatile disc.*

Eidophor – one of the earliest forms of video projector (first demonstrated in 1943) to be mass-manufactured and commercially marketed.

Electronic News Gathering (ENG) – a term used during the 1970s and 1980s to describe the growing practice of replacing 16mm film with portable video equipment for shooting location footage for use in television news broadcasts.

Element – a specific copy or *generation* of a film or videotape.

Emulsion – a photosensitive compound which is coated onto an *acetate* or *polyester base* to produce moving image film.

ENG – see *electronic news gathering.*

Exposure – the act of subjecting a photosensitive medium, such as *film* or a *charged couple device* to light. In a *camera* or *printer* the duration of an exposure can be set, whilst the *aperture* setting on a lens is adjusted to control its intensity.

Exposure Index (EI) – a scale for measuring the *speed* of film, introduced in 1941 as the American Standards Association (ASA) scale.

FCC – see *Federal Communications Commission.*

Federal Communications Commission (FCC) – the US government agency which regulates broadcasting, and which established the *National Television Standards Committee* in 1940.

Fixing – the final stage in *processing* film, which 'fixes' the *developed* image into one which cannot be altered by further *exposure* to light.

Film – the combination of a flexible, transparent *base* (such as *cellulose nitrate* or *cellulose acetate*) and a photosensitive chemical *emulsion.*

Follower – a device for reproducing a separate magnetic audio track(s) recorded on film base coated with a magnetic oxide, which is electronically synchronised to the projector, telecine or viewing device running the picture film.

Footage – a generic term referring to a specific quantity of moving image material, usually unedited rushes or stock shots as distinct from a completed production.

fps [usually written in lower case] – frames per second: a scale for measuring the speed at which film passes through a *camera* or *projector.*

Frame – an individual photograph in a moving image sequence.

Gate – an opening in the film path of a *camera, printer* or *projector* through which light is admitted to make the *exposure* or to project the image.

Gauge – the width of moving image film, for example, 16mm or 35mm.

Generation – the number of *elements* removed that a film or videotape is from the original. A copy made from a camera original film element would be termed first generation and a copy made from that copy would be second generation.

Grading – in a moving image film *laboratory,* determining the intensity and (if the film being printed is colour) the colour balance of the *exposure* received by the unexposed stock in a *printer* (known as 'timing' in North America).

Gramophone – British English term for *Phonograph.*

HD – see *Digital High Definition.*

HDTV – High Definition Television. A short-lived *analogue* television standard introduced by Sony in 1981 which used the same scanning rate as *NTSC* (60 hertz), but which had 1,125 lines of horizontal resolution and an *aspect ratio* of approximately 1:1.75.

Head – a component which makes physical contact with the magnetic tape in an audio or video recorder. In recording the head produces electromagnetic energy in response to an input signal, which modulates the magnetic oxide on the passing tape. In playback the head produces electromagnetic energy in response to the oxide, which is then amplified and reproduced (in the case of analogue audio or video tape) or passed to a computer (in the case of magnetic tape used to store digital data) for processing.

Imbibition – (from the Latin *bibere*, 'drink' or 'absorb') the method of producing *release prints* used in the original *Technicolor* process, in which organic dyes are 'imbibed' by a gelatine layer on the raw print stock.

Image DissectorTube – the name given by Philo Farnsworth to the experimental electronic television camera he developed and patented in 1927.

Interactive – an adjective which is frequently misused by those who market consumer technology and misunderstood by those who use it. In essence it refers to one of two functions: (i) the ability of a user to selectively access media content which is prerecorded (such as on a DVD) or received in a one-way transmission (e.g. a digital terrestrial television broadcast); and (ii) the ability of a user to directly command a computer in order to determine how and in what form media content is delivered (e.g. via the Internet).

Interlace – in *television* and *video*, the practice of scanning the horizontal lines which form the image alternately (such as 1-3-5 followed by 2-4-6) in order to achieve even screen illumination on early *cathode ray tube* displays. In practice this is no longer necessary, as the phosphors in most modern CRTs remain energised for long enough to support *progressive* (i.e. 1-2-3-4-5-6) scanning. But the main analogue television standards in use still broadcast and record an interlaced signal.

Intertitle – in a silent film, a card containing written text against shown against a (usually black) background for a few seconds, used to convey dialogue or information which the director is unable to express using visuals alone.

ips – [usually written in lower case] – inches per second: a scale for measuring the speed at which magnetic tape passes a recording or playback head.

Kelvin – a scale for measuring *colour temperature*.

Kinemacolor™ – the most successful and widely used mechanical *additive* film-based colour system, developed by George Albert Smith and used quite widely between 1908 and 1915.

Kinescope – North American term for *telerecording*.

Kinetoscope™ – a device for viewing moving image film developed by the Edison Company in 1891, in which an endless loop of 35mm film is viewed through a magnifying glass as it passes under a continuously rotating shutter.

Kodachrome™ – the first commercially marketed *tripack dye coupler* (or *chromogenic*) film, originally sold in *reversal* form for 16mm amateur filmmaking and 35mm still photography in the mid-1930s.

Laboratory (lab) – a facility for printing and processing moving image film.

Latent image – in photography, the chemical state of the photosensitive *emulsion* after it has been *exposed* but before it has been *processed*; the image has been recorded, but is not yet visible to the naked eye.

Legacy format – a film, videotape or data storage format which is no longer in mainstream industrial use. Legacy format media are a significant problem for archives, as the equipment and spare parts needed to keep them playable may be difficult or impossible to obtain.

Lenticular – in colour cinematography, a process in which a series of embossed indentations on monochrome film act as miniature 'lenses' for recording three colour records when exposed through red, green and blue filters. After *processing*, the film is projected through reciprocating filters and the process is reversed, thus producing a full-colour image on the screen.

Liquid Crystal Display (LCD) – the means of displaying a moving image used in some television screens, computer monitors and video projectors, which exploits the phenomenon whereby the opacity of certain organic compounds can be altered by applying an electric charge.

Lossless and **lossy compression** – see *compression*.

Macrovision – the trade name of an anti-piracy technique which works by encoding videotape and DVD recordings sold to consumers with a signal that introduces distortion into any analogue copy which is made from them.

Magnetic – in *audio* and *video*, the process of recording a moving image and/or sound using a carrier (usually tape) in which magnetically polarised particles of a metallic oxide form the *analogue* signal or *digital* data.

Masked coupler – in colour film technology, a *coupler* which itself contains a colour dye. By selectively filtering these in printing, more accurate control of the colour balance is possible in the duplicate *element*. Masked couplers are what gives modern colour negative stock its characteristic orange hue.

Microphone – in *audio*, a device which converts audible sound into electrical energy for recording or reproduction.

Microprocessor – the component within a computer which carries out mathematical data processing functions.

Monaural – see *mono*.

Mono – a sound recording which has only one *track* (from the Greek, 'one').

Monochrome – from the Greek, literally 'one colour', i.e. black-and-white.

Monopack – a version of the Technicolor system developed in the early 1940s in which 35mm Kodachrome reversal stock was exposed in a normal studio camera, from which a set of three-strip negatives was then printed. It therefore enabled origination in colour without needing the cumbersome three-strip camera, although the resulting image was not as sharp or as saturated as genuine Technicolor.

Moore's Law – a prediction made by the American computer scientist Gordon Moore that the evolution of microprocessor technology would enable a doubling of computing power available from each chip every two years.

Mosaic – see *Lenticular*.

Motion Picture Patents Company (MPPC) – an organisation founded in December 1908 in order to jointly control the use of moving image technology covered by patents owned by its members and associates, principally Thomas Edison and George Eastman. It exerted a significant influence on the economic and technical development of the US film industry for a few years, but was eventually declared to be in contravention of antitrust legislation and dissolved by a federal court order in 1917.

Moving Pictures Expert Group (MPEG) – a committee of the International Standards Organisation established in 1988 to develop encoding and compression standards for digital video. Its most widely implemented standard, MPEG-2, is used in *DVB* television broadcasting and consumer *DVD* video discs.

Multimedia – the use of computers to integrate the storage and delivery of text, still images, moving images, and sound recordings provided from a single software carrier or online source.

National Association of Theatre Owners (NATO) – a representative organisation for the cinema exhibition industry in the United States (not to be confused with the international political body whose name forms the same acronym, the North Atlantic Treaty Organisation). In the early 1990s NATO successfully campaigned for a change from acetate to polyester film stock for cinema release printing.

National Television Standards Committee (NTSC) – an agency founded by the US Government in 1940 to determine television transmission standards. Today this term is usually used to refer to a standard consisting of 525 horizontal lines and a scanning rate of 60 hertz, in which the luminance (Y) signal is amplitude modulated and the chrominance (C) signal is phase modulated.

NATO – see *National Association of Theatre Owners*.

Negative – a photographic image, usually produced by exposure in a *camera*, in which the contrast between light and shade appears as the opposite to what the naked eye would see.

NICAM (Near Instantaneously Compacted and Expanded Audio Multiplex) – a means of broadcasting two channels of *digital* sound with a *PAL* television picture, developed by the BBC and introduced in Britain in 1986.

Nipkow disc – in mechanical television, a disc with sequential perforations that spiral inward. When it is rotated at high speed, a stationary photoelectric cell or light source positioned behind the disc will scan the image as a series of horizontal lines.

Offset – the distance between the synchronisation points for picture and sound on a *combined* film element. On standard 35mm film, for example, a combined *optical* soundtrack is printed 20 frames ahead of the picture.

Nitrate – see *cellulose nitrate.*

NTSC – see *National Television Standards Committee.*

Optical – sound (either *analogue* or *digital*) which is recorded photographically onto film.

Optical printing – a method of copying film in which the image of a source element is projected through a lens onto unexposed stock. Because the quality of reproduction is lower, optical printing is generally used to make copies which cannot be produced by *contact printing*, such as on a different *gauge* from the source or to copy damaged film for archival preservation.

Orthochromatic – a black-and-white film *emulsion* which is sensitive only to the blue and green areas of the visible colour spectrum.

PAL – see *Phase Alternate Line.*

Panchromatic – a black-and-white film *emulsion* which is sensitive to the entire visible colour spectrum.

Parallax – the difference in position between the images produced or projected by two lenses mounted in proximity to each other for use simultaneously, such as the *rangefinder* form of viewfinder, or systems which use multiple strips of film exposed simultaneously such as 3D and Cinerama.

Passive conservation – attempting to preserve moving image media in an archive by storing an original (or as generationally close to the original as can be found) element in atmospheric conditions that will inhibit decomposition as much as possible.

Perforation – small holes punched out of moving image film during manufacture, which are engaged by *sprocket* teeth to transport the film through the mechanism of a camera, printer or projector.

Persistence of vision – a theory, which emerged in the early nineteenth century but which in now believed to be fundamentally flawed, whereby the human brain retains the memory of an image perceived by the eye momentarily after it has disappeared.

Phase Alternate Line (PAL) – the set of broadcast television standards originally developed by Telefunken in West Germany and adopted as the British broadcast standard to replace the original 405-line, black-and-white only system in 1967. A PAL signal has 625 lines of horizontal resolution and a scanning rate of 50 hertz. PAL remains the highest resolution system in widespread use, and at the time of writing accounts for approximately one third of the world's television broadcasting.

Phenakistiscope™ – an optical toy first demonstrated in 1833 in which the *persistence of vision* effect is induced by the viewer looking into a mirror through a rotating shutter.

Phonograph – a system for *acoustic audio* recording and reproduction first demonstrated by Thomas Edison in 1877, in which an *analogue* waveform was inscribed by a stylus into a cylinder coated with wax. This word is now used as a generic noun in US English which also refers to disc-based *analogue audio* technology.

Photosensitive – material which responds to light, either in the form of a chemical reaction (as in film), or by producing electrical energy (as in the photocells used for optical sound reproduction, or the tube in a television camera).

Piracy – the commercial exploitation of recorded moving images without the authorisation of their copyright owner (as defined by the legislation which applies in the country or countries applicable), usually by the production of unauthorised copies.

Pixel – a single 'dot', or unit of data, used in digital imaging. The more pixels an image contains, the greater its *resolution*.

Polyester – an inorganic film base which entered widespread use in the late 1980s and is now used mainly to produce *release prints*.

Polyethalene teraphalate – see *polyester*.

Positive – a photographic image, produced either by copying from a *negative* or through the *reversal* process, in which the contrast between light and shade appears as it would to the naked eye.

Primary colours – red, green and blue; in colour photography, the three colours from which all other shades in the visible spectrum are formed.

Print – a *positive* copy of a *film* which is at least one *generation* removed from the original.

Printer – a device for *exposing* images onto moving image *film* which are copied from another film *element*.

Processing – the immersion of *exposed* film in a sequence of chemical baths in order to convert the *latent image* into an image which is visible to the human eye and which cannot be altered by further exposure to light.

Progressive – in television and video, a broadcast signal or video recording which is scanned sequentially, i.e. it is not *interlaced*.

Projector – a device which displays a moving image on a flat, reflective surface.

Pulldown – the linear area of film occupied by each *frame*, which is advanced by the *intermittent* movement of a *camera*, *step printer* or *projector*.

Pulse Code Modulation (PCM) – in *digital audio*, a widely used encoding format which is not *compressed*.

Quadruplex – the first mass-produced videotape format, marketed by Ampex from 1956 and in use until the mid-1970s. It is so called because the 2-inch wide tape is scanned by four cylindrically mounted heads.

Radio Corporation of America (RCA) – a US broadcasting, telecommunications and electronics company founded in 1919 and which was a key player in the development of optical film sound and television technologies.

Rangefinder – a form of camera viewfinder which consists of an optical system entirely separate from the lens which receives the exposure used to record the actual image.

Raw stock – (i) unexposed moving image film, (ii – less commonly used) uncoated film base, i.e. no emulsion.

RCA – see *Radio Corporation of America*.

Read Only – audio-visual media which cannot be recorded by consumers. Examples include phonograph records and laserdiscs.

Reduction – in film duplication, an *intermediate* or *print* made by optical printing onto a smaller gauge than the original, such as a 16mm reduction print from a 35mm negative.

Reflex – a form of camera viewfinder which, by means of a system of mirrors, allows the photographer to see directly through the lens which receives the exposure used to record the actual image. This eliminates the possibility of *parallax* errors which can be introduced by a *rangefinder* system, especially in extreme close-up shots.

Release print – a *combined positive* copy of a film made for projection in cinemas.

Resolution – the level of perceived detail in a photographic image.

Reversal – film which can create a *positive* photographic image in a single process (i.e. without having to expose a *negative* and then copy it onto a second *generation* of film). After *exposure* the film is developed twice: once to produce a negative image, which is then 'reversed' by a second development to convert it to a positive.

S-Video – in video recording, a method whereby the video information is split into two separate signals, one carrying luminance (brightness) and the other carrying chrominance (colour) information. This is a slightly higher quality method of recording video than a *composite* signal and was used mainly in semi-professional systems such as Hi-8 and S-VHS. It does not offer the same quality as the *component* method, which is mainly found in professional studio and broadcast VTRs.

SDDS – see *Sony Dynamic Digital Sound.*

SECAM – see *Système Èlectronique Couleur avec Memoir.*

Secondary colours – the *subtractive negatives* of the three *primary colours*, i.e. magenta, cyan and yellow.

Système Èlectronique Couleur avec Memoir (SECAM) – the set of broadcast television standards developed in France and used there and in the former Soviet Union countries. SeCAM is almost identical to PAL in having 625 lines of horizontal resolution and a scanning rate of 50 hertz. The only significant difference is in the way colour information is encoded, meaning that a PAL television will display a SeCAM transmission and vice-versa, but in black-and-white only.

Shutter – a component in a *camera* or *step printer* or which interrupts the flow of light in order to facilitate the *exposure* of individual *frames* in rapid succession, or in a *projector* to display the frames individually and in rapid succession to induce the *persistence of vision* effect.

Shutter speed – in a *camera*, the duration of each *exposure*.

Single system – a form of early *optical* sound recording in which both the picture and sound record were exposed simultaneously inside the same camera.

SMPTE – Society of Motion Picture and Television Engineers (originally SMPE: 'and television' was added in 1950).

Sony Dynamic Digital Sound (SDDS) – an *optical* system, first used commercially in 1992, for encoding up to 8 channels of *digital* sound onto a *release print*, and for reproducing them in the cinema.

Sound Camera – a device which exposes an optical sound record onto raw negative film stock, independently of the camera used to expose the picture.

Speed – (i) the comparative level of sensitivity to light of different *film emulsions*; (ii) an abbreviation for *shutter speed*; (iii) the speed at which individual photographs are *exposed* and displayed as part of a moving image sequence, which is expressed in *frames* per second; (iv) the speed at which a phonograph record rotates (usually expressed in revolutions per minute) or magnetic tape passes a *heed* (usually expressed in inches per second).

Splice – a physical join between two consecutive pieces of film or magnetic tape.

Sprocket – a toothed cylinder in a camera, printer or projector which engages the *perforations* in the film as it rotates in order to transport the film through the mechanism.

Step printing – a method of *contact* or *optical* film printing in which each *frame* of the source and destination element is *exposed* to light individually whilst the two elements are stationary.

Stereo – (i) strictly speaking, an *audio* recording that has two *channels*, which are designed to be played through loudspeakers approximating a listener's left and right ear. The term is also used colloquially to describe any recording which consists of more than one channel (i.e. one which is not *mono*); (ii) an image, either still or moving, which is a combination of two separate photographs, taken of the same subject but from slightly different angles, approximating the viewer's left and right eye. When they are seen from reciprocating angles, they appear to the viewer as a single, three dimensional picture.

Strike – the act of exposing destination film stock to light in a *printer*, a term usually used in post-production and *laboratories*, e.g. 'print no. X was struck from interneg Y'.

Successive frame – in colour cinematography, the exposure of (and in some process, projection from) two or three colour records on consecutive frames of a single strip of *monochrome, panchromatic* film.

Subtractive – in colour film technology, the process of 'subtracting' combinations of the three primary colours from white light (which contains equal proportions of all three) to form a picture, as with Technicolor and colour-coupler film systems.

Subtractive negatives – the colours produced by 'subtracting' (see above), i.e. filtering out the primary colours from white light. These are red/cyan, green/magenta and blue/yellow.

SVA *Stereo Variable Area* – an optical sound record containing more than one channel.

Synchronisation – the means of ensuring that a moving image and sound are recorded and reproduced at the same time as each other.

Tape splice – a method of joining film used mainly for *release prints* in cinemas, through the application of heat resistant, transparent and adhesive tape across either or both surfaces.

Technicolor™ – a company formed in 1915, which developed the first commercially successful three-colour *subtractive* film colour system, using a special prismatic *camera* and the *imbibition* method of producing *prints*. Though neither are now used, the company still exists, providing film *laboratory* services in a number of countries.

Telecine – a machine which produces a television image from film, either for simultaneous broadcast (done mainly in the days before video) or (in more recent years) for recording onto videotape.

Telerecording – recording onto film from a television image. This was done mainly in the days before videotape, when it was the only means available of permanently recording an image produced by a television camera.

Television – an *analogue* method of encoding and representing moving images electronically, in which the picture is divided into a number of horizontal lines. Each line is scanned repeatedly by the *cathode ray tube* or *charged couple device* in a television camera or display monitor, which produces an electronic analogue waveform. This can be transmitted either by wire or broadcast as a radio signal. The number of lines and scanning interval varies according to which television standard is used: for example, in the British *PAL* system, there are 625 lines which are scanned 50 times per second.

Thin FilmTransistor (TFT) – a type of *LCD* electronic display first marketed in the late 1990s, containing microscopic transistors which, after power is applied, have the ability to remain energised (and thus visible to the naked eye) almost indefinitely. TFT displays are predominately used in laptop computers, but are already starting to be incorporated into high-quality domestic television receivers.

Three strip – in colour cinematography, a process which exposes a record of all three *primary colours* onto three separate rolls of *monochrome* film running through the camera in synchronisation.

Throw – the linear distance between the projector and the screen in a cinema auditorium.

Time Shift – a phrase invented by the Sony founder Akio Morita to describe the use of VCRs by consumers to record broadcast television programmes off-air for subsequent viewing at a more convenient time, without intending to keep the recording permanently.

Timing – North American term for *grading.*

Todd-AO™ – a widescreen system first shown in 1955 which uses film 70mm wide (twice the width of conventional 35mm) to record a larger picture frame and six channels of magnetic sound.

Tripack – in colour film technology, a *coupler emulsion* sensitised to all three *primary colours.*

Two Strip – in colour cinematography, a process which exposes a record of red and green light (but not blue) onto two separate rolls of *monochrome* film running through the camera in synchronisation.

Ultrasonic splice – a method of joining *polyester* film in which the two surfaces are subjected to intense heat and then fused through the application of ultrasonic energy.

UMatic – a videotape format launched in 1971 and used mainly for *ENG.* It was largely superseded by *Betacam* and *Betacam SP* in the 1980s.

VHS – see *Video Home System.*

Video – From the Latin verb *videre,* literally meaning 'I see': (i) a generic noun referring to the technologies used to record a *television* signal, usually on *magnetic* tape, and reproduce it on a display monitor. Both *analogue* and *digital* video formats are available; (ii) colloquial – noun describing video recording or playback hardware, i.e. 'my video is broken'; (iii) colloquial – noun describing a recorded videotape, for example, 'I have a video of *Citizen Kane*'; (iv) colloquial – verb meaning the act of making a video recording, such as 'I will video my friend's wedding'.

Video Cassette Recorder (VCR) – strictly speaking, a machine which records and reproduces video signals in which the tape is contained within a cassette (such as in the UMatic and VHS formats). Although professional as well as consumer formats are now cassette-based, this term is generally used to describe consumer hardware as distinct from the more sophisticated machines found in television studios and media production facilities.

Video CD – a consumer format which can hold approximately one hour of low resolution digital video, encoded using the MPEG-1 system, on a standard compact disc. It was very popular in Asian and Far Eastern markets in the mid-1990s, but was never used on any significant scale in Europe or North America, where DVD video superseded VHS as the primary domestic video format directly.

Video Home System (VHS) – a video cassette format developed for consumer use, which was launched in 1976. It has proven the most commercially successful format in its market sector, and was still in widespread use at the time of writing.

VideoTape Recorder (VTR) – strictly speaking, a machine which records and reproduces video signals in which the tape is mounted on open reels, which are manually threaded by the operator (such as in the 2-inch Quadruplex and 1-inch C formats). All of these formats are now obsolete; and both professional and consumer video equipment is now cassette-based. Professional recorders are still often referred to as VTRs within the broadcast and media production industries.

Vinegar syndrome – a form of chemical decomposition affecting *cellulose acetate* film bases in which acetic acid leaches from the base through hydrolysis, causing the film to shrink and become brittle. Though formally termed deacetylation, it is generally referred to as vinegar syndrome due to the characteristic smell of affected *elements*.

Visible spectrum – light which can be seen by the naked eye.

VistaVision™ – a *widescreen* process launched by Paramount in 1954 in which 35mm film is transported horizontally through the camera and projector mechanism in order to produce a frame twice the size as those produced by conventional 35mm.

VCR – see *Video Cassette Recorder.*

VTR – see *Video Tape Recorder.*

Wet Hire – the practice of an equipment supplier hiring both hardware and the technical personnel to operate and maintain it. This business model is still used for some very specialist products and services in the film and television industries (mainly in digital post-production), although it is now more common for suppliers to train and certify technical staff which are directly employed by the customer (such as Imax). The best-known example of the wet hire model was Technicolor, whose 'colour consultants' were contractually empowered to exercise a high level of control over lighting, set and costume design on any production which used the three-strip camera.

Widescreen – any *aspect ratio* wider than 1:1.33 standard initially determined by W. K. L. Dickson in 1889.

Write Once – audio-visual media which cannot be edited, erased or rerecorded after the initial recording is made. Examples include film, wax or metal masters for phonograph records and CD-R recordable compact discs.

Zoetrope – an optical toy marketed from the 1860s which exploits the *illusion of continuous movement* effect to allow a paper strip printed with twelve consecutive *frames* to be viewed as a moving image sequence.

bibliography

Abel, Richard (1994) *The Ciné Comes to Town: French Cinema, 1896–1914*. Berkeley, CA: University of California Press.

Abel, Richard and Rick Altman (eds) (2001) *The Sounds of Early Cinema*. Bloomington, IN: Indiana University Press.

Abramson, Albert (1955) 'A Short History of Television Recording', *Journal of the SMPTE*, vol. 64 (February); reprinted in Raymond Fielding (ed.) *A Technological History of Motion Pictures and Television*. Berkeley, CA: University of California Press, pp. 250–4.

____ (1976) *Electronic Motion Pictures: History of the Television Camera*, New York: Arno Press.

____ (1988) *The History of Television, 1880–1941*. Jefferson, NC: McFarland.

____ (1995) *Zworykin: Pioneer of Television*. Urbana, IL: University of Illinois Press.

____ (2002) *The History of Television, 1942–2000*. Jefferson, NC: McFarland.

Academy of Motion Picture Arts and Sciences Research Council (1938) *Motion Picture Sound Engineering*. New York: D. van Nostrand.

Adelstein, P. Z., C. L. Graham and L. E. West (1970) 'Preservation of Motion Picture Color Films Having Permanent Value', *Journal of the SMPE*, vol. 79 (November), pp. 1011–18.

Adelstein, P. Z., J. M. Reilly, D. W. Nishimura and C. J. Erbland (1992) 'Stability of Cellulose Ester Base Photographic Film: Part 1 – Laboratory Testing Procedures', *SMPTE Journal*, vol. 101, no. 5 (May), pp. 336–46.

____ (1992) 'Stability of Cellulose Ester Base Photographic Film: Part 2 – Practical Storage Considerations', *SMPTE Journal*, vol. 101, no. 5 (May), pp. 347–53.

Aldred, John (1963) *Manual of Sound Recording*. London: Fountain Press.

Alexander, Robert Charles (1999) *The Inventor of Stereo: The Life and Works of Alan Dower Blumlein*. London: Focal Press.

Allen, Jeanne Thomas (1980) 'The Industrial Context of Film Technology: Standardisation and Patents', in Teresa de Lauretis and Stephen Heath (eds) *The Cinematic Apparatus*. Basingstoke: Macmillan, pp. 25–36.

Allen, Michael, 'In the Mix: How Electrical Reproducers Facilitated the Transition to Sound in British Cinemas' in Kevin Donnelly (ed.) *Film Music: Critical Approaches*. New York: Continuum, pp. 62–87.

Allen, N. S., M. Edge, J. S. Appleyard, T. Jewitt and C. V. Horie (1990) 'Initiation of the Degredation of Cellulose Triacetate Base Motion Picture Film', *Journal of Photographic Science*, vol. 38, pp. 54–9.

Allvine, Glendon, *The Greatest Fox Of Them All*, New York, Lyle Stuart Inc. (1969).

Altman, Rick (1996) 'The Silence of the Silents', *Musical Quarterly*, vol. 80, no. 4 (Winter), pp. 648–718.

Anderson, Joseph & Barbara, 'Motion Perception in Motion Pictures', in Teresa de Lauretis & Stephen

Heath (eds), *The Cinematic Apparatus*, Basingstoke: Macmillan (1980), pp. 76–95.
Andrew, Dudley, 'The Post-War Struggle for Colour' in Teresa de Lauretis and Stephen Heath (eds) *The Cinematic Apparatus*. Basingstoke: Macmillan, pp. 61–75.
Anon, (1987) 'Nitrate Project 2000: A Race Against Time', *Historical Journal of Film, Radio and Television*, vol. 7, no. 3, pp. 325–34.
____ (1980/81) 'Colour Problem', *Sight and Sound*, vol. 50, no. 1, pp. 12–13.
Armes, Roy (1988) *On Video*. London and New York: Routledge.
Baird, Margaret (1973) *Television Baird*. Cape Town: Haum.
Baker, T. T. (1938) 'Negative-Positive Technic with the Dufay Color Process', *Journal of the SMPE*, vol. 31, pp. 240–7.
Ball, J. A. (1935) 'The Technicolor Process of Three-Color Cinematography', *Journal of the SMPE*, vol. 25, no. 2 (August), pp. 127–38.
Balshofer, Fred J. & Arthur C. Miller (1967) *One Reel A Week*. Berkeley & Los Angeles, University of California Press.
Barr, Charles (1983) '*Blackmail*: Silent and Sound', *Sight and Sound*, vol. 52, no. 2 (Spring), pp. 122–6.
Becker, Ron (2001) 'Hear and See Radio in the World of Tomorrow: RCA and the Presentation of Television at the World's Fair, 1939–40', *Historical Journal of Film, Radio and Television*, vol. 21, no. 4 (October), pp. 361–78.
Bell, Donald J. (1930) 'A Letter from Donald J. Bell', *International Photographer*, vol. 2, no. 1 (February), p.19.
Belton, John (1990) 'The Origins of 35mm Film as a Standard', *SMPTE Journal*, (August), pp. 652–61.
____ (1992) *Widescreen Cinema*. Cambridge, MA: Harvard University Press.
____ (2000) 'Getting it Right – Robert Harris on Color Restoration', *Film History*, vol. 12, no. 4, pp. 393–409.
Bernds, Edward (1999) *Mr. Bernds Goes to Hollywood: My Life and Career in Sound Recording at Columbia with Frank Capra and Others*. Lanham, MD and London: Scarecrow Press.
Bijker, Wiebe (1997) *Of Bakelites and Bulbs: Toward a Theory of Sociotechnical Change*, Cambridge, MA: MIT Press.
Bilby, Kenneth M. (1986) *The General: David Sarnoff and the Rise of the Communications Industry*. New York: Harper and Row.
Blake, Larry (1984) *Film Sound Today*. Hollywood: Reveille Press.
Boddy, William F. (1989) 'Launching Television: RCA, the FCC and the Battle for Frequency Allocations, 1940-47', *Historical Journal of Film, Radio and Television*, vol. 9, no. 1, pp. 45–57.
Bordwell, David, Janet Staiger and Kristin Thompson (1985) *The Classical Hollywood Cinema: Film Style and Mode of Production to 1960*. London: Routledge.
Borwick, John (ed.) (1976) *Sound Recording Practice*. Oxford: Oxford University Press.
Bottomore, Stephen (2002) 'A Fallen Star: Problems and Practices in Early Film Preservation', in Roger Smither and Catherine Surowiec (eds) *This Film is Dangerous: A Celebration of Nitrate Film*. Brussels: FIAF, pp. 185–90.
Brand, Neil (2002) 'Distant Trumpets – The Score to *The Flag Lieutenant* and Music of the British Silent Cinema', in Andrew Higson (ed.) *Young and Innocent? The Cinema in Britain, 1896–1930*. Exeter: University of Exeter Press, pp. 208–24.
Brendon, Piers (2000) *The Dark Valley: A Panorama of the 1930s*. London: Jonathan Cape.
Briggs, Asa (1995) *The History of Broadcasting in the United Kingdom – Competition, 1955–1974*. Oxford: Oxford University Press.
Brown, Harold (1990) *Physical Characteristics of Early Films as Aids to Identification*. Brussels: FIAF.
Brown, Simon (2003) 'Dufaycolor – The Spectacle of Reality and British National Cinema', AHRB Centre for Film and Television Studies, London; available online at http://www.bftv.ac.uk/projects/dufaycolor.htm (accessed 17 June 2003).
Brownlow, Kevin (1980) 'Silent Films – What was the Right Speed?', *Sight and Sound*, vol. 49, no. 3 (Summer), pp. 164–7.

Burch, Noël (1990) *Life to Those Shadows*. London: British Film Institute.
Burder, John (1976) *16mm Film Cutting*. London: Focal Press.
Burns, R. W. (1986) *British Television: The Formative Years*. London: Peter Peregrinus.
____ (1998) *Television: An International History of the Formative Years*, London, Institution of Electrical and Electronic Engineers.
Bustamente, Carlos (2000) 'The Bolex Motion Picture Camera' in John Fullerton and Astrid Soderbergh-Widding, *Moving Images: From Edison to Web Cam*. Eastleigh: John Libbey, pp. 59–65.
Cameron, James R. (1922) *Motion Picture Projection*, 3rd edn. New York: Technical Book Co.
____ (1929) *Motion Pictures With Sound*. Manhattan Beach, NY: Cameron Publishing Co.
Cameron, James and John F. Rider (1930) *Sound Pictures and Trouble Shooters Manual*. Manhattan Beach, Cameron Publishing Co.
Carr, Robert E. and R. M. Haynes (1988) *Wide Screen Movies*. Jefferson, NC: McFarland.
Case, Dominic (2001) *Film Technology in Post Production*, 2nd edn. Oxford: Focal Press.
Chanan, Michael (1990) *The Dream that Kicks: The Prehistory and Early Years of Cinema in Britain*. 2nd edn. London: Routledge.
Clason, W.E. (ed.), *Elsevier's Dictionary of Television and Video Recording*, Amsterdam, Elsevier Scientific Publishing Co. (1975).
Cleveland, David (2002) 'Don't Try This at Home: Some Thoughts on Nitrate Film, with Particular Reference to Home Movie Systems', in Roger Smither and Catherine Surowiec (eds) *This Film is Dangerous: A Celebration of Nitrate Film*. Brussels: FIAF, pp. 191–7.
Coe, Brian (1981) *The History of Movie Photography*. Westfield, NJ: Eastwood Editions.
Conant, Michael (1982) *Antitrust in the Motion Picture Industry*, Berkeley, University of California Press.
Cornwell-Clyne, Adrian (1951) *Colour Cinematography*, 3rd edn., London, Chapman Hall.
Coven, Frank (ed.) (1950) *The Daily Mail Television Handbook*. London: Daily Mail Publications.
Crafton, Donald (1997) *The Talkies*. Berkeley, CA: University of California Press.
Crane, Rhonda J. (1979) *The Politics of International Standards: France and the Colour TV War*. Norwood, NJ: Ablex.
Cripps, Thomas (1997) *Hollywood's High Noon*. Baltimore: Johns Hopkins University Press.
Crozier, Major T.M., *Report to the Right Honourable Secretary of State for Scotland on the Circumstances Attending the Loss of Life at the Glen Cinema, Paisley, on the 31st December, 1929*, London, HMSO (1930).
Dambacher, Paul (1996) *Digital Broadcasting*. London: Institution of Electrical and Electronic Engineers.
Davidoff, Frank (1975) 'Digital Video Recording for TV Broadcasting', *Journal of the SMPTE*, vol. 84 (July), pp. 552–5.
Davis, Darrell William (2003) 'Compact Generations: VCD Markets in Asia', *Historical Journal of Film, Radio and Television*, vol. 23, no. 2 (June), pp. 165–76.
De Forest, Lee (1941) 'Pioneering in Talking Pictures', *Journal of the SMPE*, vol. 36 (January), pp. 41–9.
____ (1942) *Television Today and Tomorrow*. New York: Dial Press.
____ (1950) *Father of Radio*. Chicago: Wilcox and Follett.
De Lauretis, Teresa and Stephen Heath (eds) (1980) *The Cinematic Apparatus*. Basingstoke: Macmillan.
Dempsey, Michael, 'Colorization', *Film Quarterly*, vol. 40, no. 2 (Winter 1986–87), pp. 2–3.
Dickson, William Kennedy Laurie (1895) *History of the Kinetograph, Kinetoscope and the Kineto-Phonograph*. New York: Albert Bunn.
Dobrow, Julia R. (ed.) (1990) *Social and Cultural Aspects of VCR Use*. Hillsdale, NJ: Lawrence Erlbaum.
Donald, James, Laura Marcus and Anne Friedberg (eds) (1998) *Close Up, 1927–1933: Cinema and Modernism*. London: Cassell.
Donnelly, Kevin (ed.) (2001) *Film Music: Critical Approaches*. New York: Continuum.
Eargle, John (2001) *The Microphone Book*. Boston: Focal Press.
Eastman Kodak Company (1927) *Tinting and Toning of Eastman Positive Motion Picture Film*. Roches-

ter: Eastman Kodak Company.
____ (1958) *How to Make Good Home Movies*. Rochester: Eastman Kodak Company.
Ede, François (1995) *Jour de fête, ou, Le couleur retrouvée*. Paris: Cahiers du Cinema Livres.
Eder, Josef Maria (1972) *History of Photography*. New York: Dover Publications.
Eisenstein, Sergei (1982) 'The Dynamic Square', in Jey Leyda (ed.) *Film Essays and a Lecture*. Princeton: Princeton University Press, pp. 48–65.
Enticknap, Leo (2002) 'The Film Industry's Conversion from Nitrate to Acetate Stocks in the Late 1940s: A Discussion of the Reasons and Consequences', in Roger Smither and Catherine Surowiec (eds) *This Film is Dangerous: A Celebration of Nitrate Film*. Brussels: FIAF, pp. 202–12.
____ (2004) 'Moving Image Technology in the Time of Mitchell and Kenyon', in Vanessa Toulmin, Simon Popple and Patrick Russell (eds) *The Lost World of Mitchell and Kenyon: Edwardian Britain on Film*. London: British Film Institute, pp. 21–30.
Eyman, Scott (1997) *The Speed of Sound: Hollywood and the Talkie Revolution, 1926–1930*. New York: Simon & Schuster.
Fairservice, Don (2001) *Film Editing: History, Theory and Practice*. Manchester: Manchester University Press.
Fielding, Raymond (ed.) (1967) *A Technological History of Motion Pictures and Television*. Berkeley, CA: University of California Press.
Fifty Years of Fuji Photo Film (1984) Fuji Photo Film Co. Ltd.
Fisher, Bob (1992) 'The Dawning of the Digital Age', *American Cinematographer*, vol. 73, no. 4 (April), pp. 70–85.
Fisher, David E. and Marshall Jon Fisher, *Tube – The Invention of Television*. San Diego, Harcourt Brace and Company (1996).
Fordyce, Charles (1948) 'Improved Safety Motion Picture Film Support', *Journal of the SMPE*, vol. 51 (October), pp. 331–50.
____ (1955) J. M. Calhoun and E. E. Moyer, 'Shrinkage Behavior of Motion Picture Film', *Journal of the SMPTE*, vol. 64 (February), pp. 62–6.
Forrester, Chris (2000) *The Business of Digital Television*. Oxford: Focal Press.
Fox, Barry (1991) 'Squeezed Sound for the Silver Screen', *New Scientist*, 13 July, p. 26.
Franklin, Harold B. (1929) *Sound Motion Pictures*. New York: Doubleday, Doran & Co.
Fullerton, John and Astrid Soderbergh-Widding (eds) (2000) *Moving Images: From Edison to Web Cam*. Eastleigh: John Libbey.
Garrity, W. E. and J. N. A. Hawkins (1941) 'Fantasound', *Journal of the SMPE*, vol. 37 (August), pp. 127–46.
Gaudreault, Andre (1985) 'The Infringement of Copyright Laws and its Effects, 1900–06', *Framework*, no. 29, pp. 3–6.
Geddes, Keith and Gordon Bussey (1986) *Television – The First 50 Years*. Bradford: National Museum of Film, Photography and Television.
Geduld, Harry (1975) *The Birth of the Talkies: From Edison to Jolson*. Bloomington: Indiana University Press.
Gelatt, Roland (1955) *The Fabulous Phonograph: The Story of the Gramophone from Tin Foil to High Fidelity*. New York: Lippincott.
Gomery, Douglas (1976) 'Warner Bros. and Sound', *Screen*, vol. 17, no. 1 (Spring), pp. 40–53.
Gosser, H. Mark (1998) 'The *Bazar de la Charité* Fire: The Reality, the Aftermath and the Telling', *Film History*, vol. 10, no. 1, pp. 70–89.
Gould, Sara and Marie-Therese Varlamoff (2000) 'The Preservation of Digitised Collections', in Ralph W. Manning (ed.) *A Reader in Preservation and Conservation*. Munich: IFLA Publications.
Graham, Gerald G. (1989) *Canadian Film Technology, 1896–1986*. London and Toronto: Associated Universities' Press.
Graham, Margaret (1986) *RCA and the Videodisc: The Business of Research*. Cambridge: Cambridge University Press.
Grosset, Philip (1959) *Making 8mm Movies*. London: Fountain Press.

Hallett, Michael (1978) *John Logie Baird and Television*. Hove: Priory.
Hampton, Benjamin B. (1931) *History of the American Film Industry*. New York: Covici Friede.
Hanson, Wesley T. Jr (1980) 'The Evolution of Motion Pictures in Color', *SMPTE Journal*, vol. 89, no. 7 (July), pp. 528–30.
Happé, L. Bernard (1975) *Basic Motion Picture Technology*, 2nd edn. London: Focal Press.
Harding, Colin and Simon Popple (1996) *In the Kingdom of Shadows: A Companion to Early Cinema*. London: Cygnus Arts.
Hartwig, Robert L. (2000) *Basic TV Technology: Digital and Analog*, 3rd edn. Boston: Focal Press.
Haver, Ronald (1986) *David O. Selznick's Gone With The Wind*. New York, Bonanza Books.
Hayward, Philip and Tana Wollen (eds) (1993) *Future Visions: New Technologies of the Screen*. London: British Film Institute.
Hendricks, Gordon (1961) *The Edison Motion Picture Myth*. Berkeley: University of California Press.
Hepworth, Cecil (1951) *Came the Dawn: Memories of a Film Pioneer*. London: Phoenix House.
Herbert, Stephen (1998) *Industry, Liberty and a Vision: Wordsworth Donisthorpe's Kinesigraph*. Hastings: The Projection Box.
____ *Persistence of Vision*, essay published online at http://www.grand-illusions.com/percept.htm (page last modified on 14 November 2004 and accessed by the author on 5 February 2005)
Hercock, R. J. and G. A. Jones (1979) *Silver by the Ton: A History of Ilford Ltd., 1879–1979*. Maidenhead: McGraw Hil.
Herzog, Charlotte (1980) *The Motion Picture Theater and Film Exhibition, 1896–1932*, unpublished PhD thesis, Northwestern University; UMI no. 8026823.
Higson, Andrew (ed.) (2002) *Young and Innocent? The Cinema in Britain, 1896–1930*. Exeter: University of Exeter Press.
Hilmes, Michele (ed.) (2003) *The Television History Book*, London, British Film Institute.
Hochheiser, Sheldon (1992) 'What Makes the Picture Talk: AT&T and the Development of Sound Motion Picture Technology', *IEEE Transactions on Education*, vol. 35, no. 4 (4 November), pp. 278–85.
Hopwood, Henry J. (1899) *Living Pictures: Their History, Production and Practical Working*. London: The Optician and Photographic Trades Review.
Houston, Penelope (1994) *Keepers of the Frame: The Film Archives*. London: British Film Institute.
Hulfish, David S. (1915) *Cyclopaedia of Motion Picture Work*, 3rd edn. New York: American Technical Society.
Hunt, R. W. G. (1975) *The Reproduction of Colour*, 3rd edn. Surbiton: Fountain Press.
Huske, Gibboney and Rick Vallières (2002) *Digital Cinema, Episode II*. London: Credit Suisse First Boston (Europe) Ltd. This publication is available in full online at www.sabucat.com/digital.pdf at the time of writing.
Huther, W., H. Tjabben and D. J. Bancroft (1999) 'Digital Film Mastering', *SMPTE Journal*, vol. 108, no. 12 (December), pp. 859–63.
Ibbetson, W. S. (1921) *The Kinema Operator's Handbook: Theory and Practice*. London: E. and F. N. Spon.
Inglis, A. F. (1992) *Video Engineering*. White Plains: Knowledge Industry Publications.
Jackson, Pat (1999) *A Retake Please: Night Mail to Western Approaches*. Liverpool: Royal Naval Museum Publications & Liverpool University Press.
Jacobs, Lewis (1939) *The Rise of the American Film*. New York: Teachers' College Press.
Jones, Bernard E. (1915) *The Cinematograph Book*. London: Cassell.
Jossé, Harald (1984) *Die Entstehung des Tonfilms*. Frieburg and Munich: Verlag Karl Alber.
Kallenberger, Richard H. and George D. Cvjetnicanin (1994) *Film into Video: A Guide to Merging the Technologies*. Boston and London: Focal Press.
Kalmus, Herbert T. (1938) 'Technicolor Adventures in Cinemaland', *Journal of the SMPE*, vol. 36, no. 6 (December), pp. 564–85.
Kamm, Anthony and Malcolm Baird (2002) *John Logie Baird – A Life*. Edinburgh: National Museums of Scotland Publishing.

Kattelle, Alan (2000) *Home Movies: A History of the American Industry, 1897–1979*. Nashua, NH: Transition Publishing.
Kearns, Paul (1998) *The Legal Concept of Art*. Oxford: Hart Publishing.
Kellogg, Edward W. (1945) 'The ABC of Photographic Sound Recording', *Journal of the SMPE*, vol. 44, no. 3 (March), pp. 151–83.
Kirk, David K. (1981) *25 Years of Video Tape Recording*. Bracknell: 3M Corporation.
Kisseloff, Jeff (1965) *The Box: An Oral History of Television, 1920–61*. New York: Penguin.
Kitsopanidou, Kira (2003) 'The Widescreen Revolution and Twentieth Century Fox's Eidophor in the 1950s', *Film History*, vol. 15, no. 1, pp. 32–56.
Kittler, Friedrich A. (1999) *Gramophone, Film, Typewriter* (trans. Geoffrey Winthrop-Young and Michael Wutz). Stanford: Stanford University Press.
Klöpfel, Don E. (1969) *Motion Picture Projection and Theater Presentation Manual*. New York: SMPTE.
Koshofer, Gert (1988) *Color – Die Farben des Films*. West Berlin: Wissenschaftverlag Volker Speiss.
Kurosawa, Akira (1982) *Something Like an Autobiography*. New York: Alfred A. Knopf.
Lardner, James (1987) *Fast Forward*. New York: W. W. Norton.
Lassally, Walter (1997) 'Alice in Format Land', *Image Technology*, vol. 79, no. 10, p. 14.
Lastra, James (2000) *Sound Technology and the American Cinema*. New York: Columbia University Press.
Levy, Mark (ed.) (1989) *The VCR Age*. Newbury Park: Sage.
Leyda, Jay (ed.) (1982) *Film Essays and a Lecture*. Princeton: Princeton University Press.
Limbacher, James L. (1969) *Four Aspects of the Film: A History of the Development of Color, Sound, 3D and Widescreen Films and their Contribution to the Art of the Motion Picture*. New York: Brussel and Brussel.
Lindgren, Ernest (1944) 'The Permanent Preservation of Cinematograph Film', *Proceedings of the British Society for International Bibliography*, vol. 5, no. 5, pp. 97–104.
Low, Rachael (1985) *Film Making in 1930s Britain*. London: George Allen and Unwin.
Lyons, Eugene (1966) *David Sarnoff: A Biography*. New York: Harper and Row.
Macnab, Geoffrey (1993) *J. Arthur Rank and the British Film Industry*. London: Routledge.
Machon, David F. P. (1981) 'The Shape of Things to Come?', *BKSTS Journal*, vol. 63, no. 12 (December), pp. 738–43.
Mannes, L. D. and L. Godowsky Jr (1935) 'The Kodachrome Process for Amateur Cinematography in Natural Color', *Journal of the SMPE*, vol. 25, pp. 24–57.
Mannoni, Laurent (2000) *The Great Art of Light and Shadow: Archaeology of the Cinema* (trans. Richard Crangle). Exeter: University of Exeter Press.
Marc, David & Robert J. Thompson (2005) *Television in the Antenna Age: A Concise History*. Oxford: Blackwell.
Mazzanti, Nicola (2002) 'Rasing the Colours (Restoring Kinemacolor)', in Roger Smither and Catherine Surowiec (eds) *This Film is Dangerous: A Celebration of Nitrate Film*. Brussels: FIAF, pp. 123–5.
McGreevey, Tom & Joanne L. Yeck, *Our Movie Heritage*, New Jersey, Rutgers University Press (1997).
McKernan, Luke, 'The Brighton School and the Quest for Natural Colour' in Vanessa Toulmin & Simon Popple (eds.), *Visual Delights 2: Audience and Reception* (qv.), pp. 205-218.
Mees, C. E. K. (1929) 'Amateur Cinematography and the Kodacolor Process', *Journal of the Franklin Institute* (January), pp. 1–17.
____ (1961) *From Dry Plates to Ektachrome Film: A Story of Photographic Research*. New York: Ziff-Davis.
Messter, Oskar (1936) *Mein weg mit dem Film*. Berlin Schöneberg: Max Hesse Verlag.
Miehling, Rudolph (1929) *Sound Projection*. New York: Mancall.
Miller, W. E. (1950) *Television In Your Home: Everything the Potential Viewer Needs to Know*. London: Iliffe and Sons Ltd.
Moore, Gordon E. (2004) 'Cramming more components onto integrated circuits', *Electronics*, vol. 38,

no. 8 (19 April 1965), republished at ftp://download.intel.com/research/silicon/moorespaper.pdf (accessed by the author on 26 October).

Moore, Jerrold Northrop (1999) *Sound Revolutions: A Biography of Fred Gaisberg, Founding Father of Commercial Sound Recording*. London: Sanctuary.

Morton, David (2000) *Off the Record: The Technology and Culture of Sound Recording in America*, New Brunswick, NJ: Rutgers University Press.

Mulvey, Laura (1999) 'Now You Has Jazz', *Sight and Sound*, vol. 9, no. 5 (May), pp. 16–18.

Murphy, Robert (1984) 'The coming of sound to the cinema in Britain', *Historical Journal of Film, Radio and Television*, vol. 4, no. 2.

Nathan, John (2000) *Sony: The Inside Story*. London: HarperCollins.

Nyquist, Harold (1928) 'Certain Topics in Telegraph Transmission Theory', *Transactions of the Institute of American Electrical Engineers*, vol. 42 (April), pp. 617–44.

Oha, Obododimma (2001) 'The Visual Rhetoric of the Ambivalent City in Nigerian Video Films', in Mark Shiel and Tony Fitzmaurice (eds) *Cinema and the City: Film and Urban Societies in a Global Context*. Oxford: Blackwell, pp. 196–205.

Ohanian, Thomas A. and Michael E. Phillips (1996) *Digital Filmmaking: The Changing Art and Craft of Making Motion Pictures*. Boston: Focal Press.

Olson, Harry (1931) 'The Ribbon Microphone', *Journal of the SMPE*, vol. 16, p. 695.

Pawley, Edward (1972) *BBC Engineering, 1922–1972*. London: BBC Publications.

Petrie, Duncan (1996) *The British Cinematographer*. London: British Film Institute.

____ (2002) 'British Low-budget Production and Digital Technology', *Journal of Popular British Cinema*, vol. 5, pp. 64–76.

Pierce, David (1997) 'The Legion of the Condemned: Why American Silent Films Perished', *Film History*, vol. 9, no. 1 (Spring), pp. 5–22.

Pohlmann, Ken C. (1989) *The Compact Disc: A Handbook of its Theory and Use*. Oxford: Oxford University Press.

Polan, Martin (1980) 'The Future of Entertainment and the Four Horsemen of Technology', *Sight and Sound*, vol. 49, no. 4, pp. 224–9.

Ramsaye, Terry (1926) *A Million and One Nights*. New York: Simon and Schuster.

Rawlence, Christopher (1990) *The Missing Reel: The Untold Story of the Lost Inventor of Moving Pictures*. London: Fontana.

Read, Paul and Mark-Paul Mayer (2000) *Restoration of Motion Picture Film*. Oxford: Butterworth-Heinemann.

Read, Walter and Leah Brodbeck Stenzel Burt (1994) *From Tinfoil to Stereo: The Tinfoil Years of the Recording Industry*. Gainsville: University Press of Florida.

Reed, Robert M. & Maxine K. Reed, *The Facts on File Dictionary of Television, Cable and Video*, New York, Facts On File Inc. (1994).

Reilly, James (2004) 'Keeping Cool and Dry: A New Emphasis in Film Preservation', available online at www.lcweb.loc.gov/film/storage.html (accessed by the author on 23 April).

Reiskind, H. I. (1941) 'Multiple-Speaker Reproducing Systems for Motion Pictures', *Journal of the SMPE*, vol. 37 (August), pp. 154–63.

Rentschler, Eric (1996) *The Ministry of Illusion: Nazi Cinema and its Afterlife*. Cambridge, MA: Harvard University Press.

Roberts, Graham & Philip M. Taylor (eds) (2001) *The Historian, Television and Television History*. Luton: University of Luton Press.

Robinson, Jack Fay (1982) *Bell and Howell Company – A 75 Year History*. Chicago: Bell and Howell.

Robson, Neil (2004) 'Living Pictures Out of Space: The Forlorn Hopes for Television in Pre-1939 London', *Historical Journal of Film, Radio and Television*, vol. 24, no. 2 (June), pp.223–32.

Rossell, Deac, *Living Pictures: The Origins of the Movies*, New York, State University of New York Press (1998).

Roud, Richard (1983) *A Passion for Films: Henri Langlois and the Cinémathèque Française*. New York: Viking Press.

Rowland, Richard (1927) 'The Speed of Projection of Film', *Transactions of the SMPE* (January), pp. 77–9.
Ryan, Roderick T. (1967) *A Study of the Technology of Color Motion Picture Processes Developed in the United States*, unpublished PhD thesis, University of Southern California.
____ (1977) *A History of Motion Picture Colour Technology*. London and New York: Focal Press.
Salt, Barry (1992) *Film Style and Technology: History and Analysis*, 2nd edn. London: Starword.
Sawyer, Martin (2002) 'The Sound of Nitrate', in Roger Smither and Catherine Surowiec (eds) *This Film is Dangerous: A Celebration of Nitrate Film*. Brussels: FIAF, pp. 136–9.
Scotland, John (1930) *The Talkies*. London: Crosby, Lockwood and Son.
Segrave, Kerry (1999) *Movies at Home: How Hollywood Came to Television*. Jefferson, NC and London: McFarland.
____ (2003) *Piracy in the Motion Picture Industry*. Jefferson, NC and London: McFarland.
Sexton, Jamie (2003) 'Televerite Hits Britain: documentary, drama and the growth of 16mm film-making in British television', *Screen*, vol. 44, no. 4 (Winter), pp. 429–44.
Sherlock, James (1997) 'Restoration', *Cinema Papers*, vol. 117 (June), pp. 18–22.
Shiel, Mark and Tony Fitzmaurice (eds) (2001) *Cinema and the City: Film and Urban Societies in a Global Context*. Oxford: Blackwell.
Shires, George (ed.) (1977) *The Technical Development of Television*. New York: Arno Press.
Slide, Anthony (1992) *Nitrate Won't Wait: A History of Film Preservation in the United States*. Jefferson, NC: McFarland.
Sloane, T. O'Conor (1922) *Motion Picture Projection*. New York: Falk.
Souto, H. Mario Raimondo (1982) *The Technique of the Motion Picture Camera*, 4th edn. London: Focal Press.
Smither, Roger and Catherine Surowiec (eds) (2002) *This Film is Dangerous: A Celebration of Nitrate Film*. Brussels: FIAF.
Sobel, Robert (1986) *RCA*. New York: Stein and Day.
Sokal, Alan, 'Transgressing the Boundaries: Toward a Transformative Hermeneutics of Quantum Gravity', *Social Text*, vol. 46, no. 47 (Spring–Summer 1996), pp. 217–52.
Sokal, Alan, & Jean Bricmont, *Intellectual Impostures: Postmodern Philosophers' Abuse of Science*, London, Profile Books (1998).
Spehr, Paul C. (1979) 'Fading: The Color Film Crisis', *American Film*, vol. 5, no.2 (November), p. 57.
____ 'Unaltered to Date: Developing 35mm Film' in John Fulerton and Astrid Soderbergh-Widding (eds) *Moving Images: From Edison to Web Cam*. Eastleigh: John Libbey, pp. 3–28.
Sponable, Earl I. (1947) 'Historical Development of Sound Films', *Journal of the SMPE*, vol. 48 (April), pp. 275–303; vol. 48 (May), pp. 407–22.
Staiger, Janet, 'Combination and Litigation: Structures of US Film Distribution, 1891–1917', *Cinema Journal*, vol. 23, no. 2 (Winter 1984), pp. 41–72.
Starrett, Bob and Josh McDaniel (2000) *The Little Audio CD Book*. Berkeley: Peachpit Press.
Sterne, Jonathan (2003) *The Audible Past: Cultural Origins of Sound Reproduction*. Duke University Press.
Swade, Doron (2001) *The Cogwheel Brain: Charles Babbage and the Quest to Build the First Computer*. London: Abacus.
Swallow, Norman (1966) *Factual Television*. London: Focal Press.
Tashiro, Charles Shiro, 'Videophilia: What Happens When You Wait For It on Video', *Film Quarterly*, vol. 45, no. 1 (Fall 1991), pp. 7–17.
Taylor, Timothy D. (2002) 'Music and the Rise of Radio in 1920s America: Technological Imperialism, Socialisation and the Transformation of Intimacy', *Historical Journal of Film, Radio and Tele-vision*, vol. 22, no. 4 (October), pp. 425–43.
Thompson, Kristin (1985) *Exporting Entertainment: America in the World Film Market, 1907–1934*. London: British Film Institute.
Thrasher, Frederic M.(1946) *Okay for Sound: How the Screen Found its Voice*. New York: Duell, Sloan and Pearce.

Toulet, Emmannuelle (1986) 'Le Cinéma à l'Exposition Universelle de 1900', in *La Revue d'Histoire moderne et contemporaine*, vol. 33 (April–June), pp. 186–8.

____ (1988) *Cinématographe, invention du siècle*. Paris: Éditions Gallimard.

Toulmin, Vanessa, Simon Popple and Patrick Russell (eds) (2004) *The Lost World of Mitchell and Kenyon: Edwardian Britain on Film*. London: British Film Institute.

Toulmin, Vanessa & Simon Popple (eds) (2005) *Visual Delights 2: Audience and Reception*. Eastleigh: John Libbey.

Truffaut, François, with Helen G. Scott (1986) *Hitchcock by Truffaut*, 2nd edn. London: Paladin.

Tümmel, Hebert (1986) *Deutsche Laufbildprojektoren: Ein Katalog*. Berlin: Stiftung Deutsches Kinemathek.

Udelson, Joseph H. (1982) *The Great Television Race: A History of the American Television Industry, 1925-41*. Urbana-Champaign: University of Alabama.

Uricchio, William (ed.) (1990) *The History of German Television, 1935–1944*, special edition of *Historical Journal of Film, Radio and Television*, vol. 10, no. 2.

Usai, Paolo Cherchi (1999) *An Introduction to the Study of Silent Cinema*. London: British Film Institute.

____ (2001) *The Death of Cinema: History, Cultural Memory and the Digital Dark Age*. London: British Film Institute.

Van Schil, George (1980) 'The Use of Polyester Film Base in the Motion Picture Industry: a Market Survey', *SMPTE Journal*, vol. 89, no. 2 (February), pp. 106–10.

Volkmann, Herbert (1965) *Film Preservation: A Report of the International Federation of Film Archives*. London: National Film Archive.

Wall, E. J. (1925) *The History of Three Colour Photography*. Boston: American Photographic Publishing.

Walls, H. J. and G. G. Attridge (1977) *Basic Photo Science: How Photography Works*, 2nd edn. London: Focal Press.

Wasser, Frederick (2000) *Veni, Vidi, Video: The Hollywood Empire and the VCR*. Austin: University of Texas Press.

Watkinson, John (1988) *The Art of Digital Audio*. London: Focal Press.

____ (1999) *MPEG-2*. Oxford: Focal Press.

____ (2001) *An Introduction to Digital Video*, 2nd edn. London: Focal Press.

Weightman, Gavin (2003) *Signor Marconi's Magic Box*. London: HarperCollins.

Wheeler, Owen (1929) *Amateur Cinematography*. London: Pitman.

White, D. R., D. J. Glass, E. Meschter and W. R. Holm (1965) 'Polyester Photographic Film Base', *Journal of the SMTPE*, vol. 64 (January), pp. 674–8.

Wilhelm, Henry and Carol Brower (1993) *The Permanence and Care of Color Photographs: Traditional and Digital Color Prints, Color Negatives, Slides and Motion Pictures*, Grinnell, IO: Preservation Publishing Co.

Winston, Brian (1996) *Technologies of Seeing: Photography, Cinematography and Television*. London: British Film Institute.

____ (1998) *Media, Technology and Society*. London: Routledge.

____ (2000) *Lies, Damn Lies and Documentaries*. London: British Film Institute.

Wood, Adrian, 'The Colour of War: A Poacher Among the Gamekeepers' in Graham Roberts & Philip M. Taylor (eds) *The Historian, Television and Television History*. Luton: University of Luton Press, pp. 45–53.

Wood, Linda (ed.) (1986) *British Films, 1927–39*. London: British Film Institute.

Zimmerman, Patricia (1995) *Reel Families: A Social History of Amateur Film*. Indianapolis, Indiana University Press.

Zworykin, Vladimir and G. A. Morton (1940) *Television: The Electronics of Image Transmission*. New York: Wiley.

____ (1954) *Television: The Electronics of Image Transmission in Colour and Monochrome*. New York: Wiley.

index

Aaton camera 39
Abramson, Albert 161, 165
Academy curve 156
Academy Ratio 51, 53, 56, 58, 65, 73, 151, 154, 175
Accelerated aging test 193
Acid detection (a/d) strip 193
Acres, Birt 24, 66, 137
Advanced Research Projects Agency (ARPA) 218
Advertising 181, 184
'AEO-light' 113
Agfa and Agfa-Gevaert – see Aktien-Gelleschaft für Analin Fabrikation (Agfa) 18–19, 22–3, 25, 91–2, 96, 195
Agfacolor 91–2
Airliners, film exhibition on 23
Akeley, Carl 36–7, 40, 44, 53
Akeley camera 37
'Akeley shot' 37
Aktien-Gelleschaft für Analin Fabrikation (Agfa) 18
'Allfex' sound effects machine 104
Altman, Rick 107, 126
Amateur cinematography 32, 66, 68
American Mutoscope and Biograph 48
American Telephone and Telegraph (AT&T) 99, 110, 113
Ampex 124, 177–8, 200, 210
Analogue 3, 27, 51–2, 74, 100, 115, 121, 127–9, 144, 157, 163, 173–6, 194, 196, 200, 202–4, 207–16, 219, 223, 229–3
Anamorphic – see Lenses, anamorphic 42–3, 47, 60, 62, 64–5, 150
Anderson, Lindsay 39
Angénieux 41
Ansco – see General Aniline and Film 92, 231
Antitrust legislation 49
Archives 23, 188–90, 192–3, 195, 201, 229
Armat, Thomas 100, 108, 135–6
Arnold, August 38
Arnold & Richter (Arri) 38
Arrhenius, Svandte 193
Arri 535 & 765 cameras 43–4
Arriflex camera 38–9, 41, 43–4
Audion Tube 109, 112–13
Aspect ratios 47–8, 50–2, 54–5, 58, 60, 62, 65, 71, 115, 146, 150, 174–5
Audio – see Sound
Automatic Gain Control (AGC) 184

Babbage, Charles 205, 208
Badische Anilin und Soda Fabrik (BASF) 124
Baird, John Logie 163–8, 176, 179, 185, 215
Bandwidth 128, 160, 163, 173–4, 176–7, 185, 214, 219–20
Bauer 139
Bausch and Lomb 42–3, 60
Bayer mask 172
Beethoven, Ludwig van 208
Belgian-American Radio Corporation (Barco) 225
Bell, Alexander Graham 108, 128, 162
Bell, Donald J. 45, 73, 107
Bell & Howell 12–14, 33, 35, 37–40, 43–6, 68–9, 93
Bell & Howell 2709 camera 142
Bell Telephone Laboratories 172, 207
Belton, John 48, 53, 126
Berliner, Emile 5, 101, 105
Berthiot 41
Berthon, Robert 83

Betacam 27, 181, 202, 210–11, 215
Betacam SP 181, 210–11
Betamax 73, 179–81, 185, 218
'Betamax case' of 1981 179–81, 218
Betts, Ernest 119
Binaural 125
Biophon 105
Birtac camera 66
Blake, E. W. 112
'Black Maria' studio 32
Blair 15
Blattnerphone 124
Blimps 43
Blumlein, Alan 125, 129, 149, 156
'Born digital' moving images 229–31
Boyerstown cinema fire 137
Brakhage, Stan 76
Brighton School 80
British Acoustic Film 119
British Board of Film Censors (BBFC) 19
British Broadcasting Corporation (BBC) 4, 124, 160–1, 165–7, 172, 174–5, 177–8, 185, 214–15
British Film Institute 39, 88, 111, 143, 188
British Marconi 165
Brown, Harold 189
Brown, Simon 85
Brownlow, Kevin 198
Burth, Willi 153
Busch (colour process) 82

Cameras, for amateur use 66–7; invention and development of 8–10, 12, 18–20, 24–54, 62–70, 124–5, 142, 168–72, 194, 224; mass-production of 30–5; for studio use 8, 13, 16, 71, 81–9, 120, 133, 171–2; television 5, 38–9, 71–6, 166–7, 171–4, 183–5, 229
Cameraphone 106
Caxton, William 182
Canal + 214
Canon 39
Capacitance Electronic Disc (CED) 179–80
Carolus cell 121
Case, Theodore W. 113–14, 118
Cathode ray tube (CRT) 164–5, 171–5, 211, 216, 223, 225
Celluloid – see Cellulose Nitrate
Cellulose 10, 15, 18
Cellulose Acetate Propionate 20
Cellulose Diacetate 124
Cellulose Nitrate 10, 15, 18, 20, 24, 31, 188–9
Cellulose Triacetate 18, 22, 70, 148, 151, 188, 190
Censorship 19–20, 183, 199
Chabrol, Claude 65
Charge Coupled Device (CCD) 171–2, 174, 222, 224–5
Chrétien, Henri 43, 60
Chromatron 171
Chronochrome 17, 82–3, 95
Chronophone 105–7
Cine-Kodak (amateur film equipment) 66–7
Cinecolor 92
Cinema buildings and auditoria, architectural design of 53; emergence of purpose built cinemas 103, 107, 140, 147; multiplexes 26, 71–2, 93, 148, 154–6, 214; 'Picture palaces' 85, 138, 141–2; sound, conversion to 52, 105–6, 112, 146–7; widescreen, adaptation for 16, 35, 38, 41–3, 53–60, 62–5, 71, 73, 115–17, 126, 147–51, 154, 158, 170, 175, 227
Cinema Digital Sound (CDS) 82, 118, 157, 227
Cinema Television Ltd. (Cintel) 174
CinemaScope 42, 44, 59–66, 73, 115, 126, 149–50, 156, 175, 227
Cinémathèque Française 173, 190
Cineon 222–3
Cinerama 53, 56–8, 60, 63–4, 71–3, 126, 149–50, 155
Clapperboards 52, 120
Closed Circuit Television (CCTV) 130
Close Up (journal) 119
Coe, Brian 75, 82, 113
Colcin (colour process) 82
Color Systems Technology 96
Colour, additive 17, 24, 79–87, 89, 95, 168–70; artificial 75, 77, 79; chromogenic 90; coupler 86, 89–93, 96, 194–6, 199; dye transfer 76, 84, 87, 90, 194; lenticular 83–4, 86, 95, 147, 156; mosaic 83; stencil 76–9, 83, 96; successive frame 80, 82–3, 86–7, 95; subtractive 79–85, 87, 89–90, 95, 168–9, 194; tinting 77–9, 94–5, 97; toning 77–9, 94–5, 97; two-colour systems 78, 81–2
Colour Reversal Intermediate 194
Colourisation 96
Columbia Broadcasting System (CBS) 169–70, 185, 215
Compact Disc 5, 100, 128, 157, 208–9
Compact Disc, recordable by consumers (CD-R) 216
Compact Disc Video (CD Video) 180
Compression (in digital media), asymmetric 208; in audio 208; lossless 208; lossy 212,

220; ratios 157, 213; in video 208–12, 230–1
Computers, data storage volumes 27, 204, 208–9, 212–13, 225–6, 231; hard discs 128, 206, 209, 220, 222; integrated circuits, use of 201, 206; microchip 206; transistors, use of 172, 205–6; valves, use of 208
Computer Generated Imagery (CGI) 202, 222–3
Contact printing – see Film, printing of
Content Scrambling System (CSS) 217
Continuity 5, 13, 107, 116
Continuous Format Migration – see Format Migration
Continuous printing – see Film, printing of
Copyright, droit d'auteur 182; infringement of 179, 218; legal principles 176, 182–3, 216
Crangle, Richard 103
Cripps, Thomas M. 115
Crosby, Bing 62, 176–7
Crosby-Mullin Video Tape Recorder 176–7

D2 (videotape format) 210
Daguerre, Louis 8
Datacine 222–5
Davy, Sir Humphrey 139
Deacetylation 190, 193, 195
Debrie Parvo camera 32, 36
Deutsches Institut für Normung (DIN) 73
Dickson, William Kennedy Laurie 1, 10, 12–14, 25, 32, 47–8, 73, 86, 99–100, 133, 219
Difference Engine 205, 208
'Digibox' TV receiver 215–16
Digital Audio Tape (DAT) 128, 211
Digital High Definition (HD) 71, 173, 224–5, 229
Digital Light Processing (DLP) 6, 225–9
Digital Millennium Copyright Act 218
Digital Screen Network 228
Digital Subscriber Line (DSL) 220
Digital Theater Systems (DTS) 157
Digital Versatile Disc (DVD) 53, 180–1, 202, 204, 209, 211, 213–17, 220–1, 227, 230–1
Digital Video Broadcasting (DVB) 214–16, 220
Director's cut 199
Dolby, A-type 156; Company 98, 126–7, 156–8; Dolby Digital (SR-D) 157; noise reduction 127; phase shifting 156; Spectral Recording (SR) 156
Dolby, Ray 126–7, 156
Domesday Disc 4, 27
Donisthorpe, Wordsworth 10
Douglas, Bill 199
Dufay-Chromex 85
Dufaycolor 24, 76, 84–6, 89, 95, 199, 229
Dufay, Louis 84–5
Du Pont 17–18
Du Pont, Eluthère Irèneé
Duplication, of film 11, 13–15, 25, 86, 101–2, 190, 194; of videotape 57, 217; of digital media 57, 217, 222
DVCAM 211, 219
DVC Pro 211

Eastman, George 1, 7, 9–10, 66, 68, 100–1
Eastman Kodak 9–10, 13, 15, 17, 20, 22, 24, 39, 45, 47–8, 66–8, 78, 83–4, 92–6, 133, 148, 157, 161, 192, 195, 222, 227
Eastmancolor 92–3, 94–5, 194–5
Ecco 198
Edison, Thomas 10, 12, 20, 31–2, 48–9, 97, 99–107, 129, 133, 149, 161, 180
Editing, using digital intermediate stages 222, 228; of film 12, 103, 106–12, 116, 120, 122, 124–5, 138, 177–8, 200, 206, 208, 210, 223, 229; of digital media 210, 223, 229
Edit Decision List (EDL) 223
Eidophor 224–5
Eisenstein, Sergei 52, 92
Electrical and Musical Industries (EMI) 129, 166–8, 185, 213, 215
Electronic Numerical Integrator and Calculator (ENIAC) 205, 208
Electronic News Gathering (ENG) 70, 178
Emitron 166
Enterprise Optical Manufacturing 139
Ernemann 139
Estar 24, 57
Evans, Mortimer 31
Exposure Index (EI) 8, 24, 88, 90–1, 93
Eyemo camera 37, 39
Eyman, Scott 98, 107

Farnsworth, Philo 164–7
Federal Communications Commission (FCC) 167, 170, 175
Fellini, Federico 195
Feminist film theory 2
Ferguson, Graeme 71–2
Ferrania 92
FIAF – see International Federation of Film Archives 189, 190, 192
Film, base 7–13, 18, 21–6, 30, 48, 69, 188–94; chemistry of 9, 14, 16, 67, 86, 132, 137, 194; CRT digital film recorders 164–5, 171–5, 211, 216, 223, 225; decomposition of 5, 11, 25,

188–98, 201, 230; digitisation of 213; duplication of 5, 11–16, 25, 47, 57, 79, 86, 101–2, 120–1, 190–6, 201, 204, 217, 222–3; emulsion 8–13, 15–19, 24, 47, 67, 73–9, 83–7, 92–4, 113, 121, 188, 191, 196–7, 222; fine grain (lavender) prints 16, 90; formats 4, 29, 31, 33, 35–41, 47–61, 66, 71–3, 219; laboratories 14; monochromatic 17, 79; orthochromatic 16–18, 80, 116; panchromatic 16–19, 80–7, 115–16, 121, 169, 196, 199; perforation 7, 12, 15, 19, 23, 29–34, 47–9, 52, 60, 62–3, 67–9, 134, 157, 191, 197–8; pre-tinted 78; processing of 14, 72, 86, 93; printing of 12–16, 34, 43, 62, 67, 71, 78, 84–90, 121, 156, 182, 191–2, 197, 203, 228; reversal 20, 66–8; safety 11, 18–20, 67–70, 117–18, 137–40, 147–8, 152–4, 185, 188, 190, 195; shrinkage of 191–2, 197; splicing of 12-13, 177; theft of 183
Filmguard 198
Filmosound projector 69
Fischer, Rudolph 90, 225
Filters, use of in cinematography 58, 79–82, 84, 196, 200, 223
Fischer, Fritz 90, 225
Flaherty, Robert 36–8, 40
Fletcher, Harvey 125
Ford, Henry 29, 86, 95
De Forest, Lee 109, 112–13, 128, 144, 162, 205
De Forest Phonofilms – see Phonofilm
Format migration (of videotape and digital media) 4, 201, 229
Format obsolescence (of videotape) 25, 27, 148, 166, 193, 215, 229–31
Fox Grandeur (widescreen process) 35, 42, 52, 54, 63, 147, 150
Fox-Talbot, Henry 8
Friese-Greene, William 10, 31
Fuji Photo Film 22–3, 25, 69, 92, 195

Gaisberg, Fred 109
Gance, Abel 53, 198
Gasparcolor 76
Gauges, single 8mm 68–9, 84; standard 8mm 68–9; super 8mm 69; 9.5mm 67–9, 73, 84, 198; 16mm 13, 20, 22, 38–41, 66–71, 83–4, 90, 125, 147–8, 175, 177–8, 199, 210; Super 16mm 23, 224; 17.5mm 41, 66–71, 73; 28mm 19, 66–8, 160; 35mm 4, 8, 12–13, 22–4, 31–45, 60, 62–75, 82, 84, 86–7, 90, 118, 124, 127, 135, 140, 147, 149–50, 125, 174–5, 181–3, 185, 195, 201–2, 219, 222–3, 223–9; 65mm 42, 44, 54, 63, 89; 70mm 35, 41, 44, 54, 63–6, 71–3, 89, 147, 150, 225
Gaumont 14, 25
Gaumont, Léon 17, 82, 95, 105
Gelatine bromide process 9
General Aniline and Film (Ansco) 92
Gevaert 23, 25, 78, 92
Glen Cinema fire 19
Godard, Jean-Luc 65, 195
Godowsky, Leopold 90
Goebbels, Joseph 91, 95
Goldmark, Peter 169–70, 215
Goldwyn, Samuel 45
Gomery, Douglas 117
Gottschalk, Robert 43
Gramophone – see Phonograph
Grandeur – see Fox Grandeur
Griffith, David Wark 77, 161

Handschiegl, Max 77, 94
Harris, Robert 89
Harman, Alfred Hugh 24
Haughten, Ken 209
Hendersons film laboratory 190
Hepworth, Cecil 13–15. 105
Herbert, Stephen 6
Hertz, Heinrich 162, 166–7, 173
Hickox, Sid 38
High Definition Television (HDTV) 71, 173
Hitchcock, Alfred 60, 89, 120
Hollerith, Herman 205
Hornbeck, Larry 225
Horner, William 7
Howell, Albert S. 12–14, 33, 37–40, 43–4, 68–9, 93, 142
Hoxie, Charles 114
Hurley, Frank 36
Hydrolysis 188, 190
Hypergonar lens 60
Hypertext Mark-Up Language (HTML) 219
Hypertext Transfer Protocol (HTTP) 219

Ibuka, Masaru 171
Iconoscope 165
I. G. Farben 124
Ilford 24, 84–5
Image Dissector Tube 164
Image Permanence Institute (IPI) 191–3
Imax 42, 60, 71–2, 155
Intermittent mechanism 13, 30–2, 72, 133–4, 136, 139, 200
International Business Machines (IBM) 205,

209
International Federation of Film Archives (FIAF) 189–90, 192
International Standards Organisation (ISO) 73
Internet 98, 184, 217–22
Internet Service Providers (ISPs) 219, 221
Ives, Herbert 168

Kalee (trademark of Kershaw Projector Manufacturing) 63, 139, 143
Kalmus, Herbert 85–6, 88
Keaton, Buster 37
Kerr, Robert 71
'Key moment' (conversion to sound) 98–9, 102, 105, 115–19, 126, 129–30
Kinarri camera 38
Kinemacolor 80–3, 95, 147, 169
Kinescope – see Telerecording
Kinesigraph 10
Kinetograph 31–2, 47, 99–100, 133
Kinetophone 107
Kinetoscope 12–13, 20, 31, 47, 100, 133–6, 158
Knopp, Leslie 149
Kodachrome 83, 90–1, 95, 195
Kodacolor 83–5, 147
Kodak – see Eastman Kodak
Koster, Henry 42, 136
Kroitor, Roman 71
Kurosawa, Akira 104

Lantern – see Magic Lantern
Laserdisc 4, 180
Latent Image 7–9, 16, 79, 84
'Latham loop' 135, 138
Latham, Woodville 32, 135, 138
Lean, David 64, 199
Lee, Frederick Marshall 80
Leevers, Norman 124
Legacy formats 200
Lenses; Anamorphic 42–3, 47, 60, 62, 64–5, 150; 'bug eye' 64; Prime 40, 43, 87; Zoom 40–1, 53
Lens turrets 34, 40
Letterboxing 175
Lighting; carbon arc 85, 139–42, 144, 151–2, 225; incandescent 12, 17–18, 85, 133, 139, 151–2; limelight 136–7, 139; mercury vapour discharge 17; in studios 29, 41, 85, 88, 90; tungsten 17–18, 68, 85, 152; xenon arc 151–4, 225
Lippmann, Gabriel 83
Liquid Crystal Display (LCD) 172, 225
Lloyd, Harold 37
Lucas, George 5, 72, 224
Lumière, Auguste and Louis 1, 12–13
Lye, Len 76

Macrovision 184, 217
Maddox, Richard Leach 9
Magic Lantern 10–11, 74, 102–3, 138
Magnafilm 54
Magnascope 53
Magnetophon 124
Maltese Cross intermittent mechanism 31, 135–6, 138
Manchester Metropolitan University 53
Mannes, Leopold 31, 135–6, 138
Marconi, Gugliemo 54
Matsushita 124
Maxwell, James Clerk 79–82, 94
Mayer, Mark-Paul 197
Mazzanti, Nicola 81
McKernan, Luke 80
McLaren, Norman 76
Mees, C. A. Kenneth 16
Messter, Oskar 105, 136
Metro-Goldwyn-Mayer (MGM) 58, 96
Microphones 2, 35, 52, 108–9, 112, 118, 120, 125, 129, 162, 203
Microsoft 49, 221
Minnesota Mining and Manufacturing (3M) 124
Mitchell 35, 40–4, 54, 87
Mitchell BNC camera 35, 43–4
Mitchell BNCR camera 35, 38
Modulation (of analogue electronic signals) 78, 109, 111, 113–14, 127, 160, 177, 181, 204
Monochrome 11, 74–6, 78–80, 84–93, 168–70, 196
Moore, Gordon 206–7, 209, 219–20
Moore's Law 206–7, 209, 220, 226, 228–9, 231
Morita, Akio 179, 208
Morse, Samuel 162
Morse Code 162
Morton, David 100, 109, 116, 126, 162
Motiograph 139
Motion Picture Experts' Coding Group (MPEG) 211–17, 220–2, 229
Motion Picture Patents (MPPC) 48–9, 62, 73, 101, 129, 161
Motion Picture Producers' and Distributors' Association of America (MPPDA) 183
Movietone 52, 54, 60, 110, 113–14, 119–20,

144
Moy and Bastie camera 81
MP3 (short for MPEG-1, layer 3 audio format) 220
MPEG – see Motion Picture Experts' Coding Group
MPEG-2 211–14, 216, 220, 222
Multimedia 4, 206
Mullin, John J. 176–7
Museum of Modern Art (MoMA) 176
My Childhood 199
Mylar 23

Nagra 127
Namias, Rodolfo 9
National Association of Theatre Owners (NATO) 24
National Film and Television Archive (UK) 75, 189–90
National Geographic 90
National Television Standards Committee (NTSC) 167, 169–75, 178, 200, 206, 210–11, 213–14, 220, 222, 225
NTSC Color 170, 173
Natural Vision (widescreen format) 54
Near Instantaneously Compacted and Expanded Audio Multiplex (NICAM) 214
Newman-Sinclair 36
Newman Sinclair Autokine camera 36, 44
Newman-Sinclair Standard camera 36
Newsreels 102, 114, 119–20
Nickelodeons 137
Nipkow disc 163–4, 166, 168–9
Nipkow, Paul163
Nitrate – see Cellulose Nitrate
Nitrocellulose – see Cellulose Nitrate
Noise reduction (in analogue audio) 121, 127, 156, 198
Norelco 150
Nyquist, Harold 207–8

O'Brien, Brian 63
Oboler, Arch 58, 60
Omnimax 72
Optical printing – see Film, printing of
Optical toys 7, 10, 134
Orwell, George 161
OrWo (Original Wolfen) 25
Oxberry 223

Paillard Bolex 39
Panaflex camera 43–4, 72
Panasonic 211, 217
Panavision 43–4, 60, 88
Pan-Cinor lens 41
Parallax 37, 42, 80–1, 87
'Paramount case' of 1948 56, 148
Paramount 54, 56, 62–5, 148, 150
Parkin, W. C. 19
Pathé [Frères] 25, 32, 48–9, 66–8, 76–7
Pathécolor 66
Pathéscope (amateur film equipment) 66–7
Pennebaker, D. A. 39
'Penny Gaffes' 137
Persistence of Vision (theory) 6–7
Perspecta 126
Petersen, Axel 119
Phase Alternate Line (PAL) 173
Phenakistiscope 6, 10
Philips 127, 139, 180, 213, 223
'Phono-Cinéma-Théâtre' sound system 105
Phonofilm 109, 113, 144
Phonograph 99–100, 105, 133
Phonovision 176, 179
Photographic chemistry 132
Photography, chemistry 132; history of 1, 7–11; still 20, 24, 48, 75, 84, 91, 133–4, 161–2
Photophone 52, 110, 114, 119–20
Piracy 182–4, 188, 216–20, 227
Plateau, Joseph 6
Polyester 11, 22–4, 69, 124, 156, 187–8, 192
Polyethylene Teraphalate – see Polyester
Polyvision 53
Porter, Edwin S. 139
Poulsen, Arnold 119, 123–4
Preservation (of moving images), chemicals, use of in 3, 13, 71, 81; 'copy to preserve' 190–1, 193, 195; of 'born digital' media 229–31; dye fading 93, 95, 194–7, 199, 20; format obsolescence 25, 193, 230–1; passive conservation 192–3, 195; preservation and restoration 187–201; separation negatives 91, 196; technical selection 197
Prestwich, John Alfred 100, 137
Prevost 139
Le Prince, Louis Augustin 10
Printing (of film), contact 13–16, 121–2, 191–2; continuous 13–14, 191; imbibition (dye transfer) 87, 89; optical 13, 15–16, 192; step 13–15
Projection, automation 153, 155; aperture plates 52, 65, 146; booths, design and layout of 20, 111, 137, 140, 145–6, 155–6, 158; of digital images 227–8; Digital Light Process-

ing (DLP) method 6, 225–9; Changeovers 140, 152; fire safety devices and precautions 22, 138, 140, 148; light sources 7, 11, 13, 17, 31, 65, 80, 113–14, 134–5, 141–2, 152; Liquid Crystal Display (LCD) method 172, 225; non-theatrical 66, 761, 147, 225; platters, use of 153–6, 209; projectionist, role of 20–1, 29, 33, 45, 52–3, 57, 65, 106, 111, 125, 140–6, 151–3, 155; shutters, design of 12, 25, 30–7, 50, 80–1, 133–4, 138–44, 164, 174–5; synchronisation of image to sound in 39, 42, 57, 72, 102, 104, 106–7
Propaganda 18, 22, 69, 81, 85, 91, 105, 124, 188, 221, 228
Propionate – see Cellulose Acetate Propionate

Quantel 223
Quicktime 221

Radio, industry 109, 162, 167, 169; Frequency Modulation (FM) 206; invention and development of 108–11, 113, 127, 154, 162–9; related technology in film sound 124, 129, 144
Radiodiffusion-Télévision Française 173
Radio Corporation of America (RCA) 52, 110, 114, 119–22, 144, 147, 164–5, 168–73, 213
Radio-Keith Orpheum (RKO) 54, 114, 119
Rank Organisation 62–3
RCA – see Radio Corporation of America 52, 110, 114, 119–22, 144, 147, 164–71, 173, 179m 213
RCA Photophone – see Photophone
Read, Paul 197
Recorded Music Industry 108–9, 111
Rectifiers 17, 140
Renovex 198
Restoration 89, 93, 187–201, 229
Reversal film 66
Reynaud, Émile 7
RKO – see Radio-Keith Orpheum
Roebuck, Alvah C. 138
Rogers, Will 17
Ryan, Roderick T. 76, 78

Salt, Barry 3, 14, 31, 76, 81, 118
Sarnoff, David 165, 167, 169–70, 185
Satie, Erik 104
Scientific American 99
Scorsese, Martin 195–6
Segrave, Kerry 183
Selsdon Report 166
Showscan 72
Simplex 54, 139, 147, 150
Sinatra, Frank 180, 183
Skall, William V. 89
Smith, George Albert 80–1, 172
Smith, Willoughby 112
Social Text 2
Society of Motion Picture [and Television] Engineers (SMP[T]E) 22, 48–9, 51, 53, 66–7, 73, 82, 140, 144, 161
Sokal, Alan 2
Sonochrome 78
Sony 27, 124, 128, 171, 178–81, 202, 224, 227
Sony Dynamic Digital Sound (SDDS) 157
Sound, acoustic 99, 101–2, 105–12, 117, 123, 129, 132; analogue 51; analogue to digital conversion 207; amplification of 105–9, 111–13, 123, 128–9, 204, 207; cassettes 127–8, 178; dialogue, recording and reproduction of 2, 103, 110, 126, 140; digital 64, 125, 128, 130, 157–8, 208, 227; digital to analogue conversion 173, 210, 215, 230–1; film 98, 102–4, 107, 109–14, 116, 122, 126–7, 147, 162; lectures, live performance of 102–3, 221; magnetic 39, 63, 69, 122–5, 149–50; music, live performance of 102, 104, 176; optical 4, 16, 18, 51–2, 69, 75, 110, 112–13, 117, 120–4, 129, 144, 146–7, 163–4; push-pull 122; sampling 207–8, 210; single system 52, 120–3; sound cameras 52, 113, 120–1, 125; sound effects 102–4; sound negatives 120–1, 124, 146; stereo 60, 62, 64–5, 73, 117, 125–9, 148–51, 208, 214; stereo variable area (SVA) 126, 193; variable area 114–15, 119, 121–2, 125, 147, 156, 158; variable density 113–16, 119, 121–2
Sovcolor 92
Speed, relative sensitivity to light of film 83; speed of film transport through camera 8, 50, 72; speed of film transport through projector 50, 72; standardisation of 19, 29–33, 44–5, 55–7, 140–2
Spehr, Paul 48
Spirit datacine 223
Spoilers 183
Sponable, Earl 84
Step printing – see Film, printing of
Stereoscopic (3-D) 42, 47
Studio lighting 29, 41, 85, 88, 90
Subtitling 45, 214
Système Électronique Couleur avec Memoir (SECAM) 173

Taiyo Yuden 216
Tati, Jacques 199
Taylor-Hobson 41
Technicolor; camera 43, 86, 88, 93, 171, 196; colour 86, 90; imbibition (dye transfer) printing process 87, 89; monopack 90–1; two-colour 78; three-strip 17, 36, 43, 55, 57, 64, 83, 88–93, 169m 194m 196
Telecine, flying spot 174; photoconductive 174; pulldown 32, 36, 49, 53–4, 63, 72, 124, 141, 175
Telefunken 173
Telegraphone 123
Telerecording 175–6, 185
Television, 405-line transmission system 166, 172–4, 200, 215; Cathode Ray Tube (CRT) 164–5, 171–2, 174–5, 211, 216, 223; chrominance ('C') signal 170, 172–3, 210–12; colour170–1, 173, 179; shadow mask 171; standards 160, 222
Texas Instruments 225
Théâtre Optique 7, 10
Thin Film Transistor (TFT) 171–2, 216
Thomas, Lowell 56
Thomsoncolor 199
'Time shifting' 177, 179, 181, 185
Todd-AO 42, 54, 63–4, 71, 126, 149–50, 155
Todd, Mike 56, 58
Tonbild-Syndikat (Tobis) 119
Transmission Control Protocol (TCP) 219
Tri-Ergon 119
Trinitron 171, 179
Truffaut, François 195
Turner, Edward 80, 94
Twentieth Century-Fox (TCF) 59–60, 62, 73, 84, 150

Uher 127
Ultrasonic splicing 23
UMatic 128, 178, 208, 210–11
United Kingdom Film Council (UKFC) 228
Universal 179
Urban, Charles 81
Usai, Paolo Cherchi 200

Valentine, Joseph 89
Varo lens 41
Vertical Helical Scan – see Video Home System (VHS) 180–5, 202, 211, 213, 215–24
Victor, Albert 66–8, 94
Victor Animatograph Corporation 66, 68
Victor Helical Scan – see Video Home System (VHS)
Vidor, King 52
Video CD (VCD) 217, 221
Videotape, cassettes 128, 178, 213; dropouts 231; editing of 178, 229; helical scan 128, 178–9, 201; quadruplex 177–8
Video Home System (VHS) 180–5, 202, 211, 213, 215–17, 221, 223, 231
Viewfinders, rangefinder 37–8; reflex 35, 37–8, 43
Vinegar Syndrome – see Deacetylation
Vitaphone 52, 57, 82, 99, 110–14, 118–20, 130, 144, 146, 161, 166, 187
Vitascope 54, 135–6
Vitaphone-Marconi 119
VistaVision 61–5, 71, 73, 82, 126, 149–50, 156
Visual Electronic Recording Apparatus (VERA) 177
Vivaphone 105–6

Wall Street Crash 55, 117
Waller, Fred 57–8
Walturdaw 106
Warner Brothers 38, 52, 54, 58, 62, 99, 110–14, 117, 130, 146
Warner, Sam 130
Watkins, Peter 39, 214
Watkinson, John 214
Western Electric 51–2, 109–11, 114, 118–22, 144, 147
Wet gate printing 192
Widescreen 16, 35, 38, 41–3, 53–8, 60, 62–5, 71, 73, 115–17, 147, 149–51, 170, 175
Wilcam-Imax camera 71
Williamson, James 32
Winchester Disc – see Computers, hard discs
Windows Media 221
Winston, Brian 39, 124, 181, 185
Wiseman, Frederick 39

Zeiss Ikon 152
Zimmerman, Patricia 56
Zoetrope 7
Zoller, H. P. 153
Zworykin, Vladimir 165–7